ELSEVIER'S DICTIONARY OF CHEMOETYMOLOGY

THE WHIES AND WHENCES OF CHEMICAL NOMENCLATURE AND TERMINOLOGY

ELSEVIER'S DICTIONARY OF CHEMOETYMOLOGY

THE WHIES AND WHENCES OF CHEMICAL NOMENCLATURE AND TERMINOLOGY

compiled by

ALEXANDER SENNING
*Department of Chemistry, Technical University of Denmark,
Kgs. Lyngby, Denmark*

ELSEVIER

Amsterdam • Boston • Heidelberg • London • New York • Oxford
Paris • San Diego • San Francisco • Singapore • Sydney • Tokyo

Elsevier
Radarweg 29, PO Box 211, 1000 AE Amsterdam, The Netherlands
The Boulevard, Langford Lane, Kidlington, Oxford OX5 1GB, UK

First edition 2007

Library of Congress Cataloging-in-Publication Data
A catalog record for this book is available from the Library of Congress

British Library Cataloguing in Publication Data
A catalogue record for this book is available from the British Library

ISBN-13: 978-0-444-52239-9
ISBN-10: 0-444-52239-5

Printed and bound in The Netherlands

07 08 09 10 11 10 9 8 7 6 5 4 3 2 1

Preface

Noting a marked lack of comprehensiveness and/or contemporaneity among typical reference works on chemical etymology[1-7] as well as a somewhat spotty coverage of chemical terms and their etymology in comprehensive dictionaries and textbooks[8-10] the present author decided to write an up-to-date desk reference on chemical etymology which would satisfy the needs of casual readers as well as those of more demanding users of etymological lore. Characteristic user-friendly features of the present work include avoidance of cumbersome abbreviations, avoidance of entries in foreign alphabets, and a broad coverage of all chemical disciplines including mineralogy. Biological, medical, geological, physical, and mathematical terms are only considered where they appear of interest to mainstream chemists.

This book does not provide definitions of terms (unless required in the etymological context) nor guidance as to the timeliness of different nomenclature systems. The typical user will from the outset be well aware of the exact meaning of the terms her or she focuses on and only require the etymological background to be used, for instance, in a lecture, in the introduction of a scientific paper, etc.

Examples of sources which have been drawn upon in the preparation of this book, apart from the extremely useful Internet resource Google, are listed below[11-25], but an exhausting enumeration would be tiresome and impractical. As a sign of the times many important sources are Internet based.

The author is especially indebted to the Wordorigins discussion forum[26] where many most subtle or awkward etymological riddles could be solved with expert help or laid to rest as inscrutable.

In every single instance where an etymology is described as unknown or unclear considerable effort has been spent to reach beyond conflicting, obscure, or apparently lacking evidence.

The author is grateful for his former Department's kind and generous hospitality as well as for our librarian's, Kirsten Randolf's, superb services.

Kgs. Lyngby, Denmark, July 2006

Alexander Senning

[1] D. Bailey and K. C. Bailey, *An Etymological Dictionary of Chemistry and Geology*, Arnold, London, UK, 1929.

[2] T. H. Savory, *The Language of Science*, Andre Deutsch, London, UK, 1967.

[3] L. Hogben, *The Vocabulary of Science*, Heinemann, London, UK, 1969.

[4] S. C. Bevan, S. J. Gregg, and A. Rosseinsky, *Concise Etymological Dictionary of Chemistry*, Applied Science Publishers Ltd, London, UK, 1976.

[5] V. Ringnes, *Navn på kjemiske stoffer*, Cappelen, Oslo, Norway, 1996.

[6] O. A. Neumüller, *Das Wörterbuch chemischer Fachausdrücke*, Dudenverlag, Mannheim, Germany, 2003.

[7] C. Giunta, Glossary of Archaic Chemical Terms http://web.lemoyne.edu/ ~giunta/archema.html

[8] P. N. Gove (Ed.), *Webster's Third New International Dictionary of the English Language Unabridged*, Merriam-Webster Inc., Springfield, MA, USA, 1993.

[9] Merck Index, 13th Ed., Merck & Co., Inc., Whitehouse Station, NJ, USA, 2001.

[10] N. Wiberg, *Holleman-Wiberg, Lehrbuch der Anorganischen Chemie*, Walter de Gruyter, Berlin-New York, 1995.

[11] J. R. Partington, *A History of Chemistry*, Vols. 1-4, Macmillan & Co. Ltd, New York etc., 1961-1970.

[12] P. van der Krogt, *Elementymology & Elements Multidict.* www.vanderkrogt.net/elements

[13] J. Andraos, Named Things in Chemistry & Physics http://www.careerchem.com/NAMED

[14] F. Kluge, *Etymologisches Wörterbuch der deutschen Sprache*, 24th Ed., Walter de Gruyter, Berlin-New York, 2002.

[15] K. Hårbøl, J. Schack, and H. Spang-Hanssen, *Fremmedordbog*, 2nd Ed., Gyldendal, Copenhagen, Denmark, 2001.

[16] V. Karpenko, The Discovery of Supposed New Elements, *Ambix* **27**, 77-102 (1980).

[17] B. Fugmann (Ed.), *Römpp-Lexikon Naturstoffe*, Georg Thieme Verlag, Stuttgart etc., 1997.

[18] D. Barthelmy, Mineralogy Database, http://webmineral.com

[19] O. Johnsen, *Photographic Guide to Minerals*, Oxford University Press, Oxford, UK, 2002

[20] U. Quattrocchi, CRC *Dictionary of Plant Names*, CRC Press, Boca Raton etc., 2000.

[21] P. Morris, *Biographies of Chemists*, Royal Society of Chemistry, http://www.chem.qmw.ac.uk/rschg/biog.html

[22] E. Farber (Ed.), *Great Chemists*, Interscience Publishers, New York, NY, USA, 1983.

[23] D. Abbott (Ed.), *The Biographical Dictionary of Scientists: Chemists*, Peter Bedrick Books, New York, NY, USA, 1983.

[24] J. Daintith, S. Mitchell, E. Tootill, D. Gjersten, *Biographical Encyclopedia of Scientists*, Institute of Physics Publishing, Bristol, UK, 1994.

[25] The Compleat Botanica, http:/www.crescentbloom.com

[26] D. Wilton, Wordorigins.org, http://wordorigins.org

A

a- (ab-, abs-)
derived from *a-, (ab-, abs-)* (Latin: away from)

-a
an arbitrary suffix for a binary metal oxide; patterned after magnesia and silica

AAS
an abbreviation for atom absorption spectroscopy

ABC
an abbreviation for atomic, biological, chemical (warfare)

Abegg's rule
named for the German chemist Richard Abegg (1869-1910)

Abel-Pensky method
named for the British chemist Sir Frederick Augustus Abel (1827-1902) and the German engineer B. Pensky (1850-1930)

abelsonite
$C_{31}H_{32}N_4Ni$, named for the US physicist Philip Hauge Abelson (1913-2004)

abeo-
a prefix derived from *abeo* (Latin: I go away), from *ire* (Latin: to go)

abietane
$C_{20}H_{36}$, derived from abiet(o)- and -an(e)

abietic acid (abietinic acid, sylvic acid)
$C_{20}H_{30}O_2$, derived from abiet(o)-

abiet(o)-
derived from the genus name *Abies* (firs), from *abies* (Latin: fir)

abikoviromycin
$C_{10}H_{11}NO$, derived from the specific epithet of the bacterial species name *Streptomyces abikoensis*, after its habitat Abiko, Japan, virus, and -mycin

abiotic
derived from a(n)- and bio-

[1]**ABR**
$C_{28}H_{24}O_9$, an abbreviation for 1-*O*-acetyl-2,3,4-tri-*O*-benzoyl-β-D-ribofuranose

[2]**ABR**
an abbreviation for acrylate butadiene synthetic rubber

abrin
a protein, coined by contraction of *Abrus* agglutinin, derived from the genus name *Abrus* (tropical vines), from *habros* (Greek: graceful, delicate), and -in(e)

abrine
$C_{12}H_{14}N_2O_2$, coined by variation of abrin

[1]**ABS**
an abbreviation for acrylonitrile butadiene styrene copolymer

[2]**ABS**
an abbreviation for alkylbenzenesulfonate

abscisic acid
$C_{15}H_{20}O_4$, derived from *abscisio* (Latin: abscission), from *abscidere* (Latin: to cut off) – referring to this compound's role as abscission-accelerating plant hormone

absinth-
derived from the specific epithet of the species name *Artemisia absinthum* L. (common wormwood), from *apsinthion* (Greek: wormwood), ultimately possibly from *aspand* (Persian: wormwood)

absinthin
$C_{30}H_{40}O_6$, derived from absinth- and -in(e)

absorb, absorption
derived from *ab* (Latin: away from) and *sorbere* (Latin: to swallow up)

abzyme
coined by contraction of antibody and enzyme

acac-
derived from the specific epithet of the species name *Robinia pseudoacacia* L. (black locust tree), ultimately from *akakia* (Greek: shittah tree), akin to *ake* (Greek: point, thorn)

acacetin
$C_{16}H_{12}O_5$, derived from acac-, -ete, and -in(e)

acacic acid
$C_{30}H_{48}O_5$, derived from acac-

acanthite (argentite)
Ag_2S, derived from acanth(o)- and -ite – referring to the thornlike shape of this mineral's crystals

acanth(o)-
derived from *akantha* (Greek: thorn, spine)

acarbose
$C_{25}H_{43}NO_{18}$, originally coined as a trademark, derived from a(n)-, carb(o)-, and -ose – referring to this compound's inhibition of α-glucosidase

acaricide
derived from acar(o)- and -cide

acar(o)-
derived from *akari* (Greek: mite), from *keirein* (Greek: to cut off)

ACC
$C_4H_7NO_2$, an abbreviation for 1-aminocyclopropanecarboxylic acid

accretium
Cf, an unsuccessfully suggested name for californium, derived from accretion and -ium

ACE
an abbreviation for angiotensin converting enzyme

ace-
derived from acetic acid

acemannan
coined by contraction of acetylated mannan

acenaphthene
$C_{12}H_{10}$, derived from ace- and naphthene

-acene
derived from anthracene

acer-
derived from the genus name *Acer* (maples), from *acer* (Latin: maple)

aceric acid
$C_6H_{10}O_6$, derived from acer-

acerin
derived from acer- and -in(e)

ACES
$C_4H_{10}N_2O_4S$, an abbreviation for *N*-(2-acetamido)-2-aminoethanesulfonic acid

acesulfame
$C_4H_5NO_4S$, originally coined as a trademark, probably derived (with contraction) from acetoacetic ester and sulfamate

acetal
coined by contraction of acet(o)- and alcohol

acetate
$C_2H_3O_2^-$, derived from acetic acid and -ate

acetic acid
$C_2H_4O_2$, derived from *acetum* (Latin: vinegar), from *acer* (Latin: sharp)

acetin
$C_5H_{10}O_4$, coined by contraction of *Acetylglycerin* (German: acetylglycerol)

acet(o)-
derived from acetic acid

acetogenin (polyketide)
derived from acet(o)-, -gen, and -in(e)

acetoin
$C_4H_8O_2$, derived from acet(o)- and -oin; patterned after [1]benzoin

acetol
$C_3H_6O_2$, coined by contraction of acetone alcohol

acetone
C_3H_6O, derived from acet(o)- and -one; patterned after margarone – referring to the formation of acetone as a condensation product of acetic acid

acetonide
derived from acetone and –ide

acetonitrile
C_2H_3N, derived from acet(o)- and nitrile

acetophenone (hypnone)
C_8H_8O, derived from acet(o)-, phene, and -one

acetoxime
C_3H_7NO, coined by contraction of acetone oxime

acetoxy
$C_2H_3O_2$–, coined by contraction of acetyloxy

aceturic acid
$C_4H_7NO_3$, derived from acet(o)- and -uric acid; patterned after hippuric acid

[1]acetyl
C_2H_3O–, derived from acet(o)- and -yl

[2]acetyl
an obsolete name for the vinyl radical C_2H_3, derived from acet(o)- and -yl

acetylcholinesterase
derived from acetyl, cholin, ester, and -ase

acetylene
C_2H_2, derived from [2]acetyl and -ene – referring to the formal derivation of C_2H_2 by hydrogen abstraction from the radical C_2H_3

AChE
an abbreviation for acetylcholinesterase

Acheson process
named for the US inventor Edward Goodrich Acheson (1856-1931)

achiral
derived from a(n)- and chiral

aci-
derived from acid

acichromism (halichromism)
derived from aci- and –chromism

acid
ultimately derived from *acidus* (Latin: sour), from *acer* (Latin: sharp)

acidophilic
derived from acidum and -philic

acidulate
derived from *acidulus*, diminutive of *acidus* (Latin: sour), from *acer* (Latin: sharp)

acidum
New Latin: acid, ultimately derived from *acidus* (Latin: sour), from *acer* (Latin: sharp)

acla- (akla-)
derived from aklavin

aclacinomycin
probably coined as a trademark, possibly derived from acla-, anthracyclinone, and -mycin

aclarubicin
$C_{42}H_{53}NO_{15}$, derived from acla-, rubi-, and -icin(e)

ACM
an abbreviation for acrylate synthetic rubber, derived from acrylate, 2-chloroethyl vinyl ether, and polymethylene

acmite (aegirine)
$NaFeSi_2O_6$, derived from *akme* (Greek: point, highest point, culmination) and -ite – referring to this mineral's characteristically pointed long prismatic crystals

aconic acid
$C_5H_4O_4$, derived from aconitic acid by contraction

aconine
$C_{25}H_{41}NO_6$, coined by variation of aconitine

aconit-
derived from the genus name *Aconitum* (monkshoods), ultimately from *akonitos* (Greek: without dust) – possibly referring to these plants' habitat on rocky ground

aconitane
$C_{18}H_{27}N$, derived from aconit- and -an(e)

aconitic acid
$C_6H_6O_6$, derived from aconit-

aconitine
$C_{34}H_{47}NO_{11}$, derived from aconit- and -in(e)

ACP
an abbreviation for acyl carrier protein

acridine
$C_{13}H_9N$, derived from acr(o)- and -idin(e)

acr(o)- derived from *acer* (Latin: sharp, acrid)

acrolein
C_3H_4O, derived from acr(o)-, *olere* (Latin: to smell), and -in(e)

acrylic acid
$C_3H_4O_2$, derived from acrolein

acrylonitrile
C_3H_3N, derived from acryl(o)-

acryl(oyl)-
C_3H_3O–, derived from acrylic acid

actaplanin
derived from the bacterial genus name *Actinoplanes*, from actin(o)- and *planes* (Greek: wanderer), and -in(e)

ACTH
an abbreviation for adrenocorticotropic hormone

actin
derived from *actus* (Latin: an act), from *agere* (Latin: to act), and -in(e)

actinide (actinoid)
derived from actinium and -ide

actinium
Ac, derived from actin(o)- and -ium – referring to this element's radioactivity

actinium emanation
Rn, an unsuccessfully suggested name for radon, derived from actinium and emanation

actino-
derived from the bacterial genus name *Actinomyces*, from actin(o)- and myc(o)- – referring to the filamentous or rod-like shape of these bacteria

[1]**actin(o)-**
derived from *aktis* (Greek: ray)

actinobolin
$C_{13}H_{20}N_2O_6$, coined by contraction of actino-, probably metabolism, and -in(c)

actinodaphnine
$C_{18}H_{17}NO_4$, derived from the genus name *Actinodaphne* (laurels), from actin(o)- and, ultimately, Daphne, a nymph in Greek mythology

actinoid (actinide)
derived from actinium and -oid

actinolite
$Ca_2(Mg,Fe)_5Si_8O_{22}(OH)_2$, derived from actin(o)- and -lite – referring to this mineral's often fibrous, radiated, or columnar crystal shape

actinomycin
derived from actino- and -mycin

actinon
Rn, unsuccessfully suggested name for radon, derived from actinium and [1]-on

actinonin
$C_{19}H_{35}N_3O_5$, derived from actino- and -in(e)

actinorhodine
$C_{32}H_{26}O_{14}$, derived from actino-, rhod(o)-, and -in(e) – referring to this antibiotic's red color

actinospectacine (spectinomycin)
$C_{14}H_{24}N_2O_7$, coined by contraction of actino-, the specific epithet of the bacterial species name *Streptomyces spectabilis*, from *spectabilis* (Latin: notable, admirable, remarkable), and -in(e)

actiphenol
$C_{15}H_{17}NO_4$, derived (with contraction) from actino- and phenol

acuminite
$SrAlF_4(OH){\cdot}H_2O$, derived from *acumen* (Latin: spear head) – referring to the characteristic crystal shape of this mincral

ACV
$C_{14}H_{25}N_3O_6S$, an abbreviation for δ-(L-α-aminoadipyl)-L-cysteinyl-D-valine

acyclic
derived from a(n)- and cyclic

acyclovir
$C_8H_{11}N_5O_3$, derived from a(n)-, cycl(o)- and virus – referring to this compound's less cyclic nature compared to the natural lead compound guanosine

acyl
derived from acid and -yl

acylal
derived from acyl and –al

acylase
derived from acyl and -ase

acyloin
derived from acyl- and -oin

ad- (ac-, af-, al-, am-, ap-, ar-, as-, at-)
derived from *ad (ac-, af-, al-, am-, ap-, ar-, as-, at-)* (Latin: to)

ADA
$C_6H_{10}N_2O_5$, an abbreviation for *N*-(2-acetamido)iminodiacetic acid

adamantane
$C_{10}H_{16}$, derived from *adamantinos* (Greek; of steel, of diamond), from *adamas* (Greek: steel, diamond), from *adamas* (Greek: indomitable), and -an(e) – referring to this hydrocarbon's diamond-like structure

adamite
$Zn_2AsO_4(OH)$, named for the French mineralogist Gilbert Joseph Adam (1795-1881)

Adams catalyst
PtO_2, named for the US chemist Roger Adams (1889-1971)

[1]adamsite (DM)
$C_{12}H_9AsClN$, named for the US chemist Roger Adams (1889-1971)

[2]adamsite
$NaY(CO_3)_2 \cdot 6H_2O$, named for the Canadian scientist Frank Dawson Adams (1859-1942)

adenine
$C_5H_5N_5$, derived from aden(o)- and -in(e) – referring to this compound's occurrence in many glandular organs

aden(o)-
derived from *aden* (Greek: gland)

adenosine
$C_{10}H_{13}N_5O_4$, derived (with contraction) from adenine, ribose, and -in(e)

adenylic acid
$C_{10}H_{14}N_5O_7P$, derived (with contraction) from adenosine

adermine (pyridoxine, vitamin B_6)
$C_8H_{11}NO_3$, coined by contraction of antidermatitis vitamin

[1]ADH
an abbreviation for alcohol dehydrogenase

[2]ADH
an abbreviation for antidiuretic hormone

ADI
an abbreviation for acceptable daily intake

adiabatic
derived from *adiabatos* (Greek: which cannot be passed), from a(n)-, dia-, and, ultimately, *bainein* (Greek: to walk)

adipic acid
$C_6H_{10}O_4$, derived from *adeps* (Latin: fat)

adip(o)-
derived from adipic acid

adipsin
derived from adiposity, from *adeps* (Latin: fat), and -in(e)

adiuretin (antidiuretin, vasopressin)
$C_{46}H_{64}N_{14}O_{12}S_2$, derived from a(n)-, diuresis, and -in(e)

Adkins catalyst
(Cr,Cu)O, named for the US chemist Homer Burton Adkins (1892-1949)

adlum-
derived from the genus name *Adlumia*

(vines), after the US viticulturist John Adlum (1759-1836)

adlumidine
$C_{20}H_{17}NO_6$, derived from adlum- and -idin(e)

adlumine
$C_{21}H_{21}NO_6$, derived from adlum- and -in(e)

admontite
$MgB_6O_{10}\cdot 7H_2O$, named after this mineral's locality Admont, Austria

adonitol (ribitol)
$C_5H_{12}O_5$, derived from the genus name *Adonis* (adonis), after the Greek mythical figure Adonis, and -itol

ADP
$C_{10}H_{15}N_5O_{10}P_2$, an abbreviation for adenosine 5'-diphosphate

adrenaline (epinephrine)
$C_9H_{13}NO_3$, derived from adrenal gland (glandula suprarenalis), from *ad* (Latin: to) and *renalis* (Latin: pertaining to the kidneys), from *ren* (Latin: kidney)

adrenochrome
$C_9H_9NO_3$, derived (with contraction) from adrenaline and –chrome

adsorb, adsorption
derived from *ad* (Latin: to) and *sorbere* (Latin: to swallow up)

adularia (orthoclase)
$KAlSi_3O_8$, named after this mineral's locality Adula Mountains, Switzerland

AEAPS
an abbreviation for Auger electron appearance spectroscopy

aegirine (acmite)
$NaFe(SiO_3)_2$, named for Ægir, the Old Norse god of the sea

aequorin
derived from the genus name *Aequorea* (jellyfish), from *aequoreus* (Latin: marine), from *aequor* (Latin: surface of the sea)

aer(o)-
derived from *aer* (Greek: air)

aerobic
derived from aer(o)- and *bios* (Greek: life)

aerogel
derived from aer(o)- and gel

aerosol
derived from aer(o)- and solution

¹AES
an abbreviation for atomic emission spectrometry

²AES
an abbreviation for Auger electron spectroscopy

affinity
derived from *affinitas* (Latin: relationship),from *affinis* (Latin: bordering on, related)

aflatoxin
coined by contraction of the microbiological species name *Aspergillus flavus,* from *aspergillum* (New Latin: short-handled brush), from *aspergere* (Latin: to sprinkle), and flav(o)-, and toxin

AFM
an abbreviation for atomic force microscopy

AFS
an abbreviation for atom fluorescence spectroscopy

agalmatolite
$Al_2Si_4O_{10}(OH)_2$, derived from agalma, from *agallein* (Greek: to adorn), and -lite – referring to this mineral's use in stone carving

agar
derived from *agar agar* (Malay: agar)

agari-
derived from the genus name *Agaricus* (mushrooms), from *agarikon* (Greek: a mushroom)

agaric acid (agaricic acid, agarinic acid, agaricin)
$C_{22}H_{40}O_7$, derived from agari-

agaritine
$C_{12}H_{17}N_3O_4$, derived from agari- and -in(e)

agarose
derived from agar and -ose

agate
SiO_2, ultimately derived from *achates* (Greek: agate), from Achates River, now Drillo River, Sicily, Italy

Agent Blue
C_2H_7AsO, an arbitrary US military code name for dimethylarsinous acid

Agent Orange
an arbitrary US military code name for a mixture of butyl (2,4-dichlorophenoxy)acetate and butyl (2,4,6-trichlorophenoxy)acetate

Agent White
$C_6H_3Cl_3N_2O_2$, an arbitrary US military code name for picloram

agglutination
derived from *agglutinare* (Latin: to adhere), from *gluten* (Latin: glue)

agglutinin
derived from agglutination and -in(e)

aglycon
derived from a(n)-, ultimately, glucose, and -on(e)

agmatine
$C_5H_{14}N_4$, derived from *agma* (Greek: fragment) and -in(e) – referring to this compound's formation from arginine in the course of putrefaction

agonist
derived from *agonistes* (Greek: combatant), from *agonizesthai* (Greek: to contend), from *agon* (Greek: contest)

agostic
derived from *agostos* (Greek: hooked up)

agr(o)-
derived from *ager* (Latin: field)

agroclavine
$C_{16}H_{18}N_2$, derived from agr(o)- and the fungal genus name *Claviceps*, from *clava* (Latin: club) and *-ceps* (Latin: -head), from *caput* (Latin: head), and -in(e)

agrocybin
$C_8H_5NO_2$, derived from the fungal genus name *Agrocybe*, from agr(o)- and *kybe* (Greek: head), and -in(e)

agropine
$C_{11}H_{20}N_2O_7$, derived from the bacterial genus name *Agrobacterium*, from agr(o)-, and opine – referring to the formation of this amino acid in crown galls caused by *Agrobacterium tumefaciens*

AIBN
$C_8H_{12}N_4$, an abbreviation for azoisobutyronitrile

-ain(e)
an arbitrary suffix patterned after cocaine

air
an archaic name for gas

ajaconine
$C_{22}H_{33}NO_3$, derived from the specific epithet of the species name *Delphinium ajacis* (rocket larkspur), from Ajax, a Greek mythical hero, possibly akin to *aiastes* (Greek: mourner), and -in(e)

ajmal-
derived from ajmaline

ajmalan
$C_{20}H_{26}N_2$, derived from ajmal- and -an(e)

ajmaline
$C_{20}H_{26}N_2O_2$, named for the Pakistani physician and Muslim leader Hakim Ajmal Khan (1863-1927)

ajoene
$C_9H_{14}OS_3$, derived from *ajo* (Spanish: garlic) and –ene

ajugarin
derived (with contraction) from the species name *Ajuga remota* (a bugleweed), from a- and *iugum* (Latin: yoke) – referring to this plant's undivided calyx – or from *abiga* (Latin: a plant inducing abortion), from *abigere* (Latin: to drive away), and -in(e)

Akabori amino acid reactions
named for the Japanese chemist Shiro Akabori (1900-1992)

åkermanite
$Ca_2MgSi_2O_7$, named for the Swedish metallurgist Anders Rikard Åkerman (1837-1922)

-akis
a suffix in Greek multiplicative numerals such as *tetrakis* (four times)

aklavin
$C_{30}H_{35}NO_{10}$, derived from its locality Aklavik, NWT, Canada, the source of the soil sample with the actinomycete which produces this antibiotic

aklavinone
$C_{22}H_{20}O_8$, derived from aklavin and -one

aksaite
$Mg[B_6O_7(OH)_6]\cdot 2H_2O$, named after this mineral's locality Aksai Valley, Kazakhstan

akuamm(a)-
derived from *akuamma* (Twi: owala tree, *Pentaclethra africana*)

akuammicine
$C_{20}H_{22}N_2O_2$, derived from akuamm(a)- and -icin(e)

akuammilan
$C_{19}H_{22}N_2$, derived from akuamm(a)-, -il, and -an(e)

akuammine
$C_{22}H_{26}N_2O_4$, derived from akuamm(a)- and -in(e)

-al
an arbitrary suffix derived from aldehyde

ALA
$C_5H_9NO_3$, an abbreviation for δ-aminolevulinic acid

alabamine
At, an unsuccessfully suggested name for astatine, after the state of Alabama, USA,

and -in(e)

alabamium
At, an unsuccessfully suggested name for astatine, after the state of Alabama, USA, and -ium

alabandite
MnS, named after the ancient town of Alabanda, Caria, Asia Minor

alabaster
$CaSO_4{\cdot}2H_2O$, derived from *alabastros* (Greek: alabaster, vase made of alabaster)

alacranite
As_4S_4, named after this mineral's locality Alacran, Pampa Larga, Chile

alane (alumane)
AlH_3, derived from aluminum and -an(e)

alanine
$C_3H_7NO_2$, derived from aldehyde – referring to this amino acid's synthesis from acetaldehyde

alanosine
$C_3H_7N_3O_4$, coined by contraction and variation of 3-(hydroxynitrosoamino)-L-alanine

alant(o)-
derived from alant, ultimately from *inula* (Latin: sneezeweed)

alantolactone
$C_{15}H_{20}O_2$, derived from alant(o)- and lactone

alazopeptin
$C_{15}H_{20}N_6O_5$, coined by contraction of alanine, az(o)-, peptide, and -in(e)

albaspidin
$C_{25}H_{32}O_8$, derived from *albus* (Latin: white) and the genus name *Aspidium* (ferns), from aspid(o)- and -ium, and -idin(e)

albite
$NaAlSi_3O_8$, derived from *albus* (Latin: white) and -ite – referring to this mineral's commonly white color

albizziin
$C_4H_9N_3O_3$, derived from the genus name *Albizzia* (Old World trees), after the Italian naturalist Filippo degli Albizzi (born 1749), and -in(e)

alb(o)-
derived from *albus* (Latin: white)

albofungin
$C_{22}H_{24}N_2O_9$, derived from the specific and subspecific epithets of the bacterial species name *Streptomyces albus* var. *fungistaticus* Solovyeva et Rudaya, from alb(o)- and *fungistaticus* (New Latin: fungistatic), and -in(e)

albomycin
derived from alb(o)- and -mycin

alborixin
derived from the specific epithet of the bacterial species name *Streptomyces albus* from alb(o)-, an unexplained name fragment, and -in(e)

albumen
derived from *album(en) ovi* (Latin: egg white), from alb(o)-

albumin
derived from albumen and -in(e)

alcalinium
Fr, an unsuccessfully suggested name for francium, derived (with contraction) from alkali metal and -ium

alcapton (homogentisic acid)
$C_8H_8O_4$, derived from alkali, hapt(o)-, and [1]-on

alchemy
derived from *al-kimiya* (Arabic: alchemy), ultimately from *chemeia* (Greek: chemistry), from *chymos* (Greek: juice), from *chein* (Greek: to pour)

alcian
as in alcian blue, probably coined by contraction (and slight alteration) of phthalocyanine

alcohol
derived from *al-kuhul* (Arabic: powder) – referring to the alchemistic procedure of reducing samples, through several purification steps, to a powder, the supposedly pure essence of the substance in question; thus alcohol is the essence (or spirit) of wine; the same word, spelled kohl, is used for powdered antimony as a cosmetic preparation

alcohol sulphuris
CS_2, an archaic New Latin name for carbon disulfide – referring to this compound's volatility and sulfur content

aldaric acid
derived from aldose and -aric acid

aldazine
derived from ald(o)- and azine

aldebaranium
Yb, a name unsuccessfully suggested for ytterbium, after the double star Aldebaran, from *al-dabaran* (Arabic: Aldebaran), from *dabara* (Arabic: to follow), and -ium

aldehyde
coined by contraction of *alcohol dehydrogenatus* (New Latin: aldehyde, literally dehydrogenated alcohol)

Alder ene reaction
named for the German chemist Kurt Alder (1902-1958)

Alder rule
named for the German chemist Kurt Alder (1902-1958)

aldimine
derived from ald(o)- and imine

alditol
derived (with contraction) from aldose and -itol

ald(o)-
derived from aldehyde

aldol
coined by contraction of aldehyde and alcohol

aldolase
derived from aldol and -ase

aldonic acid
derived from aldose and -onic acid

aldose
derived from aldehyde and -ose

aldosterone
$C_{21}H_{28}O_5$, derived from ald(o)-, sterol, and -one

aldoxime
derived from ald(o)- and oxime

aldrin
$C_{12}H_8Cl_6$, named for the German chemist Kurt Alder (1902-1958)

alembic
ultimately derived from *ambix* (Greek:

spouted cup, cap of a still)

alembroth
$(NH_4)_2HgCl_4$, an archaic name for this alchemistic preparation, used as a stimulant, of uncertain medieval origin

alendronic acid
$C_4H_{13}NO_7P_2$, originally coined as a trademark; patterned after etidronic acid

-alene
derived from naphthalene

aleuritic acid
$C_{16}H_{32}O_5$, derived from the genus name *Aleurites* (varnish tree), ultimately from *aleuron* (Greek: flour)

alexandrite (chrysoberyl)
$BeAl_2O_4$, named for the Russian emperor Alexander II (1818-1881) - this mineral was allegedly discovered on the day of this emperor's birth

alexin
derived from *alexein* (Greek: to defend) and -in(e)

Algar-Flynn-Oyamada reaction
named for the 20th century Irish chemists J. Algar and J. P. Flynn, and the 20th century Japanese chemist B. Oyamada

algaroth (powder of Algaroth)
SbOCl, an archaic name for antimony oxychloride, named for the discoverer of this compound's emetic properties, the Italian chemist Vittorio Algarotti (1553-1604)

algicide
derived from alga and -cide

algin (sodium alginate)
derived from *alga* (Latin: seaweed), and -in(e)

alginate
derived from alginic acid

alginic acid
$(C_6H_8O_6)_n$, derived from algin

alicyclic
derived from aliphatic and cyclic

aliphatic
derived from *aleiphar* (Greek: oil, fat), from *aleiphein* (Greek: to smear)

aliquot
derived from *aliquot* (Latin: a few), from *alius* (Latin: other) and *quot* (Latin: how many?)

alitame
$C_{14}H_{25}N_3O_4S$, coined by contraction of L-aspartyl-D-alanine-*N*-(2,2,4,4 tetramethylthi-etan-3-yl)amide

alite
Ca_3SiO_5, derived from the letter A and –lite

alizarin
$C_{14}H_8O_4$, ultimately derived from *al-asara* (Arabic: madder root), and -in(e)

alkahest
an archaic name of obscure etymology; first used by the Swiss-German alchemist and physician Philippus Aureolus Paracelsus (Theophrastus Bombast von Hohenheim) (1493-1541) for an ill-defined iatrochemical remedy; the Flemish alchemist Jan Batista van Helmont (1577-1644) used the same name or *ignis aqua* (New Latin: fire's water) for the ill-defined and disputed hypothetical universal solvent and universal remedy

alkali
derived from *al-qili* (Arabic: ash, the roasted)

alkali metal
derived from alkali and metal

alkaline air
NH_3, an archaic name for ammonia – referring to this compound's basicity

alkaloid
derived from alkali and -oid – referring to these compounds' basicity

alkane
derived from alkyl and -an(e)

alkannin
$C_{16}H_{16}O_5$, derived from the genus name *Alkanna* (herbs), ultimately from *al-hinna* (Arabic: henna), and -in(e)

alkazid process
derived from alkaline, az(o)-, and -id(e)

alkene
derived from alkyl and –ene

alkyl
derived from alkali and -yl – referring to a perceived chemical similarity between the ethyl radical and the alkali metals

alkyne
derived from alkyl and -yne

allanite (orthite)
$Ca(Ce,La)(Al,Fe,Cr.V)_3(Si_2O_7)(SiO_4)(O,OH)_2$, named for the British mineralogist Thomas Allan (1777-1833)

Allan-Robinson reaction
named for the British chemists J. Allan and Sir Robert Robinson (1885-1975)

allantoin
$C_4H_6N_4O_3$, derived from *allantois* (French: fetal membrane), ultimately from *allantoeides* (Greek: sausage shaped), from *allas* (Greek: sausage), and -in(e)

alleghanyite
$Mn_5(SiO_4)_2(OH)_2$, named after this mineral's locality Bald Knob, Alleghany County, NC, USA

allemontite (stibarsen)
(As,Sb), named after this mineral's locality Allemont, Département Isère, France

allene
C_3H_4, coined by contraction of allylene

allethrin
$C_{19}H_{26}O_3$, coined by contraction of allyl and pyrethrin

alli-
derived from the genus name *Allium* (garlic), from *allium* (Latin: garlic), ultimately probably from *aluka* (Sanskrit: edible root of the aroid plant *Amorphophallus campanulatus*)

allicin
$C_6H_{10}OS_2$, derived from alli- and -icin(e)

Allihn condenser
named for the German glassblower Felix Richard Allihn (1854-1915)

alliin
$C_6H_{11}NO_3S$, derived from alli- and -in(e)

allitol
$C_6H_{14}O_6$, derived from allose and –itol

all(o)-
derived from *allos* (Greek: other, different)

allochromatic
derived from all(o)- and -chromatic

allocinnamic acid
$C_9H_8O_2$, derived from all(o)- and cinnamic acid – referring to this acid's stereoisomerism with cinnamic acid

allocryptopine
$C_{21}H_{23}NO_5$, derived from all(o)- and cryptopine

allogon
derived from all(o)- and -gon

alloisoleucine
$C_6H_{13}NO_2$, derived from all(o)- and isoleucine – referring to this amino acid's stereoisomerism with isoleucin

alloisomerism (*cis-trans* isomerism)
derived from all(o)- and isomerism

allomerism
derived from all(o)- and -mer

allomone
coined by contraction of all(o)- and hormone

allomorph
derived from all(o)- and -morph

allophane
$Al_2O_3(SiO_2)_{1.3–2}\cdot(H_2O)_{2.5–3}$, derived from *allophanes* (Greek: appearing otherwise) – referring to the change this mineral undergoes in a blowpipe flame

allophanic acid
$C_2H_4N_2O_3$, derived from *allophanes* (Greek: appearing otherwise) – referring to the fact that only derivatives of this acid are isolable, not the free acid

allopurinol
$C_5H_4N_4O$, derived from all(o)- and purinol

allose
$C_6H_{12}O_6$, derived from all(o)- and -ose – referring to this monosaccharide's stereoisomerism with glucose

allosteric
derived from all(o)- and steric

allothreonine
$C_4H_9NO_3$, derived from all(o)- and threonine – referring to this amino acid's stereoisomerism with threonine

allotrope, allotropy
derived from allo- and -trope

alloxan
$C_4H_2N_2O_4$, derived (with contraction) from allantoin, oxalic acid, and -an(e)

alloxanthin
$C_{40}H_{52}O_2$, derived from all(o)-, xanth(o)-, and -in(e)

alloxanthine
$C_9H_4N_4O_2$, derived from all(o)- and xanthine

alloxantin
$C_8H_6N_4O_8$, derived from alloxan and -in(e)

alloxazine
$C_{10}H_6N_4O_2$, derived from alloxan and azine

allozyme
coined by contraction of alloenzyme

alloy
derived from *alligare* (Latin: to bind together), from *ad* (Latin: to) and *ligare* (Latin: to bind)

allyl
C_3H_5–, derived from *allium* (Latin: garlic) and -yl – referring to first discovery of allyl substituted compounds in garlic

allylene
C_3H_4, derived from allyl and -ene – referring to this alkene's formal formation by hydrogen abstraction from the allyl radical

allysine
$C_6H_{11}NO_3$, derived from all(o)- and lysine – referring to this amino acid's stereoisomerism with lysine

almandine (almandite)
$Fe_3Al_2(SiO_4)_3$, named after the ancient town of Alabanda, Caria, Asia Minor

aloe emodin (rhabarberone)
$C_{15}H_{10}O_5$, derived from aloe and the specific epithet of the species name *Rheum emodi* (Gilgiti rhubarb), from *Emodos* (Greek: Himalaya), and -in(e)

aloin
$C_{21}H_{22}O_9$, derived from aloe, ultimately from *agaru* (Sanskrit: agalloch), and -in(e)

alston-
derived from the genus name *Alstonia* (tropical trees), after the Scottish botanist Charles Alston (1683-1760)

alstonidine
$C_{22}H_{24}N_2O_4$, derived from alston- and -idin(e)

alstonine
$C_{21}H_{20}N_2O_3$, derived from alston- and -in(e)

alstophyllan
$C_{21}H_{26}N_2O$, derived from alstophylline and -an(e)

Alstophylline
$C_{22}H_{26}N_2O_3$, derived (with contraction) from the species name *Alstonia macrophylla* (batino, devil tree), from alsto- and phyll(o)-, and -an(e)

altaite
PbTe, named after this mineral's locality Ziryanovsk, Altai Mountains, Russia

altritol (talitol)
$C_6H_{14}O_6$, derived from altrose and -itol

altrose
$C_6H_{12}O_6$, derived from *alter* (Latin: other) and -ose – referring to this monosaccharide's stereoisomerism with glucose

alumane (alane)
AlH_3, derived (with contraction) from aluminum and -an(e)

alumina
Al_2O_3, derived from alum and -a

aluminon
$C_{22}H_{23}N_3O_9$, derived from aluminum and [1]-on – referring to this compound's usefulness as an analytical reagent for aluminium

aluminum
Al, derived from alum and -(i)um

alunite
$KAl_3(SO_4)_2(OH)_6$, derived from *alun* (French: alum) and –ite

alvite (hoegtveitite)
(Zr,Hf)SiO_4, named after this mineral's locality Alve, Norway

amabiline (necine)
$C_{15}H_{25}NO_4$, derived from the specific epithet of the species name *Cynoglossum amabile* (hound's tongue, Chinese forget-

me-not), from *amabilis* (Latin: loveable), from *amare* (Latin: to love), and -in(e)

Amadori rearrangement
named for the Italian chemist Mario Amadori (1886-1941)

amalgam
alloys of mercury, ultimately derived from *malagma* (Greek: soft mass), from *malassein* (Greek: to soften)

amanitin (amatoxin)
derived from the genus name *Amanita* (agarics), from *amanitai* (Greek: a kind of fungus), and -in(e)

amaranth
$C_{20}H_{11}N_2Na_3O_{10}S_3$, derived from *amaranton* (Greek: a mythical flower which never fades), from *amarantos* (Greek: immortal), also influenced by *anthos* (Greek: flower)

amaro-
derived from *amarus* (Latin: bitter)

amarogentin
$C_{29}H_{30}O_{13}$, derived from amaro-, gentian, and -in(e)

amarolide
$C_{20}H_{28}O_6$, derived from amaro- and -olide – referring to this lactone's bitter taste

amastatin
$C_{21}H_{38}N_4O_8$, coined by contraction of aminopeptidase and -statin

amatol
coined by contraction of the names of the constituents ammonium nitrate and trinitrotoluene

amazonite (microcline)
$KAlSi_3O_8$, named after this mineral's locality Amazon River, South America

amber (succinite)
$(C_{10}H_{16}O)_n$, ultimately derived from *anbar* (Arabic: ambergris)

ambergris
derived from *ambre gris* (Medieval French: ambergris, literally gray amber)

Amberlite
a trademark, derived from amber and -lite – probably referring to this solid ion exchanger's amber color

ambi-
derived from *ambo* (Latin: both)

ambident
derived from ambi- and -dent

ambient
from *ambiens* (Latin: surrounding), from *ambire* (Latin: to move around), from ambi- and *ire* (Latin: to go, to walk)

amblygonite
$(Li,Na)AlPO_4(F,OH)$, derived from *amblygonios* (Greek: blunt-angled) and -ite – referring to this mineral's blunt crystal shape

ambo-
derived from *ambo* (Latin: both)

ambrein
$C_{30}H_{52}O$, derived from *ambra* (Latin: ambergris) and -in(e)

ambrettolide
$C_{16}H_{28}O_2$, derived from ambrette (*Abelmoschus moschatus* L.) and -olide

ambrosane (pseudoguaiane)
$C_{15}H_{28}$, derived from ambros(o)- and -an(e)

ambrosin
$C_{15}H_{18}O_3$, derived from ambros(o)- and

-in(e)

ambros(o)-
derived from the genus name *Ambrosia* (ambrosia), from ambrosia, the Greek gods' immortalizing food, from *ambrotos* (Greek: immortal), from a(n)- and *brotos* (Greek: mortal)

-am(e)
an arbitrary suffix derived from lactam

amebicide
derived from amoeba and -cide

americate
AmO_5^{3-}, derived from americium and -ate

americium
Am, named for America, ultimately from the latinized name Americus Vespucius of the Italian navigator Amerigo Vespucci (1451-1512)

Ames test
named for the US biochemist Bruce N. Ames (born 1928)

amethyst
SiO_2, derived from *amethystos* (Greek: remedy against drunkenness), from a(n)- and *methystos* (Greek: drunk), from *methyskein* (Greek: to make drunk), from *methyein* (Greek: to be drunk), from *methy* (Greek: wine)

-amic acid
derived from amine and acid

amicetin
$C_{29}H_{42}N_6O_9$, a name coined for unstated reasons

amicoumacin
coined by contraction of amine, coumarane, and -acin(e)

amicron
derived from a(n)-, microscope, and [3]-on – referring to the invisibility of these particles in an optical microscope

amidase
derived from amide and -ase

amide
derived from ammonia and -ide

amidine
derived from amide and imine

amidinomycin
$C_9H_{18}N_4O$, derived from amidine and -mycin

amidogen
NH_2, derived from amide and –gen

amidoxime
derived from amide and oxime

amidrazone
coined by contraction of amide and hydrazone

amikacin
$C_{22}H_{43}N_5O_{13}$, coined by contraction of aminoglycoside and kanamycin

aminal
derived from amine and acetal

aminase
derived from amine and -ase

amine
derived from ammonia and -in(e)

amino-
derived from amine

aminoffite
$Ca_3Be_2Si_3O_{10}·2H_2O$, named for the Swedish

crystallographer and mineralogist Gregori Aminoff (1893-1947)

ammelide
$C_6H_9N_9O_3$, an arbitrary name

ammeline
$C_3H_5N_5O$, an arbitrary name

ammine
derived from ammonia and -in(e)

ammoidin (xanthotoxin)
$C_{12}H_8O_4$, derived from the genus name *Ammi* (annual herbs), from *ammi* (Greek: name of a plant), and -idin(e)

ammonia (azane)
NH_3, derived from ammoniac

ammoniac
NH_4Cl, derived from *sal ammoniacum* (Latin: salt of Ammon), named for the Old Egyptian god Amon (Old Egyptian: the hidden one)

ammoniacum (gum ammoniac)
derived from the specific epithet of the species name *Dorema ammoniacum* D. Don (ammoniacum), from *dorema* (Greek: gift) and gum ammoniac

ammonium
NH_4^+, derived from ammonia and -ium

ammono-
a prefix derived from ammonia

ammoxidation
derived (with contraction) from ammonia and oxidation

Amontons' law (Charles' law)
named for the French physicist Guillaume Amontons (1663-1705)

amorphous
derived from a(n)- and *morphe* (Greek: shape)

amosamine
$C_8H_{17}NO_4$, a name coined for unstated reasons

amosite (grünerite)
$(Al,Fe,Mg)_2[(Al,Si)_4O_{11}]_2(OH)_2$, derived from the acronym AMOS (Asbestos Mine of South Africa) and -ite

AMP
$C_{10}H_{14}N_5O_7P$, an abbreviation for adenosine 5'-monophosphate

ampel(o)-
derived from *ampelos* (Greek: grapevine)

ampelopsin
$C_{15}H_{12}O_8$, derived from the genus name *Ampelopsis* (vines), from ampel(o)- and *opsis* (Greek: appearance, resemblance), from *ops* (Greek: eye), and -in(e)

amphetamine
$C_9H_{13}N$, coined by contraction of α-methyl(phenylethylamine)

amphi-
derived from *amphi* (Greek: around, on both sides)

amphibole (hornblende)
derived from *amphibolos* (Greek: ambiguous, doubtful), from *amphi* (Greek: around, on both sides) and *ballein* (Greek: to throw) – referring to the many varieties of this mineral

amphibolite
a rock species, derived from amphibole and –ite

amphiphilic
derived from amphi- and -philic

amphiprotic
derived from amphi- and proton

amph(o)-
derived from *ampho* (Greek: both, of both kinds)

ampholysis
derived from amph(o)- and lysis

ampholyte
derived from ampholysis

amphomycin (glumamycin)
$C_{58}H_{91}N_{13}O_{20}$, derived from amph(o)- and -mycin

amphoteric
derived from *amphoteros* (Greek: in both ways)

amphotericin
derived from amphoteric and -in(e)

amsonic acid
$C_{14}H_{14}N_2O_6S_2$, coined by contraction of 2,2'-(ethene-1,2-diyl)bis(5-aminobenzenesulfonic acid)

amu
an abbreviation for atomic mass unit

amygdalin
$C_{20}H_{27}NO_{11}$, derived from amygdal(o)- and -in(e)

amygdal(o)-
ultimately derived from *amygdale* (Greek: almond)

amygdalose
$C_{12}H_{22}O_{11}$, derived from amygdal(o)- and -ose

amyl
$C_5H_{11}-$, derived from amyl alcohol

amyl alcohol
$C_5H_{12}O$, derived from amyl(o)-, -yl, and alcohol – referring to the first observation of this alcohol as a starch fermentation product

amylase
derived from amyl(o)- and -ase

amylene
C_5H_{10}, derived from amyl and -ene – referring to the formal formation of this alkene by hydrogen abstraction from the amyl radical C_5H_{11}

amylin
derived from amyloid and -in(e)

amyl(o)-
derived from *amylon* (Greek: starch, literally not ground)

amyloid
literally starch-like – referring to the color change seen upon staining of this peptide with iodine

amylopectin
derived from amyl(o)- and pectin

amylose
derived from amyl(o)- and –ose

amyrin
$C_{30}H_{50}O$, derived from the genus name *Amyris* (trees, shrubs), from *amyros* (Greek: not watery), and -in(e)

-an
derived from glucan

a(n)-
derived from *a(n)-* (Greek: non-)

ana-
derived from *ana* (Greek: up)

anabasine
$C_{10}H_{14}N_2$, derived from the genus name *Anabasis* (anabasis), from *anabasis* (Greek: a plant), from *anabasis* (Greek: act of going up), from *anabainein* (Greek: to go up), and -in(e)

anabolism
derived from *anabole* (Greek: anything thrown up), from *anaballein* (Greek: to throw up, to lift), from ana- and *ballein* (Greek: to throw)

anabsinthin
$C_{30}H_{40}O_6$, derived from analogous and absinthin – referring to this compound's isomerism with absinthin

anacardic acid
derived from the genus name *Anacardium* (trees), from ana- and *kardia* (Greek: heart) – referring to the heart-like shape of the top of these trees' fruit stem

anaerobic
derived from a(n)-, *aer* (Greek: air), and *bios* (Greek: life)

anagyrine
$C_{15}H_{20}N_2O$, derived from the genus name *Anagyris* (shrubs), from *anagyros* (Greek: bean trefoil), from ana- and gyr(o)-, and -in(e)

analcime (analcite)
$NaAlSi_2O_6{\cdot}H_2O$, derived from *analkimos* (Greek: weak), from a(n)- and *alke* (Greek: strength) – referring to the weak electrostatic charge caused by rubbing of this mineral

analysis
derived from ana- and lysis

anandamide
$C_{22}H_{37}NO_2$, derived from *ananda* (Sanskrit: bliss) and amide – referring to this compound's psychopharmacological properties

anaphoresis
coined by contraction of ana- and electrophoresis

anapterin
$C_9H_{11}N_5O_3$, derived from ana- and pterin

anatabine
$C_{10}H_{12}N_2$, derived from anabasine, *Tabak* (German: tobacco), and -in(e)

anatase
TiO_2, derived from *anatasis* (Greek: elongation), from *teinein* (Greek: to stretch) – referring to this mineral's crystal shape, resembling stretched-out octahedra

anchimeric
derived from *anchi* (Greek: near) and -meric

ancrod
coined by contraction of the species name *Agkistrodon rhodostoma* Boie (Malayan pit-viper), (irregularly) from *ankistron* (Greek: fishhook), -odon, rhod(o)-, and *stoma* (Greek: mouth)

ancylite
$Sr_3Ce_4(CO_3)_7(OH){\cdot}3H_2O$, derived from ankyl(o)- and -ite – referring to the rounded and distorted character of this mineral's crystals

-and
derived from *-andus* (Latin: to be acted upon)

andalusite
$Al_2(O)SiO_4$, named after this mineral's locality, the province of Andalusia, Spain

Anderson-Evans ion
$EM_6O_{24}^{n-8}$ (n = valency of E), named for the 20th century British chemist J. S. Anderson and the 20th century US chemist Howard T. Evans, Jr.

andesine
$(Na,Ca)(Si,Al)_4O_8$, named after this mineral's locality, the Andes Mountains, South America

andesite
a rock species, named after the Andes Mountains, South America

andorite
$PbAgSb_3S_6$, named for the Hungarian amateur mineralogist Andor von Semsey (1833-1923)

andradite
$Ca_3Fe_2(SiO_4)_3$, named for the Brazilian mineralogist José Bonifácio Andrade e Silva (1763-1838)

Andresen's acid
$C_{10}H_9NO_6S_2$, named for the German chemist Momme Andresen (1857-1951)

andr(o)-
derived from *aner* (Greek: man = male person, stamen)

androgamone
derived from andr(o)- and gamone

andrographolide
$C_{20}H_{30}O_5$, derived from the genus name *Andrographis* (creat, kariyat), from andr(o)- and *graphis* (Greek: stylus, pencil, brush), from *graphein* (Greek: to write) – referring to the shape of the stamens – and -olide

androstane
$C_{19}H_{32}$, derived from andr(o)-, sterol, and -an(e)

androsterone
$C_{19}H_{30}O_2$, derived from andr(o)-, sterol, and -one

androtermone
derived from andr(o)- and termone

Andrussow process
named for the Latvian-German chemist Leonid Andrussow (born 1896)

-an(e)
a suffix derived from *-anus* (Latin suffix denoting a relationship)

anemonin
$C_{10}H_8O_4$, derived from the genus name *Anemone* (anemone), from *anemone* (Greek: anemone), of Semitic origin, influenced by *anemos* (Greek: wind), and -in(e)

anethole
$C_{10}H_{12}O$, derived from *anethum* (Latin: dill), from *anethon* (Greek: dill), and -ol(e)

aneurin (thiamine, vitamin B_1)
$C_{12}H_{17}ClN_4OS$, derived (with contraction) from a(n)-, polyneuropathy, and vitamin – referring to this vitamin's ability to prevent beri-beri

angelic(a)-
derived from the genus name *Angelica* (angelic), from *angelicus* (Late Latin: angelic) – referring to these plants' medicinal use

angelic acid
$C_5H_8O_2$, derived from angelic(a)-

angelica lactone
$C_5H_6O_2$, derived from *Angelicalacton* (German: angelica lactone), from *Angelicasäure* (German: angelic acid) and *Lacton* (German: lactone)

Angeli-Rimini reaction
named for the Italian chemists Angelo Angeli (1864-1931) and E. Rimini (1874-1917)

Angeli's salt
$Na_2N_2O_3$, named for the Italian chemist Angelo Angeli (1864-1931)

angi(o)-
derived from *angeion*, diminutive of *angos* (Greek: vessel)

angiogenin
derived from angi(o)-, -gen, and -in(e)

angiostatin
derived from angi(o)- and -statin

angiotensin (angiotonin)
coined by contraction of angiotonin and hypertensin

angiotensinogen (hypertensinogen, pro-angiotensin)
derived from angiotensin and -gen

angiotonin (angiotensin)
derived from angi(o)- and -tonin

angiotropin
derived from angi(o)- and -tropin

anglesite
$PbSO_4$, named after this mineral's locality Anglesey Island, Wales, UK

anglium
Ar, an unsuccessfully suggested name for argon, derived from *Anglia* (New Latin: England) and -ium

anglohelvetium
At, an unsuccessfully suggested name for astatine, derived from anglo-, *Helvetia* (New Latin: Switzerland), and –ium

Ångström
Å, a unit named for the Swedish physicist Anders Jonas Ångström (1814-1874)

angularium
Ge, a jocularly suggested name for germanium, derived from angular (in the sense of causing debate) and –ium

anhalamine
$C_{11}H_{15}NO_3$, derived from anhal(o)- and amine

anhal(o)-
derived from the genus name *Anhalonium* (cacti), from a(n)- and *areola* (New Latin: areole), from *halonion*, diminutive of *halon* (Greek: threshing floor) – referring to a perceived lack of areoles in these cacti

anhalonidine
$C_{12}H_{17}NO_3$, derived from anhal(o)- and -idin(e)

anhalonine
$C_{12}H_{15}NO_3$, derived from anhal(o)- and -in(e)

anhydride
derived from a(n)-, *hydor* (Greek: water), and [1]-ide

anhydrite
$CaSO_4$, derived from a(n)-, hydr(o)-, and -ite – referring to the difference between this mineral and the water-containing gypsum

anhydro-
derived from anhydrous

anhydrous
derived from a(n)- and *hydor* (Greek: water)

anil
(an aromatic Schiff base), derived from

aniline

-anilic acid
derived from aniline

anilide
derived from aniline; patterned after amide

aniline
C_6H_7N, derived from *al-nil* (Arabic: indigo plant), ultimately from *nila* (Sanskrit: dark blue) – referring to the fact that aniline was first obtained by degradation of indigo

aniline mustard
$C_{10}H_{13}Cl_2N$, derived from aniline and mustard

anilite
Cu_7S_4, named after this mineral's locality Ani mine, Akita Prefecture, Japan

anion
derived from *anion* (Greek: going up), from *anienai* (Greek: to go up), from ana- and *ienai* (Greek: to go)

anisaldehyde
$C_8H_8O_2$, ultimately derived from *anison* (Greek: anise, dill)

anisic acid
$C_8H_8O_3$, ultimately derived from *anison* (Greek: anise, dill)

anisidine
C_7H_9NO, derived from anisole and -idin(e)

aniso-
derived from a(n)- and is(o)-

anisole
C_7H_8O, ultimately derived from *anison* (Greek: anise, dill) and -ol(e)

anisomycin
$C_{14}H_{19}NO_4$, coined by contraction of anisole and -mycin

anisotropic
derived from a(n)-, iso-, and *trope* (Greek: turn)

anisoyl
$C_8H_7O_2-$, derived from anisic acid and -oyl

ankerite (brown spar)
$CaFe(CO_3)_2$, named for the Austrian mineralogist Mathias Joseph Anker (1771-1843)

ankyl(o)-
derived from *ankylos* (Greek: crooked)

annabergite
$(Ni,Co)_3(AsO_4)_2 \cdot 8H_2O$, named after this mineral's locality Annaberg, Germany

annatto (orlean)
derived from *annoto* (Galibi: annatto tree, *Bixa orellana*)

annelation (annulation)
derived from *anulus*, diminutive of *anus* (Latin: ring)

annite
$KFe_3(Si_3Al)O_{10}(OH,F)_2$, named after this mineral's locality Cape Ann, MA, USA

annotinine
$C_{16}H_{21}NO_3$, derived from the specific epithet of the species name *Lycopodium annotinum* L. (stiff clubmoss), from *annotinus* (Latin: one year old), and -in(e)

annulene
derived from *anulus*, diminutive of *anus* (Latin: ring) and –ene

anode
derived from ana- and *hodos* (Greek: road)

anodyne
derived from a(n)- and *odyne* (Greek: pain)

anolyte
coined by contraction of anode and electrolyte

anomer
derived from *ano* (Greek: upward, above) and -mer

anorthite
$CaAl_2Si_2O_8$, derived from anortho(o)- and -ite – referring to this mineral's oblique crystals

anorth(o)-
derived from *anorthos* (Greek: not upright)

anorthoclase
$(Na,K)AlSi_3O_8$, derived from anorth(o)- and -clase – referring to the oblique fracture of this mineral's crystals

anorthosite
a rock species, derived from anorth(o)- and -ite

anot
$C_8H_9N_3O_3$, an abbreviation for 3-amino-5-nitro-*o*-toluamide

ansa compounds
derived from *ansa* (Latin: handle)

ansamycin
derived from *ansa* (Latin: handle) and -mycin

anserine
$C_{10}H_{16}N_4O_3$, derived from *anser* (Latin: goose) and -in(e)

antagonist
derived from ant(i)- and agonist

antara-
derived either from ant(i)- and *arada* (Greek: line, row) or from *antara* (Sanskrit: the other)

antarafacial
derived from antara- and face

ante-
derived from *ante* (Latin: before)

anthanthrene
$C_{22}H_{12}$, derived from [1]anthr(a)- and -anthrene

anthelminthic
derived from ant(i)- and *helmis* (Greek: worm)

antheridiol
$C_{29}H_{42}O_5$, derived from antheridium, from *antheros* (Greek: flowery), and –ol

anth(o)-
derived from *anthos* (Greek: flower)

anthocyan
derived from anth(o)- and cyan(o)-

anthocyanidin
derived from anthocyan and -idin(e)

anthocyanin
derived from anthocyan and -in(e)

anthophyllite
$(Mg,Fe)_7Si_8O_{22}(OH)_2$, derived from *anthophyllum* (New Latin: clove) and -ite – referring to this mineral's clove brown color

[1]anthr(a)-
derived from anthracene

anthr(a)-
derived from *anthrax* (Greek: coal)

[3]**anthr(a)-**
referring to carbon, derived from [2]anthr(a)-

anthracene
$C_{14}H_{10}$, derived from *anthrax* (Greek: coal) and -ene

anthracite
derived from *anthrakites* (Greek: anthracite), from *anthrax* (Greek: coal)

anthracycline
derived (with contraction) from [1]anthr(a)- and tetracycline

anthragallol (anthragallic acid)
$C_{14}H_8O_5$, derived from [1]anthr(a)-, gallic acid, and -ol

anthramycin
$C_{16}H_{17}N_3O_4$, derived (with contraction) from anthranilic acid and -mycin

anthranilic acid
$C_7H_7NO_2$, derived from [3]anthr(a)- and -anilic acid – referring to the presence of an aniline skeleton with an additional carbon atom

anthranol
$C_{14}H_{10}O$, coined by contraction of anthracenol

anthraquinone
$C_{14}H_8O_2$, derived from [1]anthr(a)- and quinone

-anthrene
derived (with contraction) from anthracene

anthrone
$C_{14}H_{10}O$, coined by contraction of anthracenone

anthryl
$C_{14}H_9-$, derived from anthacene and –yl

ant(i)-
derived from *anti* (Greek: against, opposite)

antiaromatic
derived from ant(i)- and aromatic

antibiotic
derived from ant(i)- and bio-

anti-Bredt compound
derived from ant(i)- and Bredt's rule

antidiuretin (adiuretin, vasopressin)
$C_{46}H_{64}N_{14}O_{12}S_2$, derived from ant(i)-, diuresis, and -in(e)

antigen
derived (with contraction) from antibody and -gen

anticlinal
derived from ant(i)- and *klinein* (Greek: to lean)

antigorite
$Mg_3Si_2O_5(OH)_4$, named after this mineral's locality Antigorio Valley, Italy

anti-Markovnikov addition
derived from ant(i)- and Markovnikov's rule

antimonite (stibnite)
Sb_2S_3, derived from antimony and -ite

antimony
Sb, derived from *antimonium* (Latin: antimony), ultimately from *ithmid* (Arabic: antimony)

antimycin
coined by contraction of antibiotic and -mycin

antimycotic
derived from ant(i)- and mycosis

antipain
$C_{27}H_{44}N_{10}O_6$, derived (with contraction) from ant(i)- and papain – referring to this peptide's inhibition of papain

antiperiplanar
derived from ant(i)-, peri-, and planar

antiprismo-
derived from antiprism

anti-Stokes line
derived from ant(i)- and Stokes' rule

antlerite
$Cu_3(OH)_4SO_4$, named after this mineral's locality Antler mine, Mojave County, AZ, USA

ANTU
$C_{11}H_{10}N_2S$, an abbreviation for α-naphthylthiourea

AO
an abbreviation for atomic orbital

AOX
an abbreviation for adsorbable organic halides, derived from the letter X in the sense of element symbol for halogen

ap-
derived from antiperiplanar

apamin
$C_{79}H_{131}N_{31}O_{24}S_4$, derived from api- and amine

apatite
$Ca_5(PO_4)_3(OH,Cl,F)$, derived from *apate* (Greek: deceit) and -ite – referring the fact that this mineral is readily mistaken for others such as beryl and tourmaline

-aphene
derived by anagrammatical contraction of phenanthrene

aphicide
derived from aphid and -cide

aphidicolin
$C_{20}H_{34}O_4$, derived from the specific epithet of the microbiological species name *Cephalosporium aphidicola* Petch, from *aphidicolus* (New Latin: aphid-breeding), and -in(e)

aphthitalite (aphthalose, glaserite, Vesuvian salt)
$(K,Na)_3Na(SO_4)_2$, derived from *aphthitos* (Greek: indestructible), from *phthiein* (Greek: to waste away), hal(o)-, and -ite – referring to this mineral's stability in air

aphylline
$C_{15}H_{24}N_2O$, derived (with contraction) from the species name *Anabasis aphylla* L. (anabasis), from *anabasis* (Greek: a plant), from *anabasis* (Greek: act of going up), from *anabainein* (Greek: to go up), a(n)-, and phyll(o)-, and -in(e)

api-
derived from *apium* (Latin: parsley, celery), ultimately from *apis* (Latin: bee)

apical, apico-
derived from apex

apigenin
$C_{15}H_{10}O_5$, derived from api-, -gen, and -in(e)

apiin
$C_{26}H_{28}O_{14}$, derived from api- and -in(e)

apiole
$C_{12}H_{14}O_4$, derived from api- and -ol(e)

apiose
$C_5H_{10}O_5$, derived from api- and -ose

aplasmomycin
$C_{40}H_{60}BNaO_{14}$, derived from a(n)-, the protozoan genus name *Plasmodium*, from *plasma* (Greek: form, mold), from *plassein* (Greek: to mold), and *-odes* (Greek: thing that resembles), and -mycin – referring to the specific antiprotozoan activity of this antibiotic

apo-
a prefix derived from *apo* (Greek: off, away from)

apoatropine
$C_{17}H_{21}NO_2$, derived from apo- and atropine – referring to the formal loss of one molecule of water relative to atropine

apocamphor
$C_9H_{14}O$, derived from apo- and camphor – referring to the formal loss of one methyl group relative to camphor

apocodeine
$C_{18}H_{19}NO_2$, derived from apo- and codeine – referring to the formal loss of one molecule of water relative to codeine

apocynin
$C_9H_{10}O_3$, derived from the genus name *Apocynum* (dogbanes), ultimately from *apokynon* (Greek: dogbane), from apo- and *kyon, kynos* (Greek: dog), and -in(e)

apoenzyme
derived from apo- and enzyme

apoferritin
derived from apo- and ferritin

apomorphine
$C_{17}H_{17}NO_2$, derived from apo- and morphine – referring to the formal loss of one molecule of water relative to morphine

apopain
derived (with contraction) from apoptosis and -ain(e)

apophyllite
$KCa_4(Si_4O_{10})_2(F,OH)\cdot 8H_2O$, derived from *apophylliso* (Greek: it flakes off), and -ite – referring to this mineral's tendency to flake or exfoliate when heated

apoprotein
derived from apo- and protein

apoptolidin
$C_{58}H_{96}O_{21}$, derived from apoptosis, from apo- and *ptosis* (Greek: a fall), from *piptein* (Greek: to fall), -ol, and -idin(e)

aporeine
$C_{18}H_{17}NO_2$, derived from apo-, the genus name *Rheum* (rhubarbs), from *rheon* (Greek: rhubarb), and -in(e)

aporphine
$C_{17}II_{17}N$, coined by contraction of apo-morphine

Appel's salt
$C_2Cl_3NS_2$, named for the German chemist Rolf Appel (born 1921)

aprotic
derived from a(n)- and proton

aprotinin
coined by contraction of antiprotease and -inin(e)

APS
an abbreviation for appearance potential spectroscopy

aptamer
derived from hapt(o)- and –mer

apyrase
coined by contraction of adenylpyrophosphatase

aqua
New Latin name for water, derived from *aqua* (Latin: water)

aqua fortis
HNO_3, concentrated nitric acid, (New Latin: literally strong water) – referring to the fact that this liquid can dissolve the noble metal silver

aquamarine
$Be_3Al_2Si_6O_{18}$, derived from *aqua marina* (Latin: literally sea water) – referring to this mineral's blue color

aqua regia
a mixture of nitric and hydrochloric acids, (New Latin: literally royal water) – referring to the fact that this liquid can dissolve the noble metal gold, the king of metals

aqueous
derived from *aqua* (Latin: water)

araban (arabinan)
derived from arabinose and -an

arabinan (araban)
derived from arabinose and -an

arabinitol (arabitol)
$C_5H_{12}O_5$, derived from arabinose and -itol

arabinose
$C_5H_{10}O_5$, ultimately derived from gum arabic and –ose

arabitol (arabinitol)
$C_5H_{12}O_5$, derived from arabinose and -itol

arabonic acid (arabinonic acid)
$C_5H_{10}O_6$, derived from arabinose and -onic acid

arachid-
derived from the genus name *Arachis* (herbs), ultimately from *arakis* (Greek: chickling vetch), diminutive of *arakos* (Greek: chickling vetch)

arachidic acid (arachic acid)
$C_{20}H_{40}O_2$, derived from arachid-

arachidonic acid
$C_{20}H_{32}O_2$, derived from arachid-

arachno-
derived from *arachnion* (Greek: spider web), from *arachne* (Greek: spider)

aragonite
$CaCO_3$, named after this mineral's locality, the province of Aragon, Spain

aramid
coined by contraction of aromatic polyamide

arborescin
$C_{15}H_{20}O_3$, derived from the specific epithet of the species name *Artemisia arborescens* L. (tree wormwood), from *arborescens* (Latin: becoming like a tree), from *arbor* (Latin: tree), and -in(e)

arborol
derived from *arbor* (Latin: tree) and -ol – referring to the dendrimeric structure of these polymers

arbutin
$C_{12}H_{16}O_7$, derived from the genus name *Arbutus* (strawberry trees), from *arbutus* (Latin: strawberry tree), and -in(e)

Arbuzov-Michaelis reaction (Michaelis-Arbuzov reaction)
named for the Russian chemist Aleksandr

Erminingeldovich Arbuzov (1877-1968) and the German chemist August Karl Arnold Michaelis (1847-1916)

arcanite
K_2SO_4, derived from this mineral's alchemistic name *arcanum duplicatum* (New Latin: double arcanum), from *arcanus* (Latin: closed, secret), from *arca* (Latin: chest), and -ite

archonium
a hypothetical element supposed to occur on the star γ-Argus, named for Argos, a Greek mythical watchful being with many eyes, and -ium

arec(a)-
derived from the genus name *Areca* (betel palms), from *areca* (Portuguese: betel palm), from *atekka*, *atakka* (Malayalam: betel palm)

arecaidine
$C_7H_{11}NO_2$, derived from arec(a)- and -idin(e)

arecoline
$C_8H_{13}NO_2$, derived from arec(a)-, -ol, and -in(e)

arene
derived (with contraction) from aromatic and -ene

Arens-van Dorp synthesis
named for the 20th century Dutch chemists J. F. Arens and D. A. van Dorp

arfvedsonite
$Fe_5Na_3H_2(SiO_3)_8$, named for the Swedish chemist Johan August Arfvedson (1792-1841)

argentite
Ag_2S, derived from argent(o)- and –ite

argent(o)-
derived from argentum

argentum
Ag, New Latin name for silver, derived from *argentum* (Latin: silver)

argill-
derived from *argilla* (Latin: clay)

argillite
a heterogeneous mineral, derived from argill- and -ite

arginase
derived from arginine and -ase

arginine
$C_6H_{14}N_4O_2$, derived from argentum and -in(e) – referring to this amino acid's silver salts of low solubility

argon
Ar, derived from *argos* (Greek: idle), from *a(n)-* (Greek: non-) and *ergon* (Greek: work), and [1]-on

argyr(o)-
derived from *argyros* (Greek: silver)

argyrodite
Ag_8GeS_6, derived from *argyrodes* (Greek: silver-containing) and -ite

-aric acid
derived from glucaric acid

aricine
$C_{22}H_{26}N_2O_4$, derived from the harbor town of Arica, Peru (now Chile) and -in(e) – referring to the shipping from this port to Europe of the Peruvian cinchona bark from which this alkaloid was isolated

aridium
Fe, a name suggested by the Swedish

chemist Clemens Ullgren (1811-1868) for a supposedly new element later shown to be iron, derived from Ares, the Ancient Greek god of war, *eidos* (Greek: form, shape), and -ium – referring to this hypothetical element's similarity to iron

aristolane (ferulane)
$C_{15}H_{26}$, derived from the genus name *Aristolochia* (birthwort), from *aristos* (Greek: best) and *locheia* (Greek: childbirth), and -an(e)

aristolochic acid
$C_{17}H_{11}NO_7$, derived from the genus name *Aristolochia* (birthwort), from *aristos* (Greek: best) and *locheia* (Greek: childbirth)

van Arkel and de Boer process
named for the Dutch chemists Anton Eduard van Arkel (1893-1976) and Jan Hendrik de Boer (1899-1971)

armalcolite
$(Mg,Fe)Ti_2O_5$, coined by contraction of the names of the US astronauts Neil Alden Armstrong (born 1930), Edwin Eugene Aldrin (born 1930), and Michael Collins (born 1930)

armepavine
$C_{19}H_{23}NO_3$, derived (with transposition and contraction) from the species name *Papaver armeniacum* L. (Armenian poppy), from *papaver* (Latin: poppy) and *armeniacus* (Latin: Armenian), and -in(e)

armstrongite
$CaZrSi_6O_{15}\cdot 3H_2O$, named for the US astronaut Neil Alden Armstrong (born 1930)

Armstrong's acid
$C_{10}H_8O_6S_2$, named for the British chemist Henry Edward Armstrong (1848-1937)

Arnd's alloy
(Cu,Mg), named for the 20th century German chemist T. Arnd

Arndt-Eistert synthesis
named for the German chemists Fritz Arndt (1885-1969) and Bernd Eistert (1902-1978)

arogenic acid (pretyrosine)
$C_{10}H_{13}NO_5$, derived from aromatic and -gen – referring to this compound's intermediacy in the biosynthesis of aromatic compounds

aromatase
derived from aromatization and -ase

aromatic
in the chemical sense, ultimately derived from *aromatikos* (Greek: fragrant)

aroyl
derived from aromatic acid and -oyl

Arrhenius equation
named for the Swedish chemist Svante August Arrhenius (1859-1927)

arsane (arsine)
AsH_3, derived from arsenic and -an(e)

arsanilic acid (atoxylic acid)
$C_6H_8AsNO_3$, derived from arsenic and aniline

arsenate
AsO_4^{3-}, derived from arsenic and -ate

arsenic
As, derived from *arsenikon* (Greek: yellow orpiment), ultimately from *zarnig* (Syrian: yellow orpiment), of Iranian origin

arsenicum
As, New Latin name for arsenic, derived from arsenic and –um

arsenite
AsO_3^{3-}, derived from arsenic and -ite

arsenolite
As_2O_3, derived from arsenic and -lite

arsenopyrite
FeAsS, derived from arsenic and pyrite

arsine (arsane)
AsH_3, derived from arsenic and -in(e)

arsorane
AsH_5, derived from arsenic and -orane; patterned after phosphorane

arteether
$C_{17}H_{28}O_5$, derived (with contraction) from artemisinin and ether

artemether
$C_{16}H_{26}O_5$, derived (with contraction) from artemisinin and ether

artemis-
derived from the genus name *Artemisia* (mugwort), from *artemisia* (Latin: mugwort), ultimately named after Artemis, the Greek goddess of forests and hills

artemisin
$C_{15}H_{18}O_4$, derived from artemis- and -in(e)

artemisinin
$C_{15}H_{22}O_5$, derived from artemis- and -inin(e)

artesunate
$C_{19}H_{28}O_8$, derived (with contraction) from artemisinin and succinate

arthurite
$Cu(Fe^{III})_2(AsO_4,PO_4,SO_4)_2(O,OH)_2 \cdot 4H_2O$, named for the British mineralogists Sir Arthur Edward Ian Montagu Russell (1878-1964) and Arthur William Gerald Kingsley (1906-1968)

artinite
$Mg_2CO_3(OH)_2 \cdot 3H_2O$, named for the Italian mineralogist Ettore Artini (1866-1928)

arupite
$Ni_3(PO_4)_2 \cdot 8H_2O$, named for the Danish scientist Hans Henning Arup (born 1928)

aryl
derived from arene and -yl

aryne
derived from arene and -yne

asafetida
derived from the specific epithet of the species name *Ferula assafoetida* L. (devil's dung), from *asa* (Iranian: gum), and *foetidus* (Latin: fetid)

asar-
derived from the genus name *Asarum* (snakeroot wildginger), from *asaron* (Greek: hazelwort)

asarinin
$C_{20}H_{18}O_6$, derived from asar- and -inin(e)

asarone
$C_{12}H_{16}O_3$, derived from asar- and -one

asbestos
derived from *asbestos* (Greek: inextinguishable), from a(n)- and *sbestos* (Greek: quenched), from *sbennynai* (Greek: to extinguish) - the name refers to an alchemistic belief that this material, once set afire, would be just as difficult to extinguish as it was to ignite; the modern use of the word asbestos can be traced to the Roman encyclopedist Pliny the Elder (23-79); the original meaning of the Greek word was quicklime

ascaridole
$C_{10}H_{16}O_2$, derived from from ascarid, from *askaris* (Greek: intestinal worm), and -ol(e)

ascharite (szaibelyite)
$MgHBO_3$, named after this mineral's locality Aschersleben, Germany (latinized Ascharia)

asc(o)-
derived from *ascus* (New Latin: bladder)

ascomycin
$C_{43}H_{69}NO_{12}$, derived from the fungal genus name *Ascomyces*, from asc(o)-, and –mycin

ascorbic acid
$C_6H_8O_6$, derived from a(n)- and *scorbutus* (New Latin: scurvy), ultimately from *skyr* (Old Norse: old curdled milk) and *bjugr* (Old Norse: edema) – referring to the ancient belief that scurvy was caused by the sailors' diet of old curdled milk

ascorbigen
derived from ascorbic acid and -gen

-ase
derived from diastase

asiaticoside
$C_{48}H_{78}O_{19}$, derived from the specific epithet of the species name *Centella asiatica* L. (Asiatic pennywort) and -oside

asparaginase
derived from asparagine and -ase

asparagine
$C_4H_8N_2O_3$, derived from asparagus, ultimately from *spargan* (Greek: to swell)

aspartame
$C_{14}H_{18}N_2O_5$, coined by contraction of aspartic acid and glutamic acid

aspartic acid
$C_4H_7NO_4$, derived from asparagus, ultimately from *spargan* (Greek: to swell)

aspergillic acid
$C_{12}H_{20}N_2O_2$, derived from the fungal genus name *Aspergillus*, from *aspergillum* (Latin: sprinkler for holy water), from *aspergere* (Latin: to sprinkle)

asperlicin
$C_{31}H_{29}N_5O_4$, coined by contraction of the fungal species name *Aspergillus alliaceus*, from *aspergillum* (Latin: sprinkler for holy water), from *aspergere* (Latin: to sprinkle) and *alliaceus* (Latin: having the smell of garlic), from alli-, and -in(e)

asperuloside
$C_{18}H_{22}O_{11}$, derived from the genus name *Asperula* (woodruff), from *asperulus*, diminutive of *asper* (Latin: rough), and -oside

asphalt
derived from *asphaltos* (Greek: asphalt) from a(n)- and *sphallein* (Greek: to cause to fall) – referring to the ancient use of asphalt as an adhesive in stone masonry

aspidin
$C_{25}H_{32}O_8$, derived from the genus name *Aspidosperma* (tropical trees), from aspid(o)-, and -in(e)

aspidinol
$C_{12}H_{16}O_4$, derived from the genus name *Aspidosperma* (tropical trees), from aspid(o)-, -in(e), and -ol

aspid(o)-
derived from *aspis* (Greek: shield)

aspidofractinine
$C_{19}H_{24}N_2$, derived (with contraction) from the species name *Aspidosperma refractum*

(a tropical tree), from aspidosperm- and *refractus* (Latin: broken off), and -inin(e)

aspidolite
$NaMg_3AlSi_3O_{10}(OH)_2$, derived from aspid(o)- and -lite – referring to the shield-like appearance of this mineral's crystals

aspidosperm-
derived from the genus name *Aspidosperma* (tropical trees), from aspid(o)- and *sperma* (Greek: seed, germ)

aspidospermidine
$C_{19}H_{26}N_2$, derived aspidosperm- and -idin(e)

aspidospermine
$C_{22}H_{30}N_2O_2$, derived from aspidosperm- and -in(e)

aspirin
$C_9H_8O_4$, derived (with contraction) from acetylspiraeic acid, from spiraeic acid, and -in(e)

ASS
an abbreviation for acetylsalicylic acid

asta(c)-
derived from the genus name *Astacus* (crustaceans), from *astakos* (Greek: lobster) and -in(e)

astacin
$C_{40}H_{48}O_4$, derived from asta(c)- and -in(e)

astatane
HAt, derived from astatine and -an(e)

astatine
At, derived from *astatos* (Greek: unstable), from a(n)- and *statos* (Greek: standing), and -in(e) – referring to this element's radioactivity

astatium
At, New Latin name for astatine, derived from *Astat* (German: astatine) and -ium

astaxanthin
$C_{40}H_{52}O_4$, derived from asta(c)- and xanthin

asterane
$C_{3n}H_{4n}$, derived from astr(o)- and -an(e) – referring to these compounds' star-like shape

asterium
He, name of a hypothetical element later shown to be helium, derived from astr(o)- and -ium

astrak(h)anite (bloedite)
$Na_2Mg(SO_4)_2{\cdot}4H_2O$, named after this mineral's locality Astrakhan, Russia

astr(o)-
derived from *aster, astron* (Greek: star)

astrophyllite
$(K,Na)_3(Fe,Mn)_7Ti_2Si_8(O,OH)_{31}$, derived from astr(o)-, phyll(o)-, and -ite – referring to this mineral's starlike crystal aggregates

asym-
derived from asymmetric

atacamite
$Cu_2Cl(OH)_3$, named after this mineral's locality Atacama Desert, Chile

atactic
derived from *ataktos* (Greek: untidy, disordered)

-ate
derived from *-atus* (Latin suffix denoting function)

Aten's sulfur (Engel's sulfur)
S_6, named for the 19th century Dutch chemist Adrian H. W. Aten

athamantin
$C_{24}H_{30}O_7$, derived from the genus name *Athamanta* (herbs), from *athamantikos* (Greek: bear's wort), after Athamas, son of Aeolus, Greek mythical figure, and -in(e)

atheneite
$(Pd,Hg)_3As$, derived from Pallas Athena, the Greek goddess of war, fertility, arts, and wisdom – referring to this mineral's content of palladium

athenium
Es, an unsuccessfully suggested name for einsteinium, probably derived from Pallas Athena, the Greek goddess of war, fertility, arts, and wisdom, and -ium

atidane
$C_{19}H_{31}N$, derived from atis plant (*Aconitum heterophyllum* Wall.), from *atis* (Hindi: atis plant), and -an(e)

atisane
$C_{20}H_{34}$, derived from atis(o)- and -an(e)

atisine (anthorine)
$C_{22}H_{33}NO_2$, derived from atis(o)- and -in(e)

atis(o)-
derived from atis plant (*Aconitum heterophyllum* Wall.), from *atis* (Hindi: atis plant)

atmosphere
derived from *atmos* (Greek: steam, vapor) and *sphaira* (Greek: sphere)

atom
derived from *atomos* (Greek: indivisible), from a(n)- and *temnein* (Greek: to cut)

ATP
$C_{10}H_{16}N_5O_{13}P_3$, an abbreviation for adenosine 5'-triphosphate

ATPase
an abbreviation for adenosine-5'-triphosphatase

atractyloside
$C_{30}H_{44}K_2O_{16}S_2$, derived from the genus name *Atractylis* (thistles), from *atraktos* (Greek: spindle, arrow), and -oside

atranorin (atranoric acid)
$C_{19}H_{18}O_8$, derived (with contraction) from the species name *Lecanora atra* (Huds.) (a lichen), from *lekane* (Greek: basin) and *hora* (Greek: beauty, grace), and *ater* (Latin: black), and -in(e)

atrolactic acid
$C_9H_{10}O_3$, derived (with contraction) from atropine and lactic acid

atropic acid
$C_9H_8O_2$, derived from atropine

atropine
$C_{17}H_{23}NO_3$, derived from the genus name *Atropa* (herbs), from the mythical figure *tropos* (Greek: the Inevitable, one of the Fates), and -in(e)

atropisomerism
derived from *atropos* (Greek: inevitable) and isomerism

attapulgite (palygorskite)
$(Mg,Al)_2Si_4O_{10}(OH)\cdot 4H_2O$, named after this mineral's locality Attapulgus, GA, USA

aucubin
$C_{15}H_{22}O_9$, derived from the genus name *Aucuba* (shrubs), from *aokuba* (Japanese: aucuba), from *ao* (Japanese: green), *ki*, *ko* (Japanese: tree), and *ba* (Japanese: leaf),

and -in(e)

aufbau
derived from *Aufbau* (German: build-up), from *bauen* (German: to build)

Auger electron spectroscopy
named for the French physicist Pierre V. Auger (1899-1993)

augite (fassaite)
$(Ca,Na)(Al,Fe,Mg)(Si,Al)_2O_6$, ultimately derived from *augites* (Greek: precious stone), from *auge* (Greek: brightness, luster) – referring to this mineral's luster

auramine
$C_{17}H_{21}ClN_2O$, derived from aur(o)- and amine – referring to this dye's bright yellow color

aure(o)-
derived from *aureus* (Latin: golden yellow), from *aurum* (Latin: gold)

aureolic acid (mithramycin)
$C_{52}H_{76}O_{24}$, derived from *aureolus*, diminutive of *aureus* (Latin: golden yellow) – referring to this antibiotic's yellow color

aureothin
$C_{22}H_{23}NO_6$, derived from aure(o)- and the specific epithet of the bacterial species name *Streptomyces thioluteus*, from thi(o)- and *luteus* (Latin: yellow), and -in(e)

aureothricin
$C_9H_{10}N_2O_2S_2$, derived from aure(o)-, an unexplained name fragment, and -in(e) – referring to this antibiotic's yellow color

auri-
derived from aurum

aurichalcite
$(Zn,Cu)_5(CO_3)_2(OH)_6$, derived from *aurichalcum* (Latin: yellow copper ore) and/or *oreichalkos* (Greek: mountain copper), and -ite

aurin (pararosolic acid, rosalic acid)
$C_{19}H_{14}O_3$, derived from aur(o)- and -in(e) – referring to this dye's yellow color

aur(o)-
derived from aurum

aurum
Au, New Latin name for gold, derived from *aurum* (Latin: gold, literally reddish shining metal)

ausonium
Np, an unsuccessfully suggested name for neptunium, derived from *Ausonia* (Latin: poetic name for Italy) and -ium

austenite
(Fe,Fe_3C), named for the British metallurgist Sir William Chandler Roberts-Austen (1843-1902)

austrium
At, an unsuccessfully suggested name for astatine, derived from Austria

aut(o)-
derived from *autos* (Greek: self)

autoclave
derived from aut(o)- and *clavis* (Latin: key), from *claudere* (Latin: to close, to lock)

autunite
$Ca(UO_2)_2(PO_4)_2 \cdot 10H_2O$, named after this mineral's locality Autun, Département Saône-et-Loire, France

Auwers synthesis
named for the German chemist Karl Friedrich von Auwers (1863-1939)

Auwers-Skita rule
named for the German chemist Karl Friedrich von Auwers (1863-1939) and the Austrian-German chemist Aladar Skita (1876-1953)

auxin
derived from aux(o)- and -in(e)

aux(o)-
derived from *auxe* (Greek: increase), from *auxein* (Greek: to increase)

auxochrome
derived from aux(o)- and -chrome

aventurine
SiO_2, derived from *aventur*e (French: chance) – referring to this mineral's chance discovery

avicide
derived from *avis* (Latin: bird) and –cide

avidin
derived from *avidus* (Latin: avid) and -in(e) – referring to this protein's avidity for biotin

avogadrite
$(K,Cs)BF_4$, named for the Italian chemist Lorenzo Romano Amedeo Carlo Avogadro, Conte de Quaregno e Cerreto (1776-1856)

Avogadro's law
named for the Italian chemist Lorenzo Romano Amedeo Carlo Avogadro, Conte de Quaregno e Cerreto (1776-1856)

Avogadro's number (Loschmidt's number)
named for the Italian chemist Lorenzo Romano Amedeo Carlo Avogadro, Conte de Quaregno e Cerreto (1776-1856)

axerophthal (retinal, vitamin A aldehyde)
$C_{20}H_{28}O$, derived from a(n)- and xerophthalmia, from xer(o)-, *ophthalmos* (Greek: eye, eye ball), and -al – referring to this compound's activity against xerophthalmia

axinite
$Ca_2(Mn,Fe,Mg)Al_2BSi_4O_{15}(OH)$, derived from *axine* (Greek: axhead) – referring to this mineral's acute crystal angles

az(a)-
derived from *azote* (French: nitrogen), ultimately derived from a(n)- and *zoein* (Greek: to live)

azadirachtin
$C_{35}H_{44}O_{16}$, derived from the genus name *Azadirachta* (neem tree), from *azad dirakht* (Persian: neem tree, literally free or noble tree), and -in(e)

azafrin
$C_{27}H_{38}O_4$, derived from *azafranillo* (Spanish: safflower), ultimately from *al-zafaran* (Arabic: saffron), and -in(e)

azane (ammonia)
NH_3. derived from az(a)- and -an(e)

azaserine
$C_5H_7N_3O_4$, coined by contraction of *O*-(diazoacetyl)-L-serine

azelaic acid
$C_9H_{16}O_4$, derived from az(o)- and *elaion* (Greek: olive oil) – referring to this compound's formation upon treatment of oleic acid with nitric acid

azeotrope
derived from a(n)-, *zein* (Greek: to boil), and *tropos* (Greek: turn)

-azepam
derived (with contraction) from azepine and lactam

azi-
variant of az(o)-

azide
N_3^-, derived from az(a)- and -ide

azine
derived from hydrazine

azlactone
derived from az(a)- and lactone

az(o)-
derived from *azote* (French: nitrogen), ultimately from a(n)- and *zoein* (Greek: to live)

azolitmin
derived from az(o)-, litmus, and -in(e)

azomycin
$C_3H_3N_3O_2$, derived from az(o)- and -mycin

azoproite
$(Mg,Fe^{II})_2(Fe^{III},Ti,Mg)BO_5$, named after the abbreviation AZOPRO for the Russian equivalent of International Association for the Study of Deep Zones of the Earth's Crust in commemoration of this organization's 1969 meeting at Lake Baikal, Russia

azorubin
$C_{20}H_{12}N_2Na_2O_7S_2$, derived from az(o)-, rubi-, and -in(e) – referring to this dye's azo group and red color

azoxy-
derived from az(a)- and oxy-

azoxystrobin
$C_{22}H_{17}N_3O_5$, derived (with contraction) from az(a)-, oxy-, and strobilurin

azulene
$C_{10}H_8$, derived from *azul* (Spanish: blue) and -ene – referring to this hydrocarbon's blue color

azure
a class of synthetic dyes, derived from azure

azurite
$Cu_3(CO_3)_2(OH)_2$, derived from *azura* (Latin: ultramarine), via Arabic *al-lazaward*, ultimately after the town of Lajward, Turkestan, and -ite – referring to this mineral's blue color

B

Babbitt metal
(Sn,Cu,Pb,Sb), named for the US metallurgist Isaac Babbitt (1799-1862)

Babo funnel
named for the German chemist L. H. Freiherr von Babo (1818-1899)

babingtonite
$Ca_2(Fe^{II},Mn)Fe^{III}Si_5O_{14}(OH)$, named for the Irish physicist and mineralogist William Babington (1757-1833)

bacilysin (bacilin, tetaine)
$C_{12}H_{18}N_2O_5$, probably derived from bacillus, lysis, and -in(e) – referring to this antibiotic's bacteriolytic activity

bacitracin
derived from the bacterial genus name *Bacillus*, Margaret Tracy (born about 1936), the name of a child in whose tissue these antibiotics were found, and -in(e)

-bacter
derived from bacterium

bacteri(o)-
derived from bacterium

bacteriochlorin
$C_{20}H_{18}N_4$, derived from bacteri(o)-, [1]chlor(o)-, and -in(e)

bacteriochlorophyll
derived from bacteri(o)- and chlorophyll

bacterium
ultimately derived from *bakterion*, diminutive of *bakteria* (Greek: staff), akin to *baktron* (Greek: stick)

bactobolin
$C_{14}H_{20}C_{12}N_2O_6$, coined by contraction of bacterium, probably metabolism, and -in(e)

baddeleyite
ZrO_2, named for this mineral's discoverer, the 19th century British-Ceylonese tea farmer Joseph Baddeley

Baden acid (Badische acid)
$C_{10}H_9NO_3S$, named for the chemical company Badische Anilin- & Soda-Fabrik AG, Ludwigshafen, Germany

Badger rules
named for the US chemist Richard McLean Badger (1896-1974)

Badische acid (Baden acid)
$C_{10}H_9NO_3S$, named for the chemical company Badische Anilin- & Soda-Fabrik AG, Ludwigshafen, Germany

Baeyer-Drewson indigo synthesis
named for the German chemist Johann Friedrich Wilhelm Adolf von Baeyer (1835-1917) and the chemist V. Drewson

Baeyer strain
named for the German chemist Johann Friedrich Wilhelm Adolf von Baeyer (1835-1917)

Baeyer system
named for the German chemist Johann Friedrich Wilhelm Adolf von Baeyer (1835-1917)

Baeyer tests
named for the German chemist Johann Friedrich Wilhelm Adolf von Baeyer (1835-

1917)

Baeyer-Villiger oxidation
named for the German chemist Johann Friedrich Wilhelm Adolf von Baeyer (1835-1917) and the Swiss-German chemist Victor Villiger (1868-1934)

bagasse
ultimately derived from *bac(c)a* (Latin: berry)

baicalein
$C_{15}H_{10}O_5$, derived from the specific epithet of the species name *Scutellaria baicalensis* (Baikal scullcap), from Lake Baikal, Russia, and -in(e)

bainite
(Fe,Fe_3C), named for the US metallurgist Edgar C. Bain (1891-1971)

Baizer synthesis
named for the US chemist Manuel M. Baizer (1914-1988)

bakelite
named for the Belgian-US chemist Leo Hendrik Baekeland (1863-1944)

Baker-Nathan effect
named for the British chemists John William Baker (1898-1967) and Wilfred S. Nathan

Baker-Venkatamaran rearrangement
named for the British chemist Wilson Baker (born 1900) and the Indian chemist K. Venkatamaran

BAL (dimercaprol)
$C_3H_8OS_2$, an abbreviation for British antilewisite

Baldwin's rules
named for the British chemist Sir Jack E. Baldwin (born 1938)

Balz-Schiemann reaction (Schiemann reaction)
named for the German chemists G. Balz and Günther Robert Arthur Schiemann (1899-1969)

Bamberger rearrangement
named for the German chemist Eugen Bamberger (1857-1932)

Bamford-Stevens reaction
named for the British chemists William Randall Bamford and Thomas Stevens Stevens (1900-2000)

bakankosin
$C_{16}H_{23}NO_8$, derived from the specific epithet of the species name *Strychnos bakanko* (a nightshade), erroneously believed to be the source of this alkaloid; the true source was *Strychnos vacacoua* Baill. (a nightshade)

bambermycin
derived from the specific epithet of the bacterial species name *Streptomyces bambergiensis*, after the city of Bamberg, Germany, and -mycin

baptigenin
$C_{15}H_{10}O_6$, derived from baptisin, -gen, and -in(e)

baptisin
$C_{26}H_{32}O_{14}$, derived from the genus name *Baptisia* (wild indigo), from *baptisis* (Greek: a dipping), from *baptein* (Greek: to dip) – referring to the use of indigo in vat dying, and -in(e)

barbaloin
$C_{21}H_{22}O_9$, coined by contraction of Barbados aloe and -in(e)

Barbier reactions
named for the French chemist Philippe Antoine Barbier (1848-1922)

Barbier-Wieland reaction
named for the French chemist Philippe Antoine Barbier (1848-1922) and the German chemist Heinrich Otto Wieland (1877-1957)

barbital (barbitone)
$C_8H_{12}N_2O_3$, derived from barbituric acid and the arbitrary suffix -al; patterned after chloral – referring to this drug's chloral hydrate-like hypnotic action

barbituric acid
$C_4H_4N_2O_3$, of uncertain origin, probably derived from the name Barbara, and urine

Barfoed's reagent
named for the Danish chemist Christen Thomsen Barfoed (1815-1899)

barite (barytes)
$BaSO_4$, derived from *barys* (Greek: heavy)

barium
Ba, derived from *barys* (Greek: heavy) and -ium

barrelene
C_8H_8, derived from barrel and -ene – referring to this hydrocarbon's barrel-like shape

Bart reaction
named for the 20th century German chemist H. Bart

Bartlett-Condon-Schneider reaction
named for the US chemists Paul Doughty Bartlett (1907-1997), Francis Edward Condon, and Abraham Schneider

Barton decarboxylation
named for the British-US chemist Sir Derek Harold Richard Barton (1918-1998)

Barton deoxygenation
named for the British-US chemist Sir Derek Harold Richard Barton (1918-1998)

Barton esters
named for the British-US chemist Sir Derek Harold Richard Barton (1918-1998)

Barton-Kellogg reaction
named for the British-US chemist Sir Derek Harold Richard Barton (1918-1998) and the 20th century Dutch chemist R. M. Kellogg

Barton-McCombie reaction
named for the British-US chemist Sir Derek Harold Richard Barton (1918-1998) and the British chemist S. W. McCombie

Barton reaction
named for the British-US chemist Sir Derek Harold Richard Barton (1918-1998)

Barton-Zard reaction
named for the British-US chemist Sir Derek Harold Richard Barton (1918-1998) and the 20th century French chemist Samir Z. Zard

bary-
derived from *barys* (Greek: heavy)

barysilite
$Pb_3Si_2O_7$, derived from bary-, silicon, and -ite

baryta
BaO, derived from barytes and –a

barytes (barite)
$BaSO_4$, derived from *barytes* (Greek: weight), from *barys* (Greek: heavy)

basalt
a rock species, ultimately derived from *basanos* (Greek: touchstone), from *bhnw* (Old Egyptian: touchstone)

BASF process
named for the chemical company Badische Anilin- und Soda-Fabrik AG, Ludwigshafen, Germany

base
derived from *base* (French: base), from *basis* (Greek: foundation), from *bainein* (Greek: to walk)

base metal
a name referring to the ready oxidation of these elements

Bashkirov oxidation
named for the Russian chemist A. N. Bashkirov (1903-1982)

Basolo rule
named for the US chemist Fred Basolo (born 1920)

basophilic
derived from base and -philic

bassanite
$2CaSO_4{\cdot}H_2O$, named for the Italian paleontologist Francesco Bassani (1853-1916)

bastnäsite (hamartite)
$Ce(CO_3)F$, $La(CO_3)F$, $Y(CO_3)F$, named after this mineral's locality Bastnäs mine, Riddarhyttan, Västmanland, Sweden

bathmometry
derived from *bathmos* (Greek: step) and -metry

bathochromic
derived from *bathos* (Greek: depth) and *chroma* (Greek: color)

bathy-
derived from *bathys* (Greek: deep)

batrach(o)-
derived from *batrachos* (Greek: frog)

batrachotoxin
$C_{31}H_{42}N_2O_6$, derived from batrach(o)- and toxin

batyl alcohol
$C_{21}H_{44}O_3$, derived from the genus name *Batis* (fish), from *batis* (Greek: a flat fish)

Baudisch reaction
named for the Austrian-US chemist Oskar Baudisch (1881-1950)

Baumé scale
named for the French pharmacist A. Baumé (1728-1804)

bauxite
a heterogeneous mineral, named after the town of Les Baux-de-Provence, France

bavenite
$Ca_4(Al,Be)_4Si_9O_{26}(OH)_2$, named after this mineral's locality Baveno, Italy

bayerite
$Al(OH)_3$, named for the German chemist Karl Josef Bayer (1847-1904)

Bayer process
named for the German chemist Karl Josef Bayer (1847-1904)

Bayer's acid (croceic acid)
$C_{10}H_8O_4S$, named for the chemical company Farbenfabriken Bayer AG, Leverkusen, Germany

Baylis-Hillman reaction
named for the 20th century US chemists Anthony Basil Baylis and Melville Ernest Douglas Hillman

BCS theory
named for the initials of the names of the US physicists John Bardeen (1908-1991), Leon Neil Cooper (born 1930), and John Robert Schrieffer (born 1931)

bebeerine (chondrodendrine, nectadrine, pelosine)
$C_{36}H_{38}N_2O_6$, derived from bebeeru tree (Demerara greenheart, *Nectandra rodiaei* Hook.), from *bibirú* (Spanish and Portuguese: bebeeru tree), of Carib origin, and -in(e)

Béchamp reduction
named for the French chemist Pierre Jacques Antoine Béchamp (1816-1908)

Bechgaard salt
named for the Danish chemist Klaus Bechgaard (born 1945)

Beckman pH meter
named for the US chemist Arnold Orville Beckman (1900-2004)

Beckmann fragmentation
named for the German chemist Ernst Otto Beckmann (1853-1923)

Beckmann rearrangement
named for the German chemist Ernst Otto Beckmann (1853-1923)

Beckmann thermometer
named for the German chemist Ernst Otto Beckmann (1853-1923)

behenic acid
$C_{22}H_{44}O_2$, derived from ben or behen, from *ban* (Arabic: seeds of any species of the genus *Moringa*)

beidellite (bentonite, wilkonite)
$Al_2O_3{\cdot}4SiO_2{\cdot}H_2O$, named after this mineral's locality Beidell, CO, USA

Beilstein test
named for the German-Russian chemist Friedrich Konrad Beilstein (1838-1906)

belite (larnite)
Ca_2SiO_4, derived from the letter B and -lite

Bell's rule
named for the British chemist Ronald Percy Bell (1907-1996)

Belousov-Zhabotinskii reaction
named for the Russian chemists B. P. Belousov (1893-1970) and Anatol M. Zhabotinskii (born 1938)

Benary reaction
named for the German chemist Erich Benary (born 1881)

Bence-Jones protein
named for the British physician Henry Bence Jones (1813-1873)

Bender's salt
$C_3H_5KO_2S$, named for the German chemist Friedrich August Bender (1847-1926)

Benedict's solution
named for the US chemist Stanley Rossiter Benedict (1884-1936)

benitoite
$BaTiSi_3O_9$, named after this mineral's locality San Benito River, CA, USA

Benkeser reduction
named for the US chemist Robert Anthony Benkeser (born 1920)

Benson's additivity rule
named for the US chemist Sidney William Benson (born 1918)

bentonite (beidellite, wilkonite)
$Al_2O_3{\cdot}4SiO_2{\cdot}H_2O$, named after this mineral's locality Fort Benton, WO, USA

benzal (benzylidene)
$C_7H_6<$, derived from benz(o)- and -al

benzaldehyde
C_7H_6O, derived from benz(o)- and aldehyde

benzedrine
$C_{10}H_{15}NO$, derived (with contraction) from benz(o)- and ephedrine

benzene
C_6H_6, derived from benzol and -ene

benzhydrol
$C_{13}H_{12}O$, from benz(o)- and hydrol

benzidine
$C_{12}H_{12}N_2$, derived from benz(o)- and -idin(e)

benzil
$C_{14}H_{12}O_2$, derived from benz(o)- and -il(e)

benzilic acid
$C_{14}H_{12}O_3$, derived from benzil – referring to this acid's formation by rearrangement of benzil

benzine
C_6H_6, an obsolete name for benzene, ultimately derived from [2]benzoin

benz(o)-
derived from benzene

benzoate
$C_7H_5O_2^-$, derived from benzoic acid and -ate

benzocaine
$C_9H_{11}NO_2$, derived from benz(o)- and -caine

benzoic acid
$C_7H_6O_2$, derived from [2]benzoin

benzoid
derived from benzene and -oid

[1]**benzoin**
$C_{14}H_{12}O_2$, derived from [2]benzoin

[2]**benzoin**
a resin, derived from *benzoin* (French: an odoriferous resin), ultimately from *luban jawi* (Arabic: Javanese incense)

benzol
C_6H_6, an obsolete name for benzene, ultimately from [2]benzoin and -ol(e)

benzophenone
$C_{13}H_{10}O$, derived from benz(o)-, phene, and -one

benzoquinone
$C_6H_4O_2$, derived from benz(o)- and quinone

benzoyl
C_7H_5O-, derived from [1]benzoin and -yl

benzvalene
C_6H_6, derived from benz(o)- and valence isomerism

benzyl
C_7H_7-, derived from benz(o)- and -yl

benzylidene
$C_7H_6=$, derived from benzyl, -ide, and -ene

benzyne (dehydrobenzene)
C_6H_4, derived from benzene and -yne

berbaman
$C_{32}H_{30}N_2O_2$, derived from berbamine and

-an(e)

berbamine
$C_{37}H_{40}N_2O_6$, derived (with contraction) from the genus name *Berberis* (barberry), from *barbaris* (Medieval Latin: barberry), from *barbaris* (Arabic: barberry), and amine

berb(er)-
derived from *berberis* (Latin: barberry), from *barbaris* (Arabic: barberry)

berberine
$C_{20}H_{19}NO_5$, derived from berb(er)- and -in(e)

berbine (protoberberine)
$C_{17}H_{17}N$, derived from berb(er)- and -in(e)

bergapten (bergamot camphor)
$C_{12}H_8O_4$, derived from bergamot (*Citrus bergamia* Risso), ultimately from *bey-armudu* (Turk: bergamot, literally prince's pear), and -pten(e); patterned after eleoptene

Bergius process (Bergius-Pier process)
named for the German chemist Friedrich Bergius (1884-1949)

Bergman cyclization
named for the US chemist Robert George Bergman (born 1942)

Bergmann azlactone peptide synthesis
named for the German chemist Max Bergmann (1886-1944)

Bergmann degradation
named for the German chemist Max Bergmann (1886-1944)

Bergmann-Zervas carbobenzoxy method
named for the German chemist Max Bergmann (1886-1944) and the Greek chemist Leonidas Zervas (1902-1980)

berkelium
Bk, named for the city of Berkeley, CA, USA where this element was prepared for the first time

Berlin black
named for the city of Berlin, Germany where this dye was first prepared

Berlin blue (Milori blue, Paris blue, Prussian blue)
$C_6Fe_2KN_6$, named for the city of Berlin, Germany where this dye was first prepared

Berlin brown
$C_6Fe_2N_6$, named for the city of Berlin, Germany where this dye was first prepared

berlinite
$AlPO_4$, named for the Swedish pharmacologist Nils Johan Berlin (1812-1891)

berninamycin
$C_{51}H_{51}N_{15}O_{15}S$, derived from the specific epithet of the bacterial species name *Streptomyces bernensis* Dietz, from *bernensis* (New Latin: of Berne, Switzerland), and -mycin

Bernthsen acridine synthesis
named for the German chemist August Bernthsen (1855-1931)

Berry pseudorotation
named for the US chemist Richard Steven Berry (born 1931)

Berthelot reaction
named for the French chemist Marcellin Pierre Eugène Berthelot (1827-1907)

Berthelot-Thiele reagent
named for the French chemist Marcellin Pierre Eugène Berthelot (1827-1907) and the German chemist Friedrich Karl

Johannes Thiele (1865-1918)

berthollide
non-stoichiometric compound, named for the French chemist Claude Louis Comte de Berthollet (1748-1822)

bertrandite
$Be_4Si_2O_7(OH)_2$, named for the French mineralogist Marcel Alexandre Bertrand (1847-1907)

beryl
$Be_3Al_2Si_6O_{18}$, derived from *beryllos* (Greek: beryl), ultimately from Dravidian *Velur*, name of an ancient town

beryllium
Be, derived from beryl and -ium

berzelianite
Cu_2Se, named for the Swedish chemist Jöns Jacob Berzelius (1779-1848)

berzeliite
$(Ca,Na)_3(Mg,Mn)_2(AsO_4)_3$, named for the Swedish chemist Jöns Jacob Berzelius (1779-1848)

berzelium (carolinium)
Th, a name suggested for a hypothetical element claimed to be contained, together with the hypothetical element carolinium, in thorium, chosen in honor of the Swedish chemist Jöns Jacob Berzelius (1779-1848)

Berzelius test (for arsenic)
named for the Swedish chemist Jöns Jacob Berzelius (1779-1848)

Bessemer steel
named for the British self-taught inventor Sir Henry Bessemer (1813-1898)

bestatin (ubenimex)
$C_{16}H_{24}N_2O_4$, derived from aminopeptidase B and -statin – referring to this compound's inhibition of aminopeptidase B

besyl
$C_6H_5O_2S-$, coined by contraction of benzenesulfonyl

besylate
$C_6H_5O_3S^-$, coined by contraction of benzenesulfonate

BET method
named for the US chemists Stephen Brunauer (1903-1986) and Paul Hugh Emmett (1900-1985), and the Hungarian-US physicist Edward Teller (1908-2003)

bet(a)-
derived from the genus name *Beta* (beets), from *beta* (Latin: beet), of Celtic origin

betacyanin
derived from bet(a)- , cyan(o)-, and -in(e)

betafite
$(Ca,Na,U)_2(Ti,Nb)_2O_6(O,OH,F)_7$, named after this mineral's locality Betafo, Malagasy Republic

betaine
$C_5H_{11}NO_2$, derived from bet(a)- and -in(e)

betalaine
coined by variation of betaine

betalamic acid
$C_9H_9NO_5$, derived from betalaine and -amic acid

betanin(e)
$C_{24}H_{26}N_2O_{13}$, coined by contraction of bet(a)- and cyanin

betaxanthin
derived from bet(a)-, xanth(o)-, and -in(e)

Bettendorff tests
named for the German chemist A. J. H. M. Bettendorff (1839-1902)

Betti reaction
named for the Italian chemist Mario Betti (1875-1942)

betulin (betulinol)
$C_{30}H_{50}O_2$, derived from the genus name *Betula* (birches), from *betula* (Latin: birch), of Celtic origin, and -in(e)

betweenanene
derived from between, -an(e), and -ene – referring to the bridging and shielding of a central C=C double bond in these compounds by two crossing alkane chains

beyer-
derived from the genus name *Beyeria* (shrubs), named for the Danish parish priest Seyer Mahling Beyer (1741-1840) and/or the German botanist Rudolf Beyer (1852-1932)

beyerane (hibane)
$C_{20}H_{34}$, derived from beyer- and -an(e)

beyerol
$C_{20}H_{32}O_3$, derived from beyer- and -ol

BHA
$C_{11}H_{16}O_2$, an abbreviation for *tert*-butylhydroxyanisole

BHC (HCH, lindane)
$C_6H_6Cl_6$, an abbreviation for benzene hexachloride

BHT (DBP)
$C_{15}H_{24}O$, an abbreviation for butylated hydroxytoluene

bi-
derived from *bis* (Latin: twice)

Bial reagent
named for the French physician M. Bial (1870-1908)

BiAS
an abbreviation for bismuth active substance

bibrotoxin
derived from the specific epithet of the species name *Atractaspis bibroni* (Bibron's mole viper), after the French zoologist Gabriel Bibron (1806-1848), and toxin

bicine
$C_6H_{13}NO_4$, coined by contraction of *N,N*-bis(2-hydroxyethyl)glycine

bicozamycin
$C_{12}H_{18}N_2O_7$, coined by contraction of the earlier names bicyclomycin (referring to this antibiotic's bicyclic structure) and aizumycin, derived from the specific epithet of the bacterial species name *Streptomyces aizunensis,* and -mycin

bicuculline
$C_{20}H_{17}NO_6$, derived from bi- and the specific epithet of the species name *Dicentra cucullaria* (Dutchman's-breeches), ultimately from *cucullus* (Latin: cap, hood), and -in(e)

bidentate
derived from bi-, *dens* (Latin: tooth), and -ate

bifilar
derived from bi- and *filum* (Latin: thread)

Biginelli reaction
named for the Italian chemist Pietro Biginelli (1860-1937)

biguanide (diguanide)
$C_2H_7N_5$, derived from bi-, guanidine, and -ide

bikhaconitine
$C_{36}H_{51}NO_{11}$, coined by contraction of bikh (*Aconitum spicatum* Stapf.), derived from *bikh* (Hindi: poison), aconit-, and -in(e)

bilane (bilinogen)
$C_{19}H_{20}N_4$, derived from bil(i)- and -an(e)

bilene
$C_{19}H_{18}N_4$, derived from bil(i)- and -ene

bil(i)-
derived from *bilis* (Latin: bile)

biline
$C_{19}H_{14}N_4$, derived from bil(i)- and -in(e)

biliprotein
derived from bil(i)- and protein

bilirubin
$C_{33}H_{36}N_4O_6$, derived from bil(i)-, rubi-, and -in(e)

biliverdin
$C_{33}H_{34}N_4O_6$, derived from bil(i)-, *verd* (Old French: green), and -in(e)

Billiter process
named for the Austrian chemist Jean Billiter (1877-1965)

binary
derived from *binaris* (Latin: consisting of two)

Bindschedler's green
$C_{16}H_{20}ClN_3$, named for the chemical company Bindschedler & Busch, Basel, Switzerland which developed this dye

Bingham fluid
named for the US chemist Eugene Cook Bingham (1878-1945)

bio-
derived from *bios* (Greek: life)

biocide
derived from bio- and -cide

biocytin
$C_{16}H_{28}N_4O_4S$, derived (with contraction) from biotin, cyt(o)-, and -in(e)

biopterin
$C_9H_{11}N_5O_3$, derived from bio- and pterin

biotin (coenzyme R, vitamin B$_7$, vitamin H)
$C_{10}H_{16}N_2O_3S$, derived from *bios* (Greek: life) and -in(e)

biotite
$K(Mg,Fe)_3(Si_3Al)O_{10}(OH,F)_2$, named for the French mathematician and physicist Jean-Baptiste Biot (1774-1862)

Biot's law
named for the French mathematician and physicist Jean-Baptiste Biot (1774-1862)

biphenyl
$C_{12}H_{10}$, derived from bi(s)- and phenyl

biphenylyl (xenyl)
$C_{12}H_9-$, derived from biphenyl and -yl

Birch reduction
named for the Australian chemist Arthur John Birch (1915-1995)

Birkeland-Eyde process
named for the Norwegian chemist Kristian Olaf Bernhard Birkeland (1867-1917) and the Norwegian engineer Sam Eyde (1866-1940)

bi(s)-
derived from *bis* (Latin: twice, two-fold)

bisabol-
derived from bisabol (gum resin af African trees of the genus *Commiphora*), from *bisap u ala* (Wolof: bisabol)

bisabolane
$C_{15}H_{30}$, derived from bisabol- and -an(e)

bisabolene
$C_{15}H_{24}$, derived from bisabol- and -ene

bisabolol
$C_{15}H_{26}O$, derived from bisabol- and -ol

Bischler-Möhlau indole synthesis
named for the German-Swiss chemist August Bischler (1865-1957) and the German chemist R. Möhlau

Bischler-Napieralski reaction
named for the German-Swiss chemist August Bischler (1865-1957) and the chemist Bernard Napieralski

bischofite
$MgCl_2 \cdot 6H_2O$, named for the German geologist and mineral chemist Karl Gustav Bischof (1792-1870)

bisemutum
Bi, New Latin name for bismuth, derived by latinization of bismuth

bismane
$C_5H_{11}Bi$, derived from bismuth and -an(e)

Bismarck Brown
named for the German Prince Otto Eduard Leopold von Bismarck-Schönhausen (1815-1898)

bismite
Bi_2O_3, derived from bismuth and -ite

bismorane
BiH_5, derived from bismuth and -orane; patterned after phosphorane

bismuth
Bi, derived from *Wismut* (lay German: bismuth), possibly from *wiss* (Medieval German: white) and *matte* (Medieval German: barrengrounds, deads)

bismuthane (bismuthine)
BiH_3, derived from bismuth and -an(e)

bismuthine (bismuthane)
BiH_3, derived from bismuth and -in(e)

bismuthinite (bismuthite)
Bi_2S_3, derived from bismuth, -in(e), and -ite

bisphenoid
derived from bi- and *sphenoeides* (Greek: edge-shaped)

bisphenol A
$C_{15}H_{15}O_2$, derived (with contraction) from bi(s)-, phenol, and acetone – referring to the synthesis of this compound from phenol and acetone

bittern (brine)
derived from bitter

bitumen
derived from *bitumen* (Latin: bitumen) of Celtic origin, related to *bedw* (Welsh: birch)

biuret
$C_2H_5N_3O_2$, derived from bi-, urea, and -et(e)

bixbyite
$(Fe,Mn)_2O_3$, named for the US mineralogist Maynard Bixby (1853-1935)

bixin
$C_{25}H_{30}O_4$, named after the genus name *Bixa* (bixa), ultimately derived from *bixa* (Taino: bixa)

Blaise ketone synthesis (Blaise-Maire reaction)
named for the French chemist Edmond Emile Blaise (1872-1939)

Blaise-Maire reaction (Blaise ketone synthesis)
named for the French chemists Edmond Emile Blaise (1872-1939) and M. Maire

Blaise reaction
named for the French chemist Edmond Emile Blaise (1872-1939)

Blanc reaction
named for the French chemist Gustave Louis Blanc (1872-1927)

Blanc rule
named for the French chemist Gustave Louis Blanc (1872-1927)

blasticidin
derived from blast(o)-, -cide, and -in(e)

blast(o)-
derived from *blastos* (Greek: bud, germ)

blastomycin
derived from the fungal genus name *Blastomyces*, from blast(o)- and -myces, and -in(e)

blende
derived from *blenden* (German: to dazzle, to deceive) – referring to the fact that such minerals were regarded as devoid of useful metals

bloedite (astrak(h)anite)
$Na_2Mg(SO_4)_2 \cdot 4H_2O$, named for the German chemist Karl August Bloede (1773-1820)

blomstrandine (eschynite, aeschynite, priorite)
$(Ce,Ca,Fe)(Ti,Nb)_2(O,OH)_6$, named for the Swedish chemist Christian Wilhelm Blomstrand (1826-1897)

BOD
an abbreviation for biological oxygen demand

Bodroux-Chichibabin aldehyde synthesis
named for the French chemist F. Bodroux (1873-1968) and the Russian chemist Aleksei Yevgenievich Chichibabin (1871-1945)

Bodroux reaction
named for the French chemist F. Bodroux (1873-1968)

boehmite
AlO(OH), named for the German geologist and paleontologist Johannes Böhm (1857-1938)

Bogert-Cook synthesis
named for the US chemists Marston Taylor Bogert (1868-1954) and J. W. Cook

Bohn-Schmidt reaction
named for the German chemists René Bohn (1862-1922) and Robert Emanuel Schmidt (1864-1938)

bohrium
Bh, named in honor of the Danish physicist Niels Henrik David Bohr (1885-1962)

boldine
$C_{19}H_{21}NO_4$, derived from boldo (*Peumus boldus* Molina), from *boldu* (Araucan: boldo), and -in(e)

bole
ultimately derived from *bolos* (Greek: clod, lump)

bole armoniac (Armenian bole, bolus Armena, bolus rubra, red bole)
derived from *bolus armeniacus* (Medieval Latin: Armenian bole)

bolecic acid (erythrogenic acid, isanic acid)
$C_{18}H_{26}O_2$, derived from boleko oil, from *boleko* (native African name of the tree *Ongokea klaineana*)

boléite
$KAg_9Pb_{26}Cu_{24}Cl_{62}(OH)_{48}$, named after this mineral's locality Boléo, Baja California, Mexico

Bologna phosphorus
BaS, named for the city of Bologna, Italy, the home of this phosphorescent preparation's discoverer, the Italian alchemist Vincenzo Casciorola (1571-1624)

Boltzmann's constant
named for the Austrian physicist Ludwig Eduard Boltzmann (1844-1906)

bolus alba (kaolin)
derived from *bolus alba* (New Latin: white clay)

bolus Armena (Armenian bole, bole armoniac, bolus rubra, red bole)
derived from *bolus armeniacus* (Medieval Latin: Armenian bole)

bombesin
$C_{71}H_{110}N_{24}O_{18}$, derived from the genus name *Bombina* (toads), from *bombinare*, *bombilare* (Latin: to buzz), from *bombus* (Latin: deep hollow sound), and -in(e)

bongkrekic acid
$C_{28}H_{38}O_7$, derived from *bongkrek* (an Indonesian molded coconut product)

Boord olefin synthesis
named for the 20th century US chemist Cecil E. Boord

boracite
$Mg_3B_7O_{13}Cl$, derived from borax and -ite

borane
BH_3, derived from boron and -an(e)

borate
derived from boric acid and -ate

borax (tincal)
$Na_2B_4O_7 \cdot 10H_2O$, ultimately derived from *burah* (Persian: borax, soda, literally white)

borazon
BN, derived from boron, az(ot)-, and [1]-on

Borch reduction
named for the US chemist Richard Frederic Borch (born 1941)

Bordeaux mixture
a translation of *bouillie bordelaise* (French: Bordeaux brew), named for the Bordeaux wine growing area, France where this fungicide was first used

bore
B, unsuccessfully suggested name for boron, derived from borax

boric acid
H_3BO_3, derived from boron

boride
B^{3-}, derived from boron and -ide

borium
B, New Latin name for boron, derived from *Bor* (German: boron) and -ium

bornane (camphane)
$C_{10}H_{18}$, derived from borneol and -an(e)

Borneo camphor (borneol, Malayan camphor, Sumatra camphor)
$C_{10}H_{18}O$, named after the island of Borneo, Indonesia

borneol (Borneo camphor, Malayan camphor, Sumatra camphor)
$C_{10}H_{18}O$, derived (with contraction) from Borneo camphor and -ol

Born-Haber cycle
named for the German physicist Max Born (1882-1970) and the German chemist Fritz Haber (1868-1934)

bornite
Cu_5FeS_4, named for the Austrian mineralogist Ignaz von Born (1742-1791)

Borodin reaction (Hunsdiecker reaction)
named for the Russian chemist and musical composer Alexander Porfirievich Borodin (1833-1887)

boromycin
$C_{45}H_{74}BNO_{15}$, derived from boron and -mycin

boron
B, derived (with contraction) from borax and carbon – referring to boron's perceived similarity to carbon

Borsche-Drechsel cyclization
named for the German chemists Walter Borsche (1877-1950) and E. Drechsel

bort
C_n, ultimately possibly from *burdus* (Latin: hinny)

bostrycoidin
$C_{15}H_{11}NO_5$, derived from the specific epithet of the fungal species name *Fusarium bostrycoides*, from *bostrychos* (Greek: curl), and -in(e)

boswellic acid
derived from the genus name *Boswellia* (trees), probably in honor of the Scottish lawyer and biographer James Boswell (1740-1795)

bottromycin
derived from the specific epithet of the bacterial species name *Streptomyces bottropensis*, from *bottropensis* (New Latin: of Bottrop, Germany), and -mycin

botulin
derived from the specific epithet of the bacterial species name *Clostridium botulinum*, from *botulus* (Latin: sausage), and -in(e)

Boudouard equlibrium
named for the French chemist Octave L. Boudouard (1872-1923)

Boughton system
named for the US chemist W. A. Boughton (1885-1977)

boulangerite
$Pb_5Sb_4S_{11}$, named for the French mining engineer C. L. Boulanger (1810-1849)

bourbonal (ethyl vanillin)
$C_9H_{10}O_3$, possibly derived from the French royal family of Bourbon and -al

bournonite
$CuPbSbS_3$, named for the French mineralogist Comte Jacques-Louis de Bournon (1751-1825)

Bouveault aldehyde synthesis
named for the French chemist Louis Bouveault (1864-1909)

Bouveault-Blanc reduction
named for the French chemists Louis Bouveault (1864-1909) and Gustave Louis Blanc (1872-1927)

Boyland-Sims Oxidation
named for the 20th century British chemists E. Boyland and Peter Sims

Boyle's law
named for the Irish chemist Robert Boyle (1627-1691)

Bradsher cyclization
named for the US chemist Charles K. Bradsher (1912-2000)

Bradsher reaction
named for the US chemist Charles K. Bradsher (1912-2000)
brady-
derived from *bradys* (Greek: slow)

bradykinin
$C_{50}H_{73}N_{15}O_{11}$, derived from brady- and kinin

braggite
(Pt,Pd,Ni)S, named for the British physicists Sir William Henry Bragg (1862-1942) and William Lawrence Bragg (1890-1971)

brannerite
$(U,Ca,Ce)(Ti,Fe)_2O_6$, named for the US geologist John C. Branner (1850-1922)

brass
(Cu,Zn), ultimately possibly derived from a Semitic language

brassi(ca)-
derived from the genus name *Brassica* (cabbage), from *brassica* (Latin: cabbage)

brassicasterol
$C_{28}H_{46}O$, derived from brassi(ca)- and sterol

brassidic acid
$C_{22}H_{42}O_2$, derived from brassi(ca)-

brassinolide
$C_{28}H_{48}O_6$, derived from brassi(ca)-, -in(e), and -olide

von Braun amide degradation
named for the German chemist Julius von Braun (1875-1939)

von Braun reaction
named for the German chemist Julius von Braun (1875-1939)

braunite
$Mn_7O_8(SiO_4)$, named for the German treasury official K. Braun (1890-1872)

Bray-Liebhafsky reaction
named for the US chemist William Crowell Bray (1879-1946) and the Hungarian-US chemist H. A. Liebhafsky (1905-1982)

brazilianite
$NaAl_3(PO_4)_2(OH)_4$, named after this mineral's locality Brazil

brazilin
$C_{16}H_{14}O_5$, derived from brazilwood (*Caesalpinia echinata* Lam.), ultimately named after Brazil, and -in(e)

breccia
a rock species, derived from *breccia* (Italian: (broken) stone)

Bredereck's reagent
named for the German chemist Helmut Bredereck (1904-1981)

Bredt's rule
named for the German chemist Konrad Julius Bredt (1855-1937)

brefeldin
derived from the specific epithet of the fungal species name *Penicillium brefeldianum* Dodge, after the German botanist and bacteriologist Oskar Brefeld (1839-1925), and -in(e)

breithauptite
NiSb, named for the German mineralogist Johann Friedrich August Breithaupt (1791-1873)

Bremen blue
$Cu(OH)_2$, named for the city of Bremen, Germany

Bremen green
named for the city of Bremen, Germany

brevetoxin
derived from the specific epithet of the species name *Ptychodicus brevis* Davis (a dinoflagellate), from *brevis* (Latin: short), and toxin

brevium
Pa, an unsuccessfully suggested name for protactinium; probably derived from *brevis* (Latin: short) on account of this radioactive element's short half-life

brianite
$Na_2CaMg(PO_4)_2$, named for the US scientist Brian Harold Mason (born 1917)

Briggs-Rauscher reaction
named for the 20th century US chemists Thomas S. Briggs and Warren C. Rauscher

brimstone
S, derived from *brinnen* (Middle English: to burn) and *ston* (Middle English: stone)

Brin process
named for the 19th century French-British chemists Arthur and Leon Brin

Britannia metal
(Sn,Sb,Cu,Bi,Zn), named for *Britannia* (Latin: Britain)

britholite
$(Ce,Ca)_5(SiO_4,PO_4)_3(OH,F)$, derived from *brithos* (Greek: weight) and -lite – referring to this mineral's high density and cerium content

British antilewisite (BAL, dimercaprol)
$C_3H_8OS_2$, derived from anti- and lewisite

brochantite
$Cu_4(OH)_6SO_4$, named for the French geologist and mineralogist A. J. M. Brochant de Villiers (1772-1840)

bröggerite
$(U,Th)O_2$, named for the Norwegian mineralogist Waldemar Christopher Brøgger (1851-1940)

bromal
C_2HBr_3O, coined by contraction of tri-bromoacetaldehyde

bromane
HBr, derived from bromine and -an(e)

bromargyrite
AgBr, derived from brom(o)-, argyr(o)-, and -ite

bromate
BrO_3^-, derived from bromic acid and -ate

bromatology
derived from *broma* (Greek: food)

bromcresol green
$C_{21}H_{14}Br_4O_5S$, derived from *Bromcresol* (German: bromocresol)

bromcresol purple
$C_{21}H_{16}Br_2O_5S$, derived from *Bromcresol*

(German: bromocresol)

bromelain
derived from the genus name *Bromelia* (pineapple), after the Swedish botanist Olaf Bromelius (1639-1705), and -in(e)

bromic acid
$HBrO_3$, derived from bromine

[1]**bromide**
Br^-, derived from bromine and -ide

[2]**bromide**
KBr, in the sense of sedative refers to the now obsolete use of potassium bromide as a sedative

bromine
Br, derived from *bromos* (Greek: stench) and -in(e)

bromite
BrO_2^-, derived from bromous acid and -ite

brom(o)-
Br−, derived from bromine

bromoform
$CHBr_3$, derived from brom(o)- and -form

bromohydrin
derived from brom(o)-, hydr(o)-, and -in(e)

bromous acid
$HBrO_2$, derived from bromine

bromphenol blue
$C_{19}H_{10}Br_4O_5S$, derived from *Bromphenol* (German: bromophenol)

bromthymol blue
$C_{27}H_{28}Br_2O_5S$, derived from *Bromthymol* (German: bromothymol)

bromum
Br, New Latin name for bromine, from *Brom* (German: bromine) and -um

Brönner's acid
$C_{10}H_9NO_3S$, named for the 19th century dye company Farbfabrik vorm. Brönner, Frankfurt/Main, Germany

Brønsted acid and base
named for the Danish chemist Johannes Nicolaus Brønsted (1879-1947)

Brønsted-Bjerrum equation
named for the Danish chemists Johannes Nicolaus Brønsted (1879-1947) and Niels Janniksen Bjerrum (1879-1958)

bronze
(Cu,Sn), ultimately derived from *biring* (Persian: bronze); according to another assumption from *aes brundusinum* (Latin: metal from Brundisium, now Brindisi, Italy)

bronzite (hypersthene)
$(Fe,Mg)SiO_3$, derived from bronze and -ite – referring to this mineral's bronze color

Brook rearrangement
named for the Canadian chemist Adrian Gibbs Brook (born 1924)

brookite
TiO_2, named for the British mineralogist Henry James Brooke (1771-1857)

brosyl
$C_6H_4BrO_2S-$, coined by contraction of 4-bromobenzenesulfonyl

Brown's reagent
$C_{23}H_{41}B$, named for the British-US chemist Herbert Charles Brown (born 1912)

bruceantin
$C_{28}H_{36}O_{11}$, derived (with contraction) from

the species name *Brucea antidysenterica* J. F. Mill. (a shrub), ultimately derived from the name of the Scottish explorer James Bruce (1730-1794), and anti-, and -in(e)

brucine
$C_{23}H_{26}N_2O_4$, derived from the genus name *Brucea* (shrubs) which was erroneously assumed to contain brucine, ultimately derived from the name of the Scottish explorer James Bruce (1730-1794), and -in(e)

brucite
$Mg(OH)_2$, named for the US mineralogist Archibald Bruce (1777-1818)

bry(o)-
derived from *bryon* (Greek: moss)

bryostatin
derived from the phylum name *Bryozoa* (moss animals), from bry(o)-, and -statin

BTX
$C_6H_6/C_7H_8/C_8H_{10}$, an abbreviation for benzene, toluene, xylene

Bucherer-Bergs reaction
named for the German chemists Hans Theodor Bucherer (1869-1949) and H. Bergs

Bucherer carbazole synthesis
named for the German chemist Hans Theodor Bucherer (1869-1949)

Bucherer reaction
named for the German chemist Hans Theodor Bucherer (1869-1949)

Buchner-Curtius-Schlotterbeck reaction
named for the German chemists Eduard Buchner (1860-1917), Theodor Curtius (1857-1928), and F. Schlotterbeck

Büchner funnel
named for the Dutch chemist Ernst H. Büchner (1880-1967)

Buchner method of ring enlargement
named for the German chemist Eduard Buchner (1860-1917)

Buchwald-Hartwig cross coupling reaction
named for the US chemists Stephen L. Buchwald (born 1955) and John F. Hartwig (born 1964)

buckminsterfullerene
C_{60}, named for the US architect Richard Buckminster Fuller (1895-1983) – referring to the similarity of this compound's structural symmetry to characteristic elements of Buckminster Fuller's architecture

bufagin (bufogenin)
$C_{24}H_{34}O_5$, derived (with contraction) from the species name *Bufo agua* (agua toad), from buf(o)- and *agua* (Tupi: agua toad), and -in(e)

bufalin
$C_{24}H_{34}O_4$, coined by contraction of buf(o)- and digitalin

bufanolide
$C_{24}H_{38}O_2$, derived from buf(o)-, -an(e), and -olide

buffer
an indigenous English word

buf(o)-
derived from the genus name *Bufo* (toads), from *bufo* (Latin: toad)

bufotalin
$C_{26}H_{36}O_6$, coined by contraction of buf(o)- and digitalin

bufotenine
$C_{12}H_{16}N_2O$, derived from buf(o)-, an unexplained name fragment, and -in(e)

bufotoxin
$C_{40}H_{60}N_4O_{10}$, derived from buf(o)- and toxin

bulbocapnine
$C_{19}H_{19}NO_4$, derived from the genus name *Bulbocapnos* (fumitory), from *bulbus* (Latin: onion, bulb) and *kapnos* (Greek: smoke), and -in(e)

bullvalene
$C_{10}H_{10}$, derived (with contraction) from bull, valence isomerism, and -ene – referring to the bull-like shape and valence isomerism of this molecule

buna rubber
coined by contraction of butadiene and *Natrium* (German: sodium) – referring to the sodium catalysis of butadiene polymerization

bungarotoxin
derived from the genus name *Bungarus* (krait), ultimately from *bhrugara* (Sanskrit: krait), and toxin

Bunnett-Olsen equation
named for the US chemists Joseph Frederick Bunnett (born 1921) and Frederic Phillip Olsen (born 1937)

Bunsen burner
named for the German chemist Robert Wilhelm Eberhard Bunsen (1811-1899)

Bunsen-Roscoe law
named for the German chemist Robert Wilhelm Eberhard Bunsen (1811-1899) and the British chemist Sir Henry Enfield Roscoe (1833-1915)

bunsenite
NiO, named for the German chemist Robert Wilhelm Eberhard Bunsen (1811-1899)

Bunte salt
named for the German chemist Hans Bunte (1848-1925)

buphanamine
$C_{17}H_{19}NO_4$, derived from the genus name *Buphane* (herbs), probably from *bous* (Greek: bull) and -phan(e), and amine

burette
derived from *burette*, diminutive of *buire* (French: a kind of jug)

Burgess reagent
$C_8H_{18}N_2O_4S$, named for the US chemist Edward M. Burgess (born 1934)

Burgundy mixture
named after the wine growing area Burgundy, France where this fungicide was first used

bustamite
$(Mn,Ca)_3Si_3O_9$, named for the Mexican general Anastasio Bustamente (1780-1853)

but(a)-
derived from butane

butane
C_4H_{10}, derived from butyric acid and -an(e)

butanolide (butyrolactone)
$C_4H_6O_2$, derived from butanoic acid and -olide

butein
$C_{15}H_{12}O_5$, derived from the genus name *Butea* (dhak tree), after the Scottish statesman and scholar John Stuart, 3rd Earl of Bute (1713-1792), and -in(e)

butenolide
derived from but(a)-, -enoic acid, and -olide

butirosin
$C_{21}H_{41}N_5O_{12}$, coined by variation of ambutyrosin, derived from γ-aminobutyryl, -ose, and -in(e)

butyric acid
$C_4H_8O_2$, derived from *butyron* (Greek: butter), from *bous* (Greek: bull) and *tyron* (Greek: cheese)

butyrin (tributyrin)
$C_{15}H_{26}O_6$, derived from butyr(o)- and -in(e); patterned after palmitin and stearin

butyr(o)-
derived from butyric acid

butyroin
$C_8H_{16}O_2$, derived from butyr(o)- and -oin; patterned after [1]benzoin

butyrolactone (butanolide)
$C_4H_6O_2$, derived from butyr(o)- and lactone

butyryl
C_4H_7O–, derived from butyric acid and -yl

bytownite
(Ca,Na)(Si,Al)$_4$O$_8$, named after this mineral's locality Bytown, now Ottawa, Canada

C

cabenegrin
derived from *cabeza de negro* (Spanish: annona, *Annona purpurea*), and -in(e)

cac(o)-
derived from *kakos* (Greek: bad)

[1]cacodyl
C_2H_6As, derived from *kakodes* (Greek: evil smelling) and -yl

[2]cacodyl
$C_4H_{12}As_2$, derived from *kakodes* (Greek: evil smelling) and -yl; a misnomer due to confusion with [1]cacodyl

cacodylic acid
$C_2H_7AsO_2$, derived from [1]cacodyl

cacodyl oxide (Cadet's liquid)
$C_4H_{12}As_2O$, derived from [1]cacodyl and oxide

cacosmophore
derived from cac(o)-, osm(o)-, and -phore

cactinomycin
coined by contraction of actinomycin C

cadalene
$C_{15}H_{18}$, coined by contraction of cadinene and -alene

cadaverine
$C_5H_{14}N_2$, derived from *cadaver* (Latin: cadaver), ultimately from *cadere* (Latin: to fall), and -in(e)

Cadet's liquid (cacodyl oxide)
$C_4H_{12}As_2O$, named for the French chemist and pharmacist Louis Claude Cadet de Gassicourt (1731-1799)

cadherin
coined by contraction of calcium-dependent cell adhesion molecule

cadin-
derived from cade oil, i.e. juniper tar oil, ultimately from *catanus* (Medieval Latin: cedar juniper, *Juniperus oxycedrus*)

cadinane
$C_{15}H_{30}$, derived from cadin- and -an(e)

cadinene
$C_{15}H_{24}$, derived from cadin- and -ene

Cadiot-Chodkiewicz coupling
named for the French chemist Paul Cadiot (born 1923) and the Polish chemist Wladyslaw Chodkiewicz (born 1921)

cadmium
Cd, derived from *cadmia* (Latin: calamine), ultimately from *kadmeia* (Greek: Cadmean earth, zink carbonate), from Kadmos, the mythical founder of the ancient town of Thebes, Greece, and -ium – referring to the occurrence of cadmium ores in association with calamine

cadmocene
$C_{10}H_{10}Cd$, derived from cadmium and -ocene

caerulein (cerulein, ceruletide)
$C_{58}H_{73}N_{13}O_{21}S_2$, derived from the specific epithet of the species name *Hyla caerulea* (Australian tree frog), from cerul(o)-, and -ein(e)

caeruloplasmin
derived from cerul(o)- and plasmin – referring to this glycoprotein's blue color

CAF (calpain)
an abbreviation for calcium activated factor

cafestol
$C_{20}H_{28}O_3$, derived from *café* (French: coffee) and sterol

caffe-
ultimately derived from *qahwa* (Arabic: coffee)

caffeic acid
$C_9H_8O_4$, derived from caffe-

caffeine (theine)
$C_8H_{10}N_4O_2$, derived from caffe- and -in(e)

Cahn-Ingold-Prelog rules (CIP system)
named for the British chemist Robert Sidney Cahn (1899-1981), the British chemist Sir Christopher Kelk Ingold (1893-1970), and the Yugoslav-Swiss chemist Vladimir Prelog (1906-1998)

-caine
derived from cocaine

cajeput-
derived from cajeput (*Melaleuca leucadendron*), from *kayu puteh* (Malay: cajeput, literally white wood)

cajeputene (dipentene)
$C_{10}H_{16}$, derived from cajeput- and -ene

cajeputol (cineole, eucalyptol)
$C_{10}H_{18}O$, derived from cajeput- and -ol

calamine (hemimorphite)
$Zn_4Si_2O_7(OH)_2 \cdot H_2O$, derived from *cadmia* (Latin: calamine), ultimately from *kadmeia* (Greek: Cadmean earth, zink carbonate), from Kadmos, the mythical founder of the ancient town of Thebes, Greece

calamitic
derived from *kalamites* (Greek: reedlike), from *kalamos* (Greek: reed)

calaverite
$AuTe_2$, named after this mineral's locality Carson Hill, Calaveras County, CA, USA

¹calci-
derived from *calx* (Latin: lime)

²calci-
derived from calcium

calcidiol (calcifediol)
$C_{27}H_{44}O_2$, derived (with contraction) from calciferol, di-, and -ol

calcifediol (calcidiol)
$C_{27}H_{44}O_2$, derived from calciferol, di-, and -ol

calciferol (ergocalciferol, vitamin D₂)
$C_{28}H_{44}O$, derived (with contraction) from ²calci-, *ferre* (Latin: to bring, to bear), and sterol

calcimycin
$C_{29}H_{37}N_3O_6$, derived from ²calci- and -mycin

calcination
derived from ¹calci-

calcite
$CaCO_3$, derived from ¹calci- and -ite

calcitonin (thyreocalcitonin)
derived from ²calci- and –tonin

calcitriol
$C_{27}H_{44}O_3$, derived from ²calci-, tri-, and –ol

calcium
Ca, derived from [1]calci- and -ium

caldariomycin
$C_5H_8Cl_2O_2$, derived from the fungal genus name *Caldariomyces fumago*, from *caldarius* (Latin: suitable for warming), from *calidus* (Latin: warm, hot), and -myces, and -in(e)

caliche
$CaCO_3$, from *caliche* (Spanish: pebble in a brick), ultimately from *calx* (Latin: lime)

calicheamycin
derived from the subspecific epithet of the bacterial species name *Micromonospora echinospora* ssp. *calichensis*, and -mycin

californium
Cf, named for the state of California, USA, where this element was prepared for the first time

calixarene
derived from *calix* (Latin: chalice) and arene

callase
derived from callose and -ase

callose
derived from *callus* (Latin: weal) and -ose – referring to the formation of this polysaccharide in the healing process of injured plants

calmodulin
coined by contraction of calcium, modulation, and -in(e)

cal(o)-
derived from *kalos* (Greek: beautiful)

calomel
Hg_2Cl_2, possibly derived from cal(o)- and *melas* (Greek: black) – referring to this white salt's blackening in the light by formation of elemental mercury

calotropin
$C_{29}H_{40}O_9$, derived from the genus name *Calotropis* (shrubs, trees), from cal(o)- and *tropis* (Greek: keel), akin to *trepein* (Greek: to turn), and -in(e)

calpain (CAF)
coined by contraction of calcium and papain

calsequestrin
coined by contraction of calcium, sequestration, and -in(e)

Calutron process
coined by contraction of California University Cyclotron process

Calvin cycle
named for the US chemist Melvin Calvin (1911-1997)

calycanthine
$C_{22}H_{26}N_4$, derived from the genus name *Calycanthus* (sweetshrub), from *kalyx* (Greek: chalice) and canth(o)-, and -in(e)

calyculin
derived from the specific epithet of the species name *Discodermia calyx* (a Japanese marine sponge), from *kalyx* (Greek: chalice), and -in(e)

cAMP
$C_{10}H_{12}N_5O_6P$, an abbreviation for cyclic adenosine 3',5'-monophosphate

campestane
$C_{28}H_{50}$, derived from campesterol and -an(e)

campesterol
$C_{28}H_{48}O$, derived from the specific epithet of the species name *Brassica campestris* L.

(field mustard), from *campester* (Latin: of the field), from *campus* (Latin: field), and sterol

camphane (bornane)
$C_{10}H_{18}$, derived from camphor and -an(e)

camphene
$C_{10}H_{16}$, derived from camphor and -ene

camphor
$C_{10}H_{16}O$, ultimately derived from *kapor* (Khmer: camphor)

camphorone
$C_9H_{14}O$, derived from camphor and -one

Camps quinoline synthesis
named for the 19th century German chemist R. Camps

camptothecin
$C_{20}H_{16}N_2O_4$, derived from the genus name *Camptotheca* (Chinese trees), from *kamptos* (Greek: flexible) and thec(o)-, and -in(e)

camsyl
$C_{10}H_{15}O_2S-$, coined by contraction of camphor-10-sulfonyl

canadine
$C_{20}H_{21}NO_4$, derived from the specific epithet of the species name *Hydrastis canadensis* (goldenseal), from *canadensis* (New Latin: Canadian), and -in(e)

canadol
derived from Canada and *oleum* (Latin: oil)

canava(l)-
derived from the genus name *Canavalia* (jack bean), from *kanavali* (Malabar: jackbean, *Canavalia ensiformis*), from *kanam* (Malabar: forest) and *valli* (Malabar: climber)

canavalin
derived from canava(l)- and -in(e)

canavanine
$C_5H_{12}N_4O_3$, derived from canav(al)- and -in(e)

cancrinite
$(Na,Ca)_8(Si_6Al_6)O_{24}(CO_3)_2 \cdot 2H_2O$, named for the Russian minister E. F. Kankrin (1774-1845)

candicidin
coined by contraction of candi(d)-, -cide, and -in(e)

candi(d)-
derived from the fungal genus name *Candida*, from *candidus* (Latin: white)

candidin
$C_{47}H_{71}NO_{17}$, derived from candi(d)- and -in(e) – referring to this compound's activity against candida infections

cannabidiol
$C_{21}H_{30}O_2$, derived from cannab(in)-, di-, and -ol

cannab(in)-
derived from the genus name *Cannabis* (hemp), from *kannabis* (Greek: hemp)

cannabinoid
derived from cannab(in)- and -oid

cannabinol
$C_{21}H_{26}O_2$, derived from cannab(in)- and –ol

Cannizzaro reaction
named for the Italian chemist Stanislao Cannizzaro (1826-1910)

cantharidin
$C_{10}H_{12}O_4$, derived from canth(aro)- and -idin(e)

canth(aro)-
derived from the genus name *Cantharis* (beetles), from *kantharis* (Greek: Spanish fly)

canthaxanthin
$C_{40}H_{52}O_2$, derived from the genus name *Cantharellus* (mushrooms), from *cantharellus*, diminutive of *cantharus* (Latin: drinking vessel), from *kantharos* (Greek: drinking vessel), xanth(o)-, and -in(e)

capreomycin
derived from the specific epithet of the bacterial species name *Streptomyces capreolus*, ultimately from *capreolus* (Latin: wild goat), diminutive of *caper* (Latin: goat), and -mycin

capric acid (caprinic acid)
$C_{10}H_{20}O_2$, derived from *caper* (Latin: goat)

caproic acid
$C_6H_{12}O_2$, derived from *caper* (Latin: goat)

caprolactam
$C_6H_{11}NO$, derived from caproic acid and lactam

caproyl (hexanoyl)
$C_6H_{11}O-$, derived from caproic acid and -yl

^{1}**capryl (octyl)**
$C_8H_{17}-$,derived from caprylic acid

^{2}**capryl (octanoyl)**
$C_8H_{16}O-$, derived from caprylic acid

^{3}**capryl (decanoyl)**
$C_{10}H_{19}O-$, derived from capric acid

caprylene
C_8H_{16}, derived from ^{1}capryl- and -ene

caprylic acid
$C_8H_{16}O_2$, derived from *caper* (Latin: goat) and -yl

caps(a)-
derived from the genus name *Capsicum* (hot pepper), from *kaptein* (Greek: to bite, to gulp down)

capsaicin
$C_{18}H_{27}NO_3$, derived from caps(a)- and -icin(e)

capsanthin
$C_{40}H_{56}O_3$, derived from caps(a)-, anth(o)-, and -in(e)

capsarubin
$C_{40}H_{56}O_4$, derived from caps(a)-, rubi-, and -in(e)

capside
derived from *capsa* (Latin: box, capsule) and -ide

capsomer
derived from capside and -mer

capto-dative
derived from *captus* (Latin: caught, captive), from *capere* (Latin: to catch), and *datus* (Latin: given), from *dare* (Latin: to give)

caput mortuum
(Latin: dead head), an alchemistic term for the residuum after distillation or sublimation or for dross

caput mortuum vitrioli
(New Latin: dead head of metal sulfate), an alchemistic term for an iron oxide pigment obtained by calcination of iron sulphate

car-
derived from the genus name *Carum* (aromatic herbs), ultimately from *karon* (Greek: caraway)

carane
$C_{10}H_{18}$, derived from car- and -an(e)

carba-
an unsystematic prefix indicating the replacement of a heteroatom by a carbon atom, derived from carbon

carbachol
$C_6H_{15}ClN_2O_2$, coined by contraction of carbamoylcholine chloride

carbamate
$CH_2NO_2^-$, derived from carbamic acid and -ate

carbamic acid
CH_3NO_2, coined by contraction of carbonic acid and amide

carbamide (urea)
CH_4N_2O, coined by contraction of carbonic acid and amide

carbamoyl
CH_2NO-, derived from carbamic acid and -oyl

carbane
CH_4, systematic, but discouraged name for methane, derived from carb(o)- and -an(e)

carbanil
C_7H_5NO, an obsolete name for phenyl isocyanate, derived from carb(o)- and anil

carbanilic acid
$C_7H_7NO_2$, derived from carbanil

carbanilide (1,3-diphenylurea)
$C_{13}H_{12}N_2O$, coined by contraction of carbonic acid anilide

carbanion
derived from carb(o)- and anion

carbapenem
$C_7H_7NO_3$, derived from carba- and penem – referring to the replacement of penem's sulfur atom with a methylene group

carbapenemase (β-lactamase)
derived from carbapenem and -ase

carbazic acid
$CH_4N_2O_2$, derived from carb(o)- and az(o)-

[1]**carbazide**
CH_6N_4O, coined by contraction of carbohydrazide

[2]**carbazide**
CN_6O, coined by contraction of carbonyl azide

carbazole
$C_{12}H_9N$, derived from carb(o)-, az(a)-, and -ol(e)

carbazotic acid
$C_6H_3N_3O_7$, derived from carb(o)- and az(o)-

carbene
R_2C, derived from carb(o)- and -ene

carbenium ion
R_3C^+, derived from carbene and –ium

carbenoid
derived from carbene and -oid – referring to these compounds' role as carbene precursors

carbic anhydride (nadic anhydride)
$C_9H_8O_3$, derived from carb(o)-

carbide
C^{4-}, derived from carb(o)- and -ide

carbinamine
CH_5N, an obsolete name for methylamine, derived from carbine and amine

carbine
CH_3, an obsolete name for the methyl radical, derived from carbon and -in(e)

carbinol
CH_4O, an obsolete name for methanol, derived from carbine and -ol

carb(o)-
derived from carbon

carbocation
derived from carb(o)- and cation

carbohydrase (glycosidase)
derived from carbohydrate and -ase

carbohydrate
derived from carbon and hydrate – referring to the general empirical formula $[C(H_2O)]_n$ for monosaccharides

carbohydrazide
CH_6N_4O, coined by contraction of carbonic acid hydrazide

carbolic acid
C_6H_6O, an obsolete name for phenol, derived from *carbo* (Latin: coal) and *oleum* (Latin: oil) – referring to the occurrence of phenol in coal tar

carboline
$C_{11}H_8N_2$, coined by contraction of carbazole and quinoline

carbomycin (magnamycin)
$C_{42}H_{47}NO_{16}$, derived from carb(o)- and -mycin

carbon
C, ultimately derived from *carbo* (Latin: coal)

carbonado (black diamond)
C_n, derived from *carbonado* (Portuguese: carbonated)

carbonate
CO_3^{2-}, derived from carbon and -ate

carbonatite
a rock species, derived from carbonate and -ite

carbonic acid
H_2CO_3, derived from carbon

carbonium
C, New Latin name for carbon, derived from *carbo* (Latin: coal) and -ium

carbonium ion
R_3C^+, (improperly) derived from carb(o)- and -onium

¹carbonyl
CO as ligand, derived from carbonic acid and -yl

²carbonyl
$-C(O)-$, derived from carbon and -yl

carborane
coined by contraction of carbaborane

carborundum
SiC, coined by contraction of carbon and corundum

carbostyril
C_9H_7NO, derived from carbon, styrene, and -il(e)

carboxime (isofulminic acid)
CHNO, derived from carb(o)- and oxime

carboxy
CHO_2-, coined by contraction of carbonic acid and -yl

carboxyl
CHO_2-, coined by contraction of carbonic acid and -yl

carboxylase
derived from carboxylation and -ase

carboxypeptidase
derived from carboxy and peptidase – referring to the attack of these enzymes at peptides' C-terminals

carbuncle
$Fe_3Al_2Si_3O_{12}$, derived from *carbunculus* (Latin: small piece of (glowing) coal), diminutive of *carbo* (Latin: coal)

carbylamine (isocyanide)
RNC, derived from carbon, -yl, and amine

carbyne
RC, derived from carb(o)- and -yne

carce(r)-
derived from *carcer* (Latin: prison)

carceplex
derived (with contraction) from carce(r)- and complex

carcerand
derived from carce(r)- and -and

carcin(o)-
derived from *karkinos* (Greek: crab, cancer)

carcinogen
derived from carcin(o)- and -gen

cardanol
derived (with contraction) from the genus name *Anacardium* (trees), from ana- and *kardia* (Greek: heart)

cardanolide
$C_{23}H_{36}O_2$, derived from cardenolide and -an(e)

cardenolide
$C_{23}H_{34}O_2$, coined by contraction of cardiac and butenolide

cardiotoxin
derived from cardiotoxic and -in(e)

cardol
derived (with contraction) from the genus name *Anacardium* (trees), from ana- and *kardia* (Greek: heart), and -ol

carene (isodiprene)
$C_{10}H_{16}$, derived from car- and -ene

Carius method
named for the German chemist Georg Ludwig Carius (1829-1875)

carlsbergite
CrN, named for the Carlsberg Foundation, Copenhagen, Denmark

carmellose
coined by contraction of carboxymethylcellulose

carmin-
ultimately derived from *qirmiz* (Arabic: kermes) and *minium* (Latin: cinnabar)

carminic acid
$C_{22}H_{20}O_{13}$, derived from carmin-

carmot
the substance of the mythical philosopher's stone, of uncertain origin and etymology

carnallite
$KMgCl_3{\cdot}6H_2O$, named for the German

geologist Rudolph von Carnall (1804-1874)

carnegine
$C_{13}H_{19}NO_2$, derived from the genus name *Carnegiea* (cacti), in honor of the Scottish-US industrialist and humanitarian Andrew Carnegie (1835-1919), and -in(e)

carnelian (cornelian)
SiO_2, derived from *cornelle* (Medieval French: cherry), probably influenced by *caro* (Latin: flesh) – referring to this mineral's red color

carnitine
$C_7H_{15}NO_3$, derived from *Karnin* (German: basic substance isolated from meat extract), from *caro* (Latin: flesh), -ite, and -in(e)

carnos-
derived from *carnosus* (Latin: fleshy), from *caro* (Latin: flesh)

carnosine
$C_9H_{14}N_4O_3$, derived from carnos- and -in(e)

carnosinic acid
$C_{20}H_{28}O_4$, derived from carnos- and -in(e)

carnosolic acid
$C_{20}H_{28}O_5$, derived from carnos- and -ol

carnotite
$K_2(UO_2)_2(VO_4)_2 \cdot 3H_2O$, named for the French chemist and inspector general of mines M. A. Carnot (1839-1920)

Carnot's cycle
named for the French physicist Nicolas Leonard Sadi Carnot (1796-1832)

Carnot's law
named for the French physicist Nicolas Leonard Sadi Carnot (1796-1832)

Carnot's theorem
named for the French physicist Nicolas Leonard Sadi Carnot (1796-1832)

caroate
derived from Caro's acid and -ate

carolinium (berzelium)
Th, a name suggested for a hypothetical element claimed to be contained, together with the hypothetical element berzelium, in thorium, derived from the state of North Carolina, USA, and -ium – referring to the location where this work was carried out

Caro's acid
H_2SO_5, named for the German chemist Heinrich Caro (1834-1910)

carotene
$C_{40}H_{56}$, derived from *carota* (Latin: carrot) and -ene

Carothers silk (nylon)
named for the US chemist Wallace Hume Carothers (1896-1937)

carotol
$C_{15}H_{26}O$, derived from the specific epithet of the species name *Daucus carota* L. (Queen Anne's lace), from *carota* (Latin: carrot), and –ol

carpaine
$C_{28}H_{50}N_2O_4$, derived (with contraction) from the species name *Carica papaya* L. (papaya), from *caricus* (Latin: Carian) and papaya, from *papai* (Otomac: papaya), and -in(e)

carph(o)-
derived from *karphos* (Greek: dry stalk, straw)

carpholite
$MnAl_2Si_2O_6(OH)_4$, derived from carph(o)-, sider(o)-, and -ite – referring to the straw-yellow tufty appearance of this mineral

carphosiderite
derived from carph(o)-, sider(o)-, and -ite – referring to this mineral's iron content and straw yellow color

carp(o)-
derived from *karpos* (Greek: fruit)

carrageenan
ultimately named after its locality Carragheen, Ireland and -an

Carr-Price reaction
named for the British chemists F. H. Carr (1874-1969) and E. A. Price (born 1882)

Carroll rearrangement
named for the 20th century British chemist M. F. Carroll

CARS
an abbreviation for coherent anti-Stokes Raman spectroscopy

carthamin (carthamic acid)
$C_{43}H_{42}O_{22}$, derived from the genus name *Carthamus* (safflower), ultimately from *qartam* (Arabic: safflower), and -in(e)

carubicin
$C_{26}H_{27}NO_{10}$, derived from the specific epithet of the bacterial species name *Actinomadura carminata*, ultimately from carmin-, rubi-, and -icin(e); patterned after daunorubicin and doxorubicin

carv-
derived from the specific epithet of the species name *Carum carvi* (caraway), ultimately from *carvi* (Medieval Latin: caraway)

carvacrol
$C_{10}H_{14}O$, derived from carv-, acr(o)-, and -ol

carvone
$C_{10}H_{14}O$, derived from carv- and -one

caryophyllane
$C_{15}H_{28}$, derived from caryophyll(o)- and -an(e)

caryophyllene
$C_{15}H_{24}$, derived from caryphyll(o)- and -ene

caryophyll(o)-
derived from the genus name *Caryophyllus* (clove), from *karyophyllon* (Greek: clove tree), ultimately from *karion* (Greek: grain, seed, kernel) and phyll(o)-

carzinophilin
derived from *karkinos* (Greek: cancer), phil(o)-, and -in(e)

Casale process
named for the Italian chemist Luigi Casale (1882-1927)

cascarillin
$C_{22}H_{32}O_7$, derived from cascarilla (*Croton eluteria* (L.)), from *cascarilla* (Spanish: cascarilla), diminutive of *cascara* (Spanish: bark), and -in(e)

casein
derived from *caseus* (Latin: cheese) and -in(e)

casomorphine
derived from casein and morphine

caspase
coined by contraction of cysteinylaspartic acid protease

cassa-
derived from the native African name *cassa* (African sasswood, *Erythrophleum guineense*)

cassaidine
$C_{24}H_{41}NO_4$, derived from cassa- and -idin(e)

cassaine
$C_{24}H_{39}NO_4$, derived from cassa- and -in(e)

Cassella's acid
named for the chemical company Cassella AG, Frankfurt am Main, Germany

cassic acid (rhein)
$C_{15}H_8O_6$, derived from the genus name *Cassia* (herbs, shrubs, trees), ultimately from *qesiah* (Hebrew: cassia)

cassiopeium
Lu, an unsuccessfully suggested name for lutetium, after the constellation Cassiopeia, named for Kassiopeia, in Greek mythology a queen of Ethiopia and the mother of Andromeda

cassiterite
SnO_2, derived from *kassiteros* (Greek: tin), probably ultimately from the ancient Kassite people, and -ite

Cassius purple
named for the German physician Andreas Cassius (1605-1673)

Casson fluid
named for the 20th century British physicist N. Casson

castanospermine
$C_8H_{15}NO_4$, derived from the genus name *Castanospermum* (chestnut), from *castanea* (Latin: chestnut) and *sperma* (Greek: seed), ultimately from *speirein* (Greek: to grow), and -in(e)

Castner process
named for the 19th century US chemist H. J. Castner

Castner-Kellner process
named for the 19th century US chemist H. J. Castner and the 19th century Austrian chemist K. Kellner

castorite (petalite)
$LiAlSi_4O_{10}$, named for the ancient Greek mythical figure Kastor, possibly from *kekasmai* (Greek: to shine, to excel) – referring to the common association of this mineral with pollucite

Castro-Stephens coupling
named for the US chemists Charles E. Castro (born 1931) and R. D. Stephens

cat(a)-
derived from *kata* (Greek: down, against)

catabolism
derived from *katabole* (Greek: downthrow), from *kataballein* (Greek: to throw down), from cat(a)- and *ballein* (Greek: to throw)

catalase
derived from catalysis and –ase

catalposide
$C_{22}H_{26}O_{12}$, derived from the genus name *Catalpa* (catalpa), from *kutuhlpa* (Creek Indian: head with wings), and -oside

catalysis
derived from cat(a)- and lysis

cataphoresis
coined by contraction of cat(a)- and electrophoresis

catapleiite
$Na_2Zr(SiO_3)_3{\cdot}2H_2O$, derived from cat(a)-, *pleion* (Greek: more), and -ite – referring to

other rare minerals always accompanying this mineral

catechin (catechol)
$C_{15}H_{14}O_6$, derived from catech(u)- and -in(e)

catechol
$C_6H_6O_2$, derived from catech(u)- and -ol

catecholamine
$C_8H_{11}NO_2$, derived from catechol and amine

catecholase
derived (with contraction) from catechol and oxidase

catech(u)-
derived from the specific epithet of the species name *Acacia catechu* (catechu, Jerusalem thorn), from *catechu* (Malay: catechu), a constituent of this tree's heartwood

catechuic acid
$C_{15}H_{14}O_6$, derived from catech(u)-

catena-
derived from *catena* (Latin: chain), from *cassis* (Latin: net, web)

catenane
derived from catena- and -an(e)

catharanthine
$C_{21}H_{24}N_2O_2$, derived from the genus name *Catharanthus* (tropical shrubs), ultimately from *katharos* (Greek: pure) and anth(o)-, and -in(e)

cathepsin
derived from *kathepsein* (Greek: to boil down, to digest), from kata- and *hepsein* (Greek: to boil), and -in(e)

cathinone
$C_9H_{11}NO$, derived from the genus name *Catha* (khat), from *qat* (Arabic: khat), -in(e), and -one

cathode
derived from cat(a)- and *hodos* (Greek: road)

cathyl
$C_3H_5O_2$–, an obsolete name for ethoxycarbonyl coined by contraction of carboethoxyl

cation
derived from cat(a)- and ion

caul(o)-
derived from *kaulos* (Greek: stem, stalk)

caulophylline
$C_{12}H_{16}N_2O$, derived from the genus name *Caulophyllum* (cohosh), from caul(o)- and phyll(o)-, and -in(e)

caustic
ultimately derived from *kaustikos* (Greek: burning, corrosive), from *kaiein* (Greek: to burn)

cavansite
$Ca(VO)Si_4O_{10}\cdot 4H_2O$, coined by contraction of calcium, vanadium, silicon, and -ite

caviplex
derived (with contraction) from cavitand and complex

cavitand
derived from *cavum* (Latin: cavity, hole) and -and

CCC
$C_6H_{13}Cl_2N$, an abbreviation for chlorocholine chloride

CCK
an abbreviation for cholecystokinin

cCMP
$C_9H_{12}N_3O_7P$, an abbreviation for cyclic cytidine 3',5'-monophosphate

CD
an abbreviation for circular dichroism

cDNA
an abbreviation for complementary deoxyribonucleic acid

CDP
$C_9H_{15}N_3O_{11}P_2$, an abbreviation for cytidine 5'-diphosphate

CE
an abbreviation for capillary electrophoresis

ceanothic acid (emmolic acid)
$C_{30}H_{46}O_5$, derived from the genus name *Ceanothus* (American vines, shrubs, small trees), from *keanothos* (Greek: a thistle)

cecropin
derived from the specific epithet of the species name *Hyalophora cecropia* (cecropia silkmoth), ultimately from *kekropios* (Greek: Athenian), from Kekrops, mythical first king of Athens, and -in(e)

cedrane
$C_{15}H_{26}$, derived from cedr(o)- and -ol

cedr(o)-
ultimately derived from *kedros* (Greek: cedar)

cedrol
$C_{15}H_{26}O$, derived from cedr(o)- and -ol

celadonite
$(K,Na)(Fe,Al,Mg)_2(Si,Al)_4O_{10}(OH)_2$, derived from celadon green, named for Céladon, Astrée's lover in Honoré d'Urfé's romance *L'Astrée* (1610), and -ite

celest-
derived from *caelum* (Latin: sky)

celesticetin
$C_{24}H_{36}N_2O_9S$, derived from the specific epithet of the bacterial species name *Streptomyces caelestis*, from celest- – referring to this microorganism's blue color – and -icetin(e)

celestin blue
$C_{17}H_{18}ClN_3O_4$, derived from celest- and -in(e)

celestine (celestite)
$SrSO_4$, derived from *celestino* (Italian: sky blue) and -in(e) – referring to this mineral's blue color

¹celite (brownmillerite)
Ca_2AlFeO_5, derived from the letter C and -lite

²celite
SiO_2, coined as a trademark, of unknown etymology

cell(o)-
derived from cellulose

cellobiase
derived from cellobiose and -ase

cellobionic acid
$C_{12}H_{22}O_{12}$, derived from cellobiose

cellobiose
$C_{12}H_{22}O_{11}$, derived from cell(o)-, bi-, and -ose

cellocidin
$C_4H_4N_2O_2$, derived from cell(o)-, -cide, and -in(e)

cellone
$C_2H_2Cl_4$, coined as a trademark, of unknown etymology

cellophane
derived from cellulose and *diaphanes* (Greek: translucent)

cellosolve
$C_4H_{10}O_2$, coined as a trademark, of unknown etymology

cellulase
derived from cellulose and -ase

celluloid
coined as a trademark, derived from cellulose and -oid

cellulose
derived from *cellule* (French: living cell), from *cellula* (New Latin: living cell), diminutive of *cella* (Latin: small room), and -ose

celsian
$Ba(AlSiO_4)_2$, named for the Swedish astronomer and natural scientist Anders Celsius (1701-1744)

Celsius temperature scale
named for the Swedish astronomer and natural scientist Anders Celsius (1701-1744)

celtium
Hf, unsuccessfully suggested name for hafnium, in honor of the Celts (Latin *Celtae*)

cembrene
$C_{20}H_{32}$, derived from the specific epithet of the species name *Pinus cembra* (Swiss pine), ultimately from *zimbar* (Old High German: wood), and -ene

cement
ultimately derived from *caementum* (Latin: stone chippings), from *caedere* (Latin: to chop)

cementite
Fe_3C, derived from cement and -ite

centaurein
$C_{24}H_{26}O_{13}$, derived from the genus name *Centaurea* (centaury), ultimately named for the centaur Chiron who, according to Greek myths, discovered these plants

centrifuge
derived from *centrifugus* (New Latin: fleeing from the center), from *centrum* (Latin: center) and *fugere* (Latin: to flee)

cephaeline
$C_{28}H_{38}N_2O_4$, derived from the genus name *Cephaelis* (tropical shrubs, trees), from cepha(lo)- and *eilein* (Greek: to compress), and -in(e)

cephalin
derived from cephal(o)- and -in(e)

cephal(o)-
derived from *kephale* (Greek: head)

cephalonic acid
$C_{25}H_{36}O_4$, derived from the fungal genus name *Cephalosporium*, from cephal(o)-, and -one

cephalosporanic acid
$C_{10}H_{11}NO_5S$, derived from cephalosporin and -an(e)

cephalosporin
derived from the fungal genus name *Cephalosporium*, from cephal(o)- and *spora* (Greek: seed) from *speirein* (Greek: to sow, to strew), and -in(e)

cephalotaxine
$C_{18}H_{21}NO_4$, derived from the genus name *Cephalotaxus* (plum yew), from cephal(o)- and taxo-, and -in(e)

cepham
C_6H_9NOS, derived (with contraction) from cephalosporanic acid and lactam

cephamycin
derived from cephalosporanic acid and -mycin

cepharantine
$C_{37}H_{38}N_2O_6$, derived from the specific epithet of the species name *Stephania cephalantha* Hayata (a weed), from cephal(o)- and anth(o)-, and -in(e)

cephem
C_6H_7NOS, derived from cepham and -ene

ceramic
derived from *keramikos* (Greek: pottery, earthenware), from *keramos* (Greek: clay), possibly akin to *kerannymi* (Greek: to mix)

ceramide
derived (with contraction) from cerebroside and amide

cerane (hexacosane)
$C_{26}H_{54}$, derived from ^{2}cer(o)- and -an(e)

cerase
derived from ^{2}cer(o)- and -ase

cerberin
$C_{32}H_{48}O_9$, derived from the genus name *Cerbera* (dogbanes), from *Kerberos*, a Greek mythical three-headed watchdog, and -in(e)

cerebr(o)-
derived from *cerebrum* (Latin: brain)

cerebrose
$C_6H_{12}O_6$, an obsolete name for galactose, derived from cerebr(o)- and -ose

cerebroside
derived from cerebrose and -ide

ceresin
derived from ^{2}cer(o)- and -in(e)

ceri-
derived from cerium

ceria
Ce_2O_3, derived from cerium and -a

ceric
Ce^{4+}, derived from cerium

cerite
$(Ca,Mg)_2X_8(SiO_4)_7$, X = lanthanoid, derived from cerium and -ite

cerium
Ce, named for the asteroid Ceres, ultimately derived from the Roman goddess Ceres, from the root *ker* (Indo-European: to grow)

cermet
coined by contraction of ceramic and metal

^{1}cer(o)-
derived from cerium

^{2}cer(o)-
ultimately derived from *keros* (Greek: wax)

cerotic acid
$C_{26}H_{52}O_2$, derived from *keroton* (Greek: pomade), from ^{2}cer(o)-

cerotin (hexacosan-1-ol)
$C_{26}H_{54}O$, derived from cerotic acid

cerous
Ce^{3+}, derived from cerium and –ous

cerulenin
$C_{12}H_{17}NO_3$, derived from the specific epithet of the fungal species name *Cephalosporium caerulens*, from *caerulens* (Latin: bluish), from cerul(o)- – referring to the bluish color of this microorganism – and -in(e)

ceruletide (caerulein, cerulein)
$C_{58}H_{73}N_{13}O_{21}S_2$, derived from the specific epithet of the species name *Hyla caerulea* (Australian tree frog), from cerul(o)-, -ete, and -ide

cerul(o)-
derived from *caeruleus* (Latin: dark blue)

cerulosplasmin
derived from cerul(o)-, plasma, and -in(e)

cerussite
$PbCO_3$, derived from *cerussa* (Latin: white lead), possibly of Greek origin, and -ite

cervantite
Sb_2O_4, named after this mineral's locality Cervantes, Spain

cervicarcin
$C_{19}H_{20}O_9$, derived (with contraction) from cervical cancer and -in(e)

ceryl
$C_{26}H_{53}-$, derived from ceryl alcohol

ceryl alcohol
$C_{26}H_{54}O$, derived from ^{2}cer(o)- and -yl

cesium
Cs, derived from *caesius* (Latin: bluish gray) – referring to the color of this element's characteristic spectral lines

cetane (hexadecane)
$C_{16}H_{34}$, derived from cet(o)- and -an(e)

cet(o)-
derived from *cetus* (Latin: whale)

cetrimonium
$C_{19}H_{42}N^+$, coined by contraction of cetyltrimethylammonium

cetyl
$C_{16}H_{33}-$, derived from cetyl alcohol

cetyl alcohol
$C_{16}H_{34}O$, derived from ^{2}cet(o)- and -yl

ceva-
derived from *cebadilla* (Spanish: sabadilla, *Schoenocaulon officinale*), diminutive of *cebada* (Spanish: barley), ultimately from *cibus* (Latin: food, meal)

cevadine
$C_{32}H_{49}NO_9$, derived from ceva- and -idin(e)

cevane
$C_{27}H_{45}N$, derived from ceva- and -an(e)

cevine
$C_{27}H_{43}NO_8$, derived from ceva- and -in(e)

ceylonite (pleonaste)
$Al_2(Fe,Mg)O_4$, derived from this mineral's locality Ceylon and -ite

cGMP
$C_{10}H_{12}N_5O_7P$, an abbreviation for cyclic guanosine 3',5'-monophosphate

chabazite (chabasite)
$CaAl_2Si_4O_{12}\cdot 6H_2O$, derived from *chabazios*, a common misspelling of *chalazios* (Greek: a precious stone resembling a hailstone), from *chalaza* (Greek: hailstone), and -ite

chalcanthite
$CuSO_4\cdot 5H_2O$, derived from chalc(o)-, anth(o)-, and -ite – referring to this mineral's copper content and flowery

appearance

chalcedony
SiO_2, possibly named after the ancient Greek town of Chalcedon in Asia Minor

chalc(o)-
derived from *chalkos* (Greek: copper)

chalcocite (chalcosine)
Cu_2S, derived from *chalcosine* (French: chalcocite) and -ite – referring to this mineral's content of copper

chalcogen
derived from chalc(o)- and -gen – referring to the occurrence of these elements in copper ores

chalcomenite
$CuSeO_3{\cdot}2H_2O$, derived from chalc(o), *mene* (Greek: moon), and -ite – referring to this mineral's content of selenium, which has been named for the ancient Greek moon goddess Selene

chalcomycin
$C_{35}H_{56}O_{14}$, derived from chalc(o)- and -mycin

chalcone
derived from chalc(o)- and -one – referring to these compounds' often yellow color

chalcopyrite
$CuFeS_2$, derived from chalc(o)- and pyrite

chalcose (lankavose)
$C_7H_{14}O_4$, derived from chalcomycin and -ose

chalcosine (chalcocite)
Cu_2S, derived from chalc(o)- and -in(e) – referring to this mineral's content of copper

chalcostibite (wolfsbergite)
$CuSbS_2$, derived from chalc(o)-, stibium, and -ite – referring to this mineral's content of copper and antimony

chalcotrichite
Cu_2O, derived from chalc(o)-, trich(o)-, and -ite – referring to this mineral's content of copper and hairy appearance

chalk
$CaCO_3$, ultimately derived from *chalix* (Greek: stone)

chalone
coined by contraction of *chalasis* (Greek: a slackening), from *chalan* (Greek: to slacken), and hormone

chamazulene
$C_{14}H_{16}$, derived from the specific epithet of the species name *Matricaria chamomilla* L. (chamomile), from *chamaimelon* (Greek: chamomile), from *chamai* (Greek: on the ground) and *melon* (Greek: apple) – referring to the apple-like smell of this plant – and azulene

chamosite
$(Fe^{II},Mg,Fe^{III})_5Al(Si_3Al)O_{10}(OH,O)_8$, named after this mineral's locality Chamoson, Switzerland

champacol
$C_{15}H_{26}O$, derived from champac (*Michelia champaca*), from *campaka* (Sanskrit: champac), of Dravidian origin, and -ol

chanoclavine (secaclavine)
$C_{16}H_{20}N_2O$, derived (with latinization) from chain, clavi(c)-, and -in(e) – referring to the opened D ring relative to the parent compound ergolene

chaoite
C_n, named for the Chinese-US petrologist Edward C. T. Chao (born 1919)

chaperonin
derived from chaperon, from *chaperon* (French: chaperon, literally hood), and -in(e)

Chapman mechanism
named for the British mathematician and physicist Sydney Chapman (1888-1970)

Chapman rearrangement
named for the British chemist Arthur William Chapman (born 1898)

Chardonnet silk
named for the French chemist and physicist L.-M.-H. Bernigand Comte de Chardonnet de Grange (1839-1924)

Charles' law (Amontons' law)
named for the French physicist Jacques Alexandre César Charles (1746-1823)

charoite
$(K,Na)_5(Ca,Ba,Sr)_8(Si_6O_{15})_2(Si_6O_{16})(OH,F)\cdot nH_2O$, most likely derived from *chary* (Russian: magic, delusion) and -ite

chartreusin
$C_{32}H_{32}O_{14}$, derived from the specific epithet of the bacterial species name *Streptomyces chartreusis*, from the color of Chartreuse liqueur, ultimately from the Carthusian monastery La Grande Chartreuse, France, and -in(e)

charybdotoxin (CTX)
$C_{176}H_{277}N_{57}O_{55}S_7$, derived from Charybdis, from the ancient tale of Odysseus' dangerous passage through the channel between the headland Scylla and the whirlpool Charybdis, and toxin – referring to this toxin's action on cellular potassium channels

chaulmoogric acid
$C_{18}H_{32}O_2$, derived from *caulmugra* (Bengal: chaulmoogra), from *caul* (Bengal: rice) and *mugra* (Bengal: bowstring hemp, *Sansevieria zeylanica*)

chavibetol
$C_{10}H_{12}O_2$, derived (with contraction) from the species name *Chavica betle* (betel pepper), from *cavika* (Sanskrit: a pepper) and *betle* (Portuguese: betel), ultimately from *verrilai* (Tamil: *Piper betle*) , and -ol

chavi(c)-
derived from the genus name *Chavica* (peppers), from *cavika* (Sanskrit: a pepper)

chavicic acid
$C_{12}H_{10}O_4$, derived from chavi(c)-

chavicine
$C_{17}H_{19}NO_3$, derived from chavi(c)- and -in(e)

chavicol
$C_9H_{10}O$, derived from chavi(c)- and -ol

chelate
from chel(e)- and -ate

chel(e)-
from *chele* (Greek: claw)

chelerythrine
$C_{21}H_{18}NO_4^+$, derived from chel(idon)-, erythr(o)-, and -in(e)

cheletropic (chelotropic)
derived from chel(e)- and -tropic

chelidamic acid
$C_7H_5NO_5$, coined by contraction of chelidonic acid and amide

chel(idon)-
derived from the genus name *Chelidonium* (celandine), from *chelidonios* (Greek: of the swallow), from *chelidon* (Greek: a swallow)

chelidonic acid (jerva acid, jervaic acid, jervasic acid)
$C_7H_4O_6$, derived from chel(idon)-

chelidonine (stylophorine)
$C_{20}H_{19}NO_5$, derived from chel(idon)- and -in(e)

chemiluminescence
derived from chemical and luminescence

chemisorption
derived from chemical and (ad)sorption

chemistry
derived from *chemeia* (Greek: chemistry), from *chymos* (Greek: juice), from *chein* (Greek: to pour)

chemo-
derived from chemistry

chemometry
derived from chemo- and -metry

chen(o)-
derived from *chen* (Greek: goose)

chenodesoxycholic acid
$C_{24}H_{40}O_4$, derived from chen(o)- and desoxycholic acid

chenodiol
$C_{24}H_{40}O_4$, derived from chen(o)-, di-, and -ol

chert
SiO_2, a name of unknown origin

Chevrel phase
named for the French chemist R. A. P. Chevrel (born 1944)

Chevreul's salt
$Cu_3(SO_3)_2{\cdot}2H_2O$, named for the French chemist Eugène Chevreul (1786-1889)

chiast(o)-
derived from *chiastos* (Greek: cross)

chiastolite
$Al_2(O)SiO_4$, derived from chiast(o)- and -lite – referring to cross-shaped inclusions of carbon in this mineral

chiavennite
$CaMnBe_2(Si,Al)_5O_{13}(OH)_2{\cdot}2H_2O$, named after this mineral's locality Chiavenna, Italy

Chicago blue
$C_{34}H_{24}N_6Na_4O_{16}S_4$, named for the city of Chicago, IL, USA

Chichibabin's hydrocarbon
$C_{38}H_{28}$, named for the Russian chemist Aleksei Yevgenievich Chichibabin (1871-1945)

Chichibabin pyridine synthesis
$C_{38}H_{28}$, named for the Russian chemist Aleksei Yevgenievich Chichibabin (1871-1945)

Chichibabin reaction
named for the Russian chemist Aleksei Yevgenievich Chichibabin (1871-1945)

chimaphilin
$C_{12}H_{10}O_2$, derived from the genus name *Chimaphila* (prince's pine), from chim(o)- and phil(o)-, and -in(e)

chim(o)-
derived from *cheima* (Greek: winter)

chimonanthine
$C_{22}H_{26}N_4$, derived from the genus name

Chimonanthus (Asiatic shrubs), from chim(o)- and anth(o)-, and -in(e)

chimyl alcohol
$C_{19}H_{40}O_3$, derived from the genus name *Chimaera* (fish), from *chimaira* (Greek: she-goat, mythical monster), and -yl

chiolite
$Na_5Al_3F_{14}$, derived from chi(ono)- and -lite – referring to this mineral's white color

chi(ono)-
derived from *chion* (Greek: snow)

chiral
derived from *cheir* (Greek: hand)

chiron
coined by contraction of chiral synthon

chiroptical
coined by contraction of chiral and optical

chitin
derived from *chiton* (Greek: tunic), of Semitic origin, and -in(e)

chitinase
derived from chitin and -ase

chit(o)-
derived from chitin

chitobiose
$C_{12}H_{24}N_2O_9$, derived from chit(o)-, bi-, and -ose

chitosamine (glucosamine)
$C_6H_{13}NO_5$, derived from chit(o)-, -ose, and -amine

chitosan
derived from chit(o)-, -ose, and -an

chitose
$C_6H_{10}O_5$, derived from chit(o)- and -ose

chkalovite
$Na_2BeSi_2O_6$, possibly named for the Russian aviator Valerii Pavlovich Chkalov (1904-1938)

chloanthite (nickel-skutterudite)
$(Ni,Co)As_{3-x}$, derived from *chloanthes* (Greek: budding, pale), from *chloos* (Greek: light green color) and *anthes* (Greek: blooming) – referring to this mineral's light green color

chloral
C_2HCl_3O, coined by contraction of trichloroacetaldehyde

α-chloralose
$C_8H_{11}Cl_3O_6$, coined by contraction of 1,2-*O*-[(1*R*)-2,2,2-trichloroethylidene]-α-D-glucofuranose

chloramine-B
$C_6H_5ClNNaO_2S$, derived from chlorine, amine, and benzenesulfonic acid

chloramine-T
$C_7H_7ClNNaO_2S$, derived from chlorine, amine, and *p*-toluenesulfonic acid

chloramphenicol
$C_{11}H_{12}Cl_2N_2O_5$, coined by contraction of chlorine, amide, phenyl-, nitro-, and glycol

chlorane
HCl, derived from chlorine and -an(e)

chloranil
$C_6Cl_4O_2$, derived from chlorine and aniline – referring to this compound's synthesis from aniline

chloranilic acid
$C_6H_2Cl_2O_4$, derived from chloranil

chlorapatite
$Ca_5Cl(PO_4)_3$, derived from ^{2}chlor(o)- and apatite

chlorargyrite (cerargyrite, horn silver)
AgCl, derived from ^{2}chlor(o)-, argyr(o)-, and –ite

chlorate
ClO_3^-, derived from chloric acid and -ate

chlordan(e)
$C_{10}H_6Cl_8$, derived from chlorine and indane

chlorendic anhydride
$C_9H_2Cl_6O_3$, coined by contraction of 1,4,5,6,7,7-hexachloro-*endo*-5-norbornene-2,3-dicarboxylic anhydride

chloric acid
$HClO_3$, derived from chlorine

chloride
Cl^-, derived from chlorine and -ide

chlorin (dihydroporphine)
$C_{20}H_{16}N_4$, derived from ^{1}chlor(o)- and -in(e)

chlorine
Cl, derived from ^{1}chlor(o)- and -in(e) – referring to the green color of Cl_2

^{1}chlorite
ClO_2^-, derived from chlorous acid and -ite

^{2}chlorite
$(Mg,Fe)_5Al(Si_3O_{10})(OH)_8$, derived from ^{1}chlor(o)- and -ite – referring to this mineral's green color

chloritoid
$(Fe,Mg,Mn)Al_2SiO_5(OH)_2$, derived from ^{2}chlorite and -oid – referring to this mineral's resemblance of chlorite minerals

^{1}chlor(o)-
derived from *chloros* (Greek: yellow green)

^{2}chlor(o)-
derived from chlorine

chlorocalcite
$CaKCl_3$, derived from ^{2}chlor(o)-, calcium, and -ite – coined in the erroneous belief that this mineral only consists of these two elements

chloroform
$CHCl_3$, derived from ^{2}chlor(o)- and -form

chlorogenic acid
$C_{16}H_{18}O_9$, derived from ^{1}chlor(o)- and -genic – referring to the fact that an aerated solution of this acid in aqueous ammonia turns green with time

chlorogenin
$C_{27}H_{44}O_4$, derived from the genus name *Chlorogalum* (soap plant), from ^{1}chlor(o)- and *gala* (Greek: milk), -gen, and -in(e)

chlorohydrin
derived from ^{2}chlor(o)-, ^{1}hydr(o)-, and -in(e)

chlorophyll
derived from ^{1}chlor(o)- and *phyllon* (Greek: leaf)

chloropicrin
CCl_3NO_2, derived from ^{2}chlor(o)-, picric acid, and -in(e) – referring to a synthesis of this compound by exhaustive chlorination of picric acid

chloroprene
C_4H_5Cl, derived from ^{2}chlor(o)- and isoprene

chlorothricin
$C_{50}H_{63}ClO_{16}$, derived from ^{2}chlor(o)-, an unexplained name fragment, and -in(e)

chlorous acid
$HClO_2$, derived from chlorine

chlorum
Cl, New Latin name for chlorine, from ^{1}chlor(o)- and –um

cholamine
C_2H_7NO, coined by contraction of choline and amine

cholane
$C_{24}H_{42}$, derived from chol(e)- and -an(e)

cholanic acid
$C_{24}H_{40}O_2$, derived from chol(e)- and -an(e)

cholanthrene
$C_{20}H_{14}$, derived from chol(e)- and -anthrene

chol(e)-
derived from *chole* (Greek: bile)

cholecalciferol (vitamin D_3)
$C_{27}H_{44}O$, derived from chol(e)- and calciferol

cholecystokinin (CCK)
derived from chol(e)-, cyst(o)-, and kinin

choleinic acid
$C_{24}H_{38}O_5$, derived (with contraction) from desoxycholic acid and -ein(e)

cholestane
$C_{27}H_{48}$, derived from chol(e)-, sterol, and -an(e)

cholesteric
derived from cholesterin

cholesterin
$C_{27}H_{46}O$, an obsolete name for cholesterol, derived from chol(e)-, *stereos* (Greek: solid), and -in(e)

cholesterol
$C_{27}H_{46}O$, derived from chol(e)-, *stereos* (Greek: solid), and -ol

cholic acid
$C_{24}H_{40}O_5$, derived from chol(e)-

choline
$C_5H_{14}NO^+$, derived from chol(e)- and -in(e)

cholinesterase
derived from choline ester and –ase

chondrillasterol
$C_{29}H_{48}O$, derived from the genus name *Chondrilla* (sponges), from *chondrilla*, diminutive of *chondrile* (Greek: plant which exudes a gum), and sterol

chondr(o)-
derived from *chondros* (Greek: cartilage, grain)

chondrocurine (chondocurine)
$C_{36}H_{38}N_2O_6$, derived from the genus name *Chondrodendron* (South American vines), from chondr(o)- and dendr(o)-, curare, and -in(e)

chondrodite
$(Mg,Fe,Ti)_5(SiO_4)_2(F,OH)_2$, derived from chondr(o)- and -ite – referring to this mineral's grainy crystal habit

chondrofoline
$C_{37}H_{40}N_2O_6$, derived from the genus name *Chondrodendron* (South American vines), from chondr(o)- and dendr(o)-, fol-, and -in(e)

chondroitin sulfate (chondroitinsulfuric acid, chondroitic acid)
derived from chondr(o)- and sulfate

chondrosamine
$C_6H_{13}NO_5$, derived (with contraction) from chondroitin sulfate, -ose, and amine

chondrosine
$C_{12}H_{21}NO_{11}$, derived (with contraction) from chondroitin sulfate, -ose, and -in(e)

chori(o)-
derived from chorion, from *chorion* (Greek: skin)

choriogonadotropin
derived from chori(o)-, gonad, and –tropin

choriomammotropin
derived from chori(o)-, *mamma* (Latin: female breast), and -tropin

chorismic acid
$C_{10}H_{10}O_6$, derived from chorism, from *chorizein* (Greek: to separate) – coined to describe chorismic acid's position at a bifurcation of the biosynthetic 'tree' leading to anthranilic acid in one and prephenic acid in the other direction

chroman
$C_9H_{10}O$, derived from [1]chrom(o)- and -an(e)

-chromatic
derived from chromat(o)-

chromatin
derived from chromat(o)- and -in(e)

chromat(o)-
derived from *chroma* (Greek: color)

chromatography
derived from chromat(o)- and *graphein* (Greek: to write)

chromatophore
derived from chromat(o)- and -phore

chrome
Cr, an obsolete name for chromium, derived from [1]chrom(o)-

-chrome
derived from *chroma* (Greek: color)

chrome diopside
$(Ca,Cr)MgSi_2O_6$, derived from chrome and diopside

chromene
C_9H_8O, derived from chroman and –ene

chromia
Cr_2O_3, derived from chromium and -a

-chromism
derived from -chrome

chromite
$FeCr_2O_4$, derived from chromium and -ite

chromium
Cr, derived from [1]chrom(o)- and -ium – referring to the fact that all chromium compounds are colored

[1]chrom(o)-
derived from *chroma* (Greek: color)

[2]chrom(o)-
derived from chromium

chromocene
$C_{10}H_{10}Cr$, derived from [2]chrom(o)- and -ocene

chromogen
derived from [1]chrom(o)- and -gen

chromomycin
derived from [1]chrom(o)- and -mycin – referring to these antibiotics' yellow color

chromone
$C_9H_6O_2$, derived from chroman and -one

chromophore
derived from [1]chrom(o)- and -phore

chromoprotein
derived from [1]chrom(o)- and protein

chromotrope
derived from [1]chrom(o)- and -trope

chromotropic acid
$C_{10}H_8O_8S_2$, derived from chromotrope

chron(o)-
derived from *chronos* (Greek: time)

chrysamine
derived from chrys(o)- and amine

chrysammic acid (chrysamminic acid)
$C_{14}H_4N_4O_{12}$, derived (with contraction) from chrysazin and a name fragment of obscure etymology

chrysanthe(ma)-
derived from the genus name *Chrysanthemum* (plants), from chrys(o)- and anth(o)-

chrysanthemaxanthin
$C_{40}H_{56}O_3$, derived from chrysanthe(ma)- and xanthin

chrysanthemic acid (chrysanthemumic acid)
$C_{10}H_{16}O_2$, derived from chrysanthe(ma)-

chrysanthemin (asterin)
$C_{21}H_{20}O_{11}$, derived from chrysanthe(ma)- and -in(e)

chrysanthenone
$C_{10}H_{14}O$, derived from chrysanthe(ma)- and -one

chrysarobin
derived from chrys(o)- and the specific epithet of the species name *Andira araroba* Aguiar (araroba), from *arariba* (Tupi: araroba)

chrysazin
$C_{14}H_8O_4$, coined by contraction of chrys(o)- and alizarin

chrysene
$C_{18}H_{12}$, derived from chrys(o)- and –ene

chrysergonic acid
$C_{32}H_{30}O_{14}$, derived from chrys(o)- and ergot

chrysin
$C_{15}H_{10}O_4$, derived from chrys(o)- and -in(e)

chrys(o)-
derived from *chrysos* (Greek: gold)

chrysoberyl (alexandrite, cat's eye, cymophane)
$BeAl_2O_4$, derived from chrys(o)- and beryl – referring to this mineral's golden color

chrysocolla
$CuSiO_3{\cdot}2H_2O$, derived from chrys(o)- and *kolla* (Greek: glue) – referring to the use of this mineral to solder gold

chrysoidine
$C_{12}H_{13}ClN_4$, derived from *chrysoeides* (Greek: gold-like) and -in(e)

chrysolite (olivine, peridot)
$(Fe,Mg)_2SiO_4$, derived from chrys(o)- and -lite – referring to this mineral's yellow color

chrysophanic acid
$C_{15}H_{10}O_4$, derived from *chrysophanes* (Greek: shining like gold), from chrys(o)- and -phan(e)

chrysoprase
SiO_2, derived from chrys(o)- and prase(o)- – referring to this mineral's green color

chrysotile
$Mg_3Si_2O_5(OH)_4$, derived from chrys(o)- and *tilos* (Greek: anything plucked, hair, fiber), from *tillein* (Greek: to pluck) – referring to this mineral's yellow color and fibrous appearance

Chugaev reaction
named for the Russian chemist Lev Aleksandrovich Chugaev (1871-1945)

Chugaev's reagent (dimethylglyoxime)
$C_4H_8N_2O_2$, named for the Russian chemist Lev Aleksandrovich Chugaev (1871-1945)

chym(o)-
derived from *chymos* (Greek: juice)

chymopapain
derived from chym(o)- and papain

chymosin (rennin)
derived from chym(o)- and -in(e)

chymotrypsin
derived from chym(o)- and trypsin

[1]CI
an abbreviation for chemical ionization

[2]CI
an abbreviation for configuration interaction

C.I.
an abbreviation for Colour Index

Ciamician-Dennstedt rearrangement
named for the Armenian-Italian chemist Giacomo Luigi Ciamician (1857-1922) and the German chemist M. Dennstedt

cichoriin
$C_{15}H_{16}O_9$, derived from the genus name *Cichorium* (chicory), ultimately from *kichora* (Greek: chicory), and -in(e)

cicutoxin
$C_{17}H_{22}O_2$, derived from the genus name *Cicuta* (water hemlock), from *cicuta* (Latin: poison hemlock, probably *Conium maculatum*), and toxin

-cide
derived from *-cida* (Latin: murderer), from *caedere* (Latin: to fell)

CIDEP
an abbreviation for chemically induced dynamic electronic polarization

CIDNP
an abbreviation for chemically induced dynamic nuclear polarization

ciguatoxin
derived from *ciguatera* (American Spanish: sea snail poisoning), from *cigua* (Taino: sea snail)

-cillin
derived from penicillin

cimigenol
$C_{30}H_{48}O_5$, derived from the genus name *Cimicifuga* (bugbane), from the genus name *Cimex* (bedbugs), from *cimex* (Latin: bedbug), and *fugare* (Latin: to put to flight), -gen, and -ol

cIMP
$C_{10}H_{11}N_4O_7P$, an abbreviation for cyclic inosine 3',5'-monophosphate

cinchamidine
$C_{19}H_{24}N_2O$, coined by contraction of cinch(ona)- and amidine

cinch(ona)-
derived from the genus name *Cinchona* (tropical trees), named for Doña Francisca Henriquez de Ribera, Countess of Chinchón, vicereine of Peru (deceased 1641)

cinchomeronic acid
$C_7H_5NO_4$, derived from cinch(ona)-, -mer, and -one

cinchonamine
$C_{19}H_{24}N_2O$, derived from cinch(ona)- and amine – referring to this alkaloid's isomerism with certain cinchona alkaloids

cinchonan
$C_{19}H_{22}N_2$, derived from cinch(ona)- and -an(e)

cinchonidine
$C_{19}H_{22}N_2O$, derived from cinch(ona)- and -idin(e)

cinchonine
$C_{19}H_{22}N_2O$, derived from cinch(ona)- and -in(e)

cinchotoxin
$C_{19}H_{22}N_2O$, derived from cinch(ona)- and toxin

cinenic acid
$C_9H_{16}O_3$, coined by variation of cineolic acid

cineol (eucalyptol)
$C_{10}H_{18}O$, derived from the specific epithet of the species name *Artemisia cina* (santonica, levant wormwood), from cino-, and -ol

cineolic acid
$C_{10}H_{16}O_5$, derived from cineol

cinerin
$C_{20}H_{28}O_3$, derived from ciner(o)- and -in(e)

ciner(o)-
derived from the specific epithet of the species name *Chrysanthemum cinerariifolium* (Dalmatian pyrethrum), ultimately from *cinis* (Latin: ashes)

cinerolone
$C_{10}H_{14}O_2$, derived from ciner(o)-, -ol, and -one

cinerubin
derived from the specific epithet of the bacterial species name *Streptomyces cinereoruber*, from ciner(o)- and *ruber* (Latin: dark red), and -in(e)

***cine*-substitution**
derived from kin(e)-

cinnabar (cinnabarite, paragite)
HgS, ultimately derived from *zanzifrah* (Persian: cinnabar)

cinnabarine
$C_{14}H_{10}N_2O_5$, derived from the specific epithet of the species name *Trametes cinnabarina* Jacq. (cinnabar polypore), from cinnabar, and -in(e)

cinnamic acid
$C_9H_8O_2$, derived from cinnamon, ultimately from *qinnamon* (Hebrew: cinnamon)

cinnamol
C_8H_8, an obsolete name for styrene, derived from cinnamic acid and -ol(e) – referring to the formation of this liquid hydrocarbon by decarboxylation of cinnamic acid

cinnoline
$C_8H_6N_2$, coined by alteration of *Chinolin* (German: quinoline)

cino-
derived from *chinensis* (New Latin: Chinese)

cinobufotalin
$C_{26}H_{34}O_7$, derived from cino- and bufotalin

CIP system
an abbreviation for Cahn-Ingold-Prelog system

circulin
derived from the specific epithet of the bacterial species name *Bacillus circulans*, from *circulans* (Latin: circulating), from *circulare* (Latin: to circulate), and -in(e)

circum-
derived from *circum-* (Latin: all around)

cis-
derived from *cis* (Latin: on this side)

cisoid
derived from cis- and -oid

citicoline
$C_{14}H_{26}N_4O_{11}P_2$, coined by contraction of cytidine diphosphate choline ester

citraconic acid
$C_5H_6O_4$, derived from citr(o)- and aconic acid

citral (geranial)
$C_{10}H_{16}O$, derived from citr(o)- and -al

citramalic acid
$C_5H_8O_5$, derived from citr(o)- and malic acid

citraurin
derived from citr(o)-, aur(o)-, and -in(e)

citrazinic acid
$C_6H_5NO_4$, derived from citric acid, az(o)-, and -in(e)

citreoviridin
$C_{23}H_{30}O_6$, derived from the specific epithet of the fungal species name *Penicillium citreo-viride*, from citr(o)- and virid(o)-, and -in(e)

citric acid
$C_6H_8O_7$, derived from citr(o)-

citrine
SiO_2, derived from citr(o)- and -in(e) – referring to this mineral's light olive color

citrinin
$C_{13}H_{14}O_5$, derived from the specific epithet of the fungal species name *Penicillium citrinum*, from citr(o)-, and -in(e)

citr(o)-
derived from the genus name *Citrus* (trees, shrubs), from *citrus* (Latin: citron tree, *Tetraclinis articulata*)

citromycetin
$C_{14}H_{10}O_7$, derived from citr(o)- and -mycetin

citronell(o)-
derived from citronella oil

citronellal
$C_{10}H_{18}O$, derived from citronell(o)- and -al

citronellic acid
$C_{10}H_{18}O_2$, derived from citronell(o)-

citronellol (cephrol)
$C_{10}H_{20}O$, derived from citronell(o)- and -ol

citrovorum factor (folinic acid)
$C_{20}H_{23}N_7O_7$, derived from the specific epithet of the bacterial species name *Leuconostoc citrovorum*, from citr(o)- and -*vorus* (Latin: -devouring), from *vorare* (Latin: to devour)

citrulline
$C_6H_{13}N_3O_3$, derived from citrull(o)- and -in(e)

citrull(o)-
derived from the genus name *Citrullus* (watermelon), from *citrullus* (Medieval Latin: a kind of cucumber), ultimately from citr(o)-

citrullol
$C_{22}H_{38}O_4$, derived from citrull(o)- and -ol

civetone
$C_{17}H_{30}O$, derived from civet and -one

CLA
an abbreviation for conjugated linoleic acid(s)

Claisen adapter
named for the German chemist Ludwig Claisen (1851-1930)

Claisen condensation
named for the German chemist Ludwig Claisen (1851-1930)

Claisen flask
named for the German chemist Ludwig Claisen (1851-1930)

Claisen rearrangement
named for the German chemist Ludwig Claisen (1851-1930)

Claisen-Ireland rearrangement
named for the German chemist Ludwig Claisen (1851-1930) and the US chemist Robert Ellsworth Ireland (born 1929)

Claisen-Schmidt condensation
named for the German chemists Ludwig Claisen (1851-1930) and J. Gustav Schmidt

Claisen-Schmidt rearrangement
named for the German chemists Ludwig Claisen (1851-1930) and J. Gustav Schmidt

-clase
derived from *klasis* (Greek: breaking), from *klan* (Greek: to break)

clathrand
derived from clathrate and -and

clathrasil
SiO_2, derived from clathrate and silicon

clathrate
derived from *clathratus* (Latin: furnished with lattice), ultimately from *kleiein* (Greek: to shut)

Claude process
named for the French chemist Georges Claude (1870-1960)

claudetite
As_2O_3, named for the 19th century French chemist F. Claudet

Claus benzene
C_6H_6, named for the German chemist Adolph Claus (1840-1900)

Claus process
named for the German-British chemist Carl Friedrich Claus (born 1885)

Clausius-Clapeyron equation
named for the German physicist Rudolf Julius Emanuel Clausius (1822-1888) and the French civil engineer Bénoit Paul Emile Clapeyron (1799-1864)

Clausius cycle
named for the German physicist Rudolf Julius Emanuel Clausius (1822-1888)

clausthalite
PbSe, named after this mineral's locality Clausthal, Germany

clavacin
$C_7H_6O_4$, derived from the specific epithet of the fungal species name *Aspergillus clavatus*, from *clavatus* (Latin: furnished with a club or clubs), from *clava* (Latin: club), and -cin(e)

clavam
$C_5H_7NO_2$, derived (with contraction) from clavulanic acid and lactam

clavi(c)-
ultimately derived from *clavis* (Latin: key)

clavulanic acid
$C_8H_9NO_5$, derived from the specific epithet of the bacterial species name *Streptomyces clavuligerus*, from *clavuligerus* (New Latin: bearing little clubs), from *clavula*, diminutive of *clava* (Latin: club), and -an(e)

clay
a rock species, an indigenous Germanic word

cleavelandite (clevelandite)
$NaAlSi_3O_8$, named for the US chemist and mineralogist Parker Cleaveland (1780-1858)

Cleland rules
named for the US chemist William Wallace Cleland (born 1930)

Clemmensen reduction
named for the Danish-US chemist Erik Christian Clemmensen (1876-1941)

Clerici solution
named for the Italian geologist and engineer E. Clerici (1862-1938)

cleveite
U_3O_8, named for the Swedish chemist Per Theodor Cleve (1840-1905)

Cleve's acids
$C_{10}H_9NO_3S$, named for the Swedish chemist Per Theodor Cleve (1840-1905)

cliffordite
UTe_3O_9, named for the US mineralogist Clifford Frondel (1907-2002)

clin(o)-
derived from *kline* (Greek: couch), from *klinein* (Greek: to bend, to cause to slope)

clinochlore
$(Mg,Fe)_5Al(Si_3O_{10})(OH)_8$, derived from clin(o)- and ^{1}chlor(o)- – referring to this mineral's oblique crystal shape and green color

clinochrysotile
$Mg_3Si_2O_5(OH)_4$, derived from clin(o)- and chrysotile

clinoenstatite
$Mg_2Si_2O_6$, derived from clin(o)- and enstatite

clinoferrosilite
$FeSiO_3$, derived from clin(o)- and ferrosilite

clinohumite
$(Mg,Fe,Ti)_9(SiO_4)_4(F,OH)_2$, derived from clin(o)- and humite

clinoptilolite
$(Na,K,Ca)_{2-3}(Si_{15}Al_3)O_{36}\cdot 11H_2O$, derived from clin(o)- and ptilolite

clinozoisite
$Ca_2Al_3(SiO_4)_3(OH)$, derived from clin(o)- and zoisite

clintonite (xanthophyllite)
$Ca(Mg,Al)_3(Al,Si)_4O_{10}(OH,F)_2$, named for the US statesman De Witt Clinton (1769-1828)

clito-
derived from *kleitos* (Greek: splendid, excellent)

closo- (clovo-)
derived (incorrectly) from *klobos* (New Greek: cage)

closyl
$C_6H_4ClO_2S$–, coined by contraction of chlorobenzenesulfonyl

clovene
$C_{15}H_{24}$, derived from clove and -ene

clup(e)-
derived from *clupea* (Latin: a small river fish)

clupanodonic acid
$C_{22}H_{34}O_2$, derived from the genus name *Clupanodon* (fish), from clup(e)- and *anodon* (Greek: toothless), from a(n)- and *odon* (Greek: tooth)

clupeine
derived from the genus name *Clupea* (herring), from clup(e)-, and -in(e)

Clusius-Dickel column
named for the German chemists Klaus Paul Alfred Clusius (1903-1963) and Gerhard Dickel (born 1913)

[1]CMC
an abbreviation for carboxymethylcellulose

[2]CMC
an abbreviation for critical micelle concentration

CMP
$C_9H_{14}N_3O_8P$, an abbreviation for cytidine 5'-monophosphate

CN
C_8H_7ClO, an arbitrary US military code name for ω-chloroacetophenone

CNDO
an abbreviation for complete neglect of differential overlap

cnicin
$C_{20}H_{26}O_7$, derived from the genus name *Cnicus* (thistles), ultimately from *knekos* (Greek: safflower, thistle), and -in(e)

co- (col-, con-, com-, cor-)
derived from *co- (col-, con-, com-, cor-)* (Latin: together, with)

cobalamine
$C_{62}H_{88}CoN_{13}O_{14}P$, derived from cobalt and amine

cobalt
Co, derived from *Kobold* (German: gnome, goblin) – referring to this metal's appearance in silver ores where it was believed in the Middle Ages to have been placed by silver-stealing goblins

cobaltic
Co^{3+}, from cobalt and -ic

cobaltite
CoAsS, derived from cobalt and -ite

cobalt(o)-
derived from cobalt

cobaltocene
$C_{10}H_{10}Co$, derived from cobalt(o)- and -ocene

cobaltous
Co^{2+}, derived from cobalt and -ous

cobaltum
Co, New Latin name for cobalt, derived from cobalt and -um

cobamamide (coenzyme B_{12})
$C_{72}H_{100}CoN_{18}O_{17}P$, derived from cobalt, amine, and amide

cobamic acid
$C_{53}H_{74}CoN_5O_{21}P$, derived from cobalt, amine, and acid

cobinic acid
$C_{48}H_{66}CoN_5O_{14}$, derived from cobalt, -in(e), and acid

cobrotoxin
derived from cobra (*Naja*) and toxin

cobyrinic acid
$C_{45}H_{59}CoN_4O_{14}$, derived from cobalt

cocarboxylase (thiamine diphosphate)
$C_{12}H_{19}ClN_4O_7P_2S$, derived (with contraction) from coenzyme and carboxylase

cocaine
$C_{17}H_{21}NO_4$, derived from the specific epithet of the species name *Erythroxylon coca* (coca), ultimately from *kuka* (Quechua: coca), and -in(e)

cocarcinogen
derived from co(n)- and carcinogen

coclaurine
$C_{17}H_{19}NO_3$, derived (with contraction) from the species name *Cocculus laurifolius* D.C. (laurel leaf cocculus), from *cocculus* (Latin: small berry) and *laurus* (Latin: laurel), and -in(e)

COD
an abbreviation for chemical oxygen demand

codamine
$C_{20}H_{25}NO_4$, derived from codeine and amine

codecarboxylase (pyridoxal 5'-phosphate)
$C_8H_{10}NO_6P$, derived (with contraction) from coenzyme and decarboxylase

[1]**codehydrogenase (NAD, NAD^+, coenzyme I)**
$C_{21}H_{27}N_7O_{14}P_2$, derived (with contraction) from coenzyme and dehydrogenase

[2]**codehydrogenase (NADH, $NADH_2$)**
$C_{21}H_{29}N_7O_{14}P_2$, derived (with contraction) from coenzyme and dehydrogenase

[3]**codehydrogenase (NADP, $NADP^+$, coenzyme II)**
$C_{21}H_{28}N_7O_{17}P_3$, derived (with contraction) from coenzyme and dehydrogenase

[4]**codehydrogenase (NADPH, $NADPH_2$)**
$C_{21}H_{30}N_7O_{17}P_3$, derived (with contraction) from coenzyme and dehydrogenase

codeine
$C_{18}H_{21}NO_3$, derived from *kodeia* (Greek: poppy head) and -in(e)

coenzyme
derived from co(n)- and enzyme

coesite
SiO_2, named for the US chemist Loring Coes, Jr. (1915-1973)

coenzyme
derived from co(n)- and enzyme

coerulein (cerulein)
derived from *caeruleus* (Latin: sky blue) and -ein(e)

coffinite
$U[SiO_4,(OH)_4]$, named for the US geologist Reuben Clare Coffin (1886-1972)

cohenite
$(Fe,Ni,Co)_3C$, named for the German mineralogist Emil Wilhelm Cohen (1842-1905)

coherin
derived from coherent and -in(e)

cohumulone
$C_{20}H_{28}O_5$, derived from co(n)- and humulone – referring to the fact that this compound is a minor hop constituent compared to humulone

colamine
C_2H_7NO, coined by contraction of alcohol and amine

colchiceine
$C_{21}H_{23}NO_6$, coined by variation of colchicine

colchicine
$C_{22}H_{25}NO_6$, derived from the genus name *Colchicum* (Old World cormous herbs), ultimately from Colchis, an ancient country on the eastern shore of the Black Sea

colemanite
$Ca_2B_6O_{11}\cdot 5H_2O$, named for the US mine owner William Tell Coleman (1824-1893)

cole(o)-
derived from *koleos, koleon* (Greek: sheath)

colfosceril palmitate
$C_{40}H_{80}NO_8P$, coined by contraction of choline, phosphorus, glyceryl, and palmitate

colforsin (forskolin)
$C_{22}H_{34}O_7$, derived (with contraction) from the species name *Coleus forskohlii* Briq. (makandi), from cole(o)-, named for the Swedish-Danish botanist Peter Forskål (1736-1763), and -in(e)

colicin
derived from the specific epithet of the bacterial species name *Escherichia coli*, from *kolon* (Greek: intestine), and -icin(e)

colistin
derived from the subspecific epithet of the bacterial species name *Bacillus polymyxa* var. *colistinus*, and -in(e)

coll(a)-
derived from *kolla* (Greek: glue)

collagen
derived from coll(a)- and -gen

collagenase
derived from collagen and -ase

collidine
$C_8H_{11}N$, derived from coll(a)- and -idin(e)

collinomycin
$C_{27}H_{20}O_{12}$, derived from the specific epithet of the bacterial species name *Streptomyces collinus*, from *collinus* (Latin: of or pertaining to a hill; New Latin: living on hills), from *collis* (Latin: hill), and -mycin

Collins oxidation
named for the US chemist Joseph Charles Collins, Jr. (born 1931)

Collins' reagent
$CrO_3\cdot(C_5H_5N)_2$, named for the US chemist Joseph Charles Collins, Jr. (born 1931)

Collman's reagent
$Na_2Fe(CO)_4$, named for the US chemist

James Paddock Collman (born 1932)

collodion
derived from *kollodes* (Greek: glue-like), from *kolla* (Greek: glue), and [1]-on

colloid
derived from *kolla* (Greek: glue) and -*oeides* (Greek: -like)

col(o)-
derived from *kolon* (Greek: large intestine)

colocynthin
$C_{38}H_{54}O_{13}$, derived from the specific epithet of the species name *Citrullus colocynthis* Schrad. (colocynth), from col(o)-, cynth-, and -in(e)

coloradoite
HgTe, named after this mineral's locality Smuggler mine, Ballerat district, Boulder County, CO, USA

colostrokinin
derived from colostrum and kinin

colubrine
$C_{22}H_{24}N_2O_3$, derived from the genus name *Colubrina* (nakedwood), from *colubrinus* (Latin: snake-like), from *colubra* (Latin: snake), and -in(e)

columb-
derived from calumba (*Jatrorrhiza palmata* Miers), from *kalumba* (Hausa: a small tree)

columbamine
$C_{20}H_{20}NO_4^+$, derived from columb- and amine

columbin
$C_{20}H_{22}O_6$, derived from columb- and -in(e)

columbite
(Fe,Mn)(Nb,Ta)$_2$O$_6$, named for *Columbia* (New Latin: USA), named for the latinized name Christophorus Columbus of the Italian-Spanish discoverer Cristoforo Colombo (1451-1506)

columbium
Nb, an obsolete name for niobium, derived from columbite and -ium

colupone
$C_{30}H_{44}O_4$, derived from co- and lupone – referring to the fact that this compound is a minor hop constituent compared with lupone

comanic acid
$C_6H_4O_4$, anagrammatically derived from meconic acid

Combes quinoline synthesis
named for the 19th century French chemist A. Combes

combretastatin
derived from the genus name *Combretum* (bushwillow tree), from *combretum* (Latin: a kind of plant, cattail), and -statin

comenic acid
$C_6H_4O_5$, anagrammatically derived from meconic acid

commo-
derived from *communis* (Latin: common)

comonomer
derived from co(n)- and monomer

compound
ultimately from *componere* (Latin: to put together)

COMT
an abbreviation for catecholamine *O*-methyltransferase

co(n)-
derived from *cum* (Latin: with)

conanine
$C_{22}H_{37}N$, derived from conessi bark, of unknown etymology, -an(e), and -in(e)

concanamycin
$C_{46}H_{75}NO_{14}$, derived from concanavalin A and -mycin – referring to this antibiotic's inhibition of the proliferation of murine splenic lymphocytes stimulated by concanavalin A

concanavalin
derived from co(n)- and canavalin

conchagen (conchiolin)
derived from *concha* (Latin: shell), from *konche* (Greek: shell), and -gen

condis crystal
coined by contraction of conformational disorder crystal

condurangin
$C_{37}H_{60}O_{16}$, derived from the specific epithet of the species name *Marsdenia condurango* Reichb. (eagle vine), from *kunturanko* (Quechua: condor vine), and -in(e)

conessine (neriine, roquessine, wrightine)
$C_{24}H_{40}N_2$, derived from conessi bark, of unknown etymology, and -in(e)

Congo red
$C_{32}H_{22}N_6Na_2O_6S_2$, named for Congo, Africa

congressane (diamantane)
$C_{14}H_{20}$, derived from congress and -an(e) – referring to the fact that the structure of this then hypothetical hydrocarbon was used as the logo of a IUPAC congress

conformer
derived from conformation and –mer

conhydrine
$C_8H_{17}NO$, derived from the genus name *Conium* (hemlock), ultimately from *koneion* (Greek: hemlock), hydr(o)-, and -in(e)

Conia reaction
named for the 20th century French chemist Jean Marie Conia

coniceine
$C_8H_{15}N$, coined by variation of conhydrine

coniferin
$C_{16}H_{22}O_8$, derived from conifer and -in(e)

coniferyl alcohol
$C_{10}H_{12}O_3$, derived from coniferin and -yl

coniine
$C_8H_{17}N$, derived from the genus name *Conium* (hemlock), ultimately from *koneion* (Greek: hemlock), and -in(e)

conjuncto-
derived from *coniunctus* (Latin: joined), from *coniungere* (Latin: to join)

conotoxin
derived from the genus name *Conus* (cone snail), from *conus* (Latin: cone), and toxin

conquinamine
$C_{19}H_{24}N_2O_2$, coined anagrammatically from cinch(ona)- and amine

Conrad-Limpach cyclization
named for the German chemists M. Conrad and Leonhard Limpach (1853-1933)

conrotatory
derived from co(n)- and rotation

contortin
$C_{22}H_{26}O_8$, derived from the specific epithet of the species name *Psoroma contortum* (a

lichen), from *contortus* (Latin: twisted), from *torquere* (Latin: to twist), and -in(e)

contra-
derived from *contra* (Latin: against)

convalla(r)-
derived from the genus name *Convallaria* (lily of the valley), from *convallis* (Latin: enclosed valley), from *vallis* (Latin: valley)

convallarin
derived from convalla(r)- and -in(e)

convallatoxin
$C_{29}H_{42}O_{10}$, derived from convalla(r)- and toxin

convicine
$C_{10}H_{15}N_3O_8$, derived from co(n)-, the genus name *Vicia* (vetch), from *vicia* (Latin: vetch), and -in(e)

cooperite
(Pt,Pd,Ni)S, named for the 20th century South African scientist R. A. Cooper

Cooper pair
named for the US physicist Leon Neil Cooper (born 1930)

copaene
$C_{15}H_{24}$, derived from copaiba balsam oil, from *copaiba* (Tupi: copaiba), and -ene

Cope elimination
named for the US chemist Arthur Clay Cope (1909-1966)

Cope rearrangement
named for the US chemist Arthur Clay Cope (1909-1966)

copolymer
derived from co(n)- and polymer

copper
Cu, derived from *cuprum* (Late Latin: copper), ultimately from *aes cyprium* (Latin: metal of Cyprus); *Kypros*, the Greek name of Cyprus, has been linked to *kyparissos* (Greek: cypress), a loan from a non-Indoeuropean language

copperas
$FeSO_4{\cdot}7H_2O$, derived from *coperose*, *couperose* (Medieval French: copperas), ultimately from cuprum and *rosa* (Latin: rose)

copr(o)-
derived from *kopros* (Greek: dung)

coproergostane (pseudoergostane)
$C_{28}H_{50}$, derived from copr(o)- and ergostane

coprogen
$C_{35}H_{53}FeN_6O_{13}$, derived from copr(o)- and -gen

coproporphyrin
derived from copr(o)- and porphyrin

coprostane (pseudocholestane)
$C_{27}H_{48}$, coined by contraction of copr(o)- and cholestane

coprosterol (stercorin)
$C_{27}H_{48}O$, derived from copr(o)- and sterol

coptisine
$C_{19}H_{14}NO_4^{+}$, derived from the genus name *Coptis* (goldthread), from *koptein* (Greek: to cut off), and -in(e)

corand (coronand)
coined by contraction of coronand

coraplex
coined by contraction of corand and complex

cordierite (dichroite, iolite)
$Mg_2Al_4Si_5O_{18}$, named for the French mining engineer and geologist Pierre Louis A. Cordier (1777-1861)

cordycepin
$C_{10}H_{13}N_5O_3$, derived from the fungal genus name *Cordyceps*, from *kordyle* (Late Greek: club) and *-ceps* (Latin: headed), from *caput* (Latin: head), and -in(e)

Corey aldehyde
$C_{14}H_{24}O_4Si$, named for the US chemist Elias James Corey (born 1928)

Corey-Bakshi-Shibata reduction
named for the US chemist Elias James Corey (born 1928) and the chemists S. Bakshi and R. K. Shibata

Corey-Kim oxidation
named for the US chemist Elias James Corey (born 1928) and the chemist C. U. Kim

Corey-Pauling-Koltun molecular model
named for the US chemists Elias James Corey (born 1928), Linus Carl Pauling (1901-1994), and Walter Lang Koltun (born 1928)

[1]**Corey's reagent**
$C_5H_6ClCrNO_3$, named for the US chemist Elias James Corey (born 1928)

[2]**Corey's reagent**
C_3H_8OS, named for the US chemist Elias James Corey (born 1928)

Corey-Winter olefin synthesis
named for the US chemist Elias James Corey (born 1928) and the Estonian-US chemist Roland Arthur Edwin Winter (born 1935)

coriamyrtin
$C_{15}H_{18}O_5$, derived (with contraction) from the species name *Coriaria myrtifolia* L. (myrtle-leaved coriaria, Currier's sumach), from *coriarius* (Latin: useful for tanning leather), from *corium* (Latin: leather), and *myrtus* (Latin: myrtle), and -in(e)

Cori ester
$C_6H_{13}O_9P$, named for the Czech-US biochemists Carl Ferdinand Cori (1896-1984) and Gerty Theresa Cori (1896-1957)

Cornforth reagent
$CrO_3 \cdot C_5H_5N \cdot H_2O$, named for the Australian-British chemist Sir John Warcup Cornforth (born 1917)

Cornforth rearrangement
named for the Australian-British chemist Sir John Warcup Cornforth (born 1917)

coron-
derived from *corona* (Latin: crown)

coronand (corand)
derived from coron- and -and – referring to crown compounds

coronate
derived from coron- and -ate – referring to crown compounds

coronene
$C_{22}H_{12}$, derived from coron- and -ene

coronium
a hypothetical element inferred from certain lines in the spectrum of the Sun's corona (later related to iron), derived from corona, from coron-, and –ium

corphyrin (corrin)
$C_{19}H_{22}N_4$, coined by contraction of core (of vitamin B_{12}) and porphyrin

corr-
derived from core (of vitamin B_{12})

corrin (corphyrin)
$C_{19}H_{22}N_4$, derived from corr- and -in(e)

corrinoid
derived from corrin and -oid

corrol
$C_{19}H_{14}N_4$, derived from corr- and -ol(e)

cortico-
derived from adrenal cortex, from *cortex* (Latin: bark)

corticoliberin (corticotropin-releasing factor, corticotropin-releasing hormone, CRF, CRH)
derived from cortico- and liberin

corticosteroid
derived from cortico- and steroid

corticosterone
$C_{21}H_{30}O_4$, derived from cortico-, sterol, and -one

corticotrophin (corticotropin)
derived from cortico- and -trophin

corticotropin (corticotrophin)
derived from cortico- and -tropin

cortisol
$C_{21}H_{30}O_5$, coined by contraction of cortico-sterol

cortisone
$C_{21}H_{28}O_5$, coined by contraction of cortico-sterone

cortol
$C_{21}H_{36}O_5$, derived (with contraction) from hydrocortisone and –ol

cortolone
$C_{21}H_{34}O_5$, derived (with contraction) from hydrocortisone, -ol, and -one

corubin
Al_2O_3, derived from corundum, ruby, and -in(e) – referring to the formation of small rubies during the manufacture of this abrasive

corundum
Al_2O_3, derived from *kurundam* (Tamil: ruby)

corybulbine
$C_{21}H_{25}NO_4$, derived (with contraction) from the species name *Corydalis bulbosa* (bulbous corydalis), from coryd(al)- and *bulbosus* (Latin: bulbous), and -in(e)

coryd(al)-
derived from the genus name *Corydalis* (fumewort), from *korydallis* (Greek: crested lark)

corydaline
$C_{22}H_{27}NO_4$, derived from coryd(al)- and -in(e)

corydine
$C_{20}H_{23}NO_4$, derived from coryd(al)- and -in(e)

corynan (corynanthean)
$C_{19}H_{26}N_2$, derived (with contraction) from corynanth(e)- and -an(e)

corynanth(e)-
derived from the genus name *Corynanthe* (flowering plants), from coryn(o)- and anth(o)-

corynantheine
$C_{22}H_{26}N_2O_3$, derived from corynanth(e)- and -ein(e)

corynanthine
$C_{21}H_{26}N_2O_3$, derived from corynanth(e)- and -in(e)

coryn(o)-
derived from *koryne* (Greek: club)

corynoxan
$C_{19}H_{28}N_2$, derived from corynanth(e)-, ox(a)-, and -an(e)

corypalmine
$C_{20}H_{23}NO_4$, derived from coryd(al)-, palm, and -in(e)

corytuberine
$C_{19}H_{21}NO_4$, derived (with contraction) from the species name *Corydalis tuberosa* (butterflyweed), ultimately from *korydallis* (Greek: crested lark) and *tuber* (Latin: hump, knob, tumor, truffle, tuber)

cosmene
$C_{10}H_{14}$, derived from the genus name *Cosmos* (herbs), from *kosmos* (Greek: order, ornament, universe), and -ene

COT
C_8H_8, an abbreviation for cycloocta-1,3,5,7-tetraene

cotarnine
$C_{12}H_{15}NO_4$, derived anagrammatically from narcotine

cotinine
$C_{10}H_{12}N_2O$, derived anagrammatically from nicotine

Cotton effect
named for the French physicist A. A. Cotton (1869-1951)

Cotton-Mouton effect
named for the French physicists A. A. Cotton (1869-1951) and H. J. D. Mouton (1869-1935)

Cottrell process
named for the US chemist F. G. Cottrell (1877-1948)

coulometry
derived from Coulomb, named for the French physicist Charles Augustin de Coulomb (1736-1806), and -metry

coumalic acid
$C_6H_4O_4$, derived from coumar- and malic acid

coumalin (pyrone)
C_5H_6O, derived from coumalic acid and -in(e)

coum(ar)-
derived from coumarou, ultimately from *cumarú* (Tupi: tonka-bean tree, *Dipteryx odorata*)

coumaran
C_8H_8O, derived from coum(ar)- and -an(e)

coumaric acid
$C_9H_8O_3$, derived from coum(ar)-

coumarin
$C_9H_6O_2$, derived from coum(ar)- and -in(e)

coumarone (cumarone)
C_8H_6O, derived from coum(ar)- and -one

coumermycin
derived from coumarin and -mycin

coumestrol
$C_{15}H_8O_5$, derived from coum(ar)-, estr(o)-, and –ol

coumingine
$C_{29}H_{47}NO_6$, derived from the specific epithet of the species name *Erythrophleum*

coumingа Baillon (an African tree), and -in(e)

Couper-Brown-Butlerov structure
named for the British chemists A. S. Couper (1831-1892) and A. C. Brown (1838-1922) and the Russian chemist Aleksandr M. Butlerov (1828-1886)

covellite
CuS, named for the Italian mineralogist Niccolò Covelli (1790-1829)

COX
an abbreviation for cyclooxygenase

cozymase (NAD, NAD$^+$)
derived (with contraction) from coenzyme and zymase

CPK
an abbreviation for Corey-Pauling-Koltun

CR
$C_{13}H_9NO$, an arbitrary US military code name for dibenzo[*b,f*][1,4]oxazepin

Crabtree's catalyst
named for the British chemist Robert H. Crabtree (born 1948)

Craig method
named for the US chemist Lyman Creighton Craig (1906-1974)

Craig partition
named for the US chemist Lyman Creighton Craig (1906-1974)

Cram's rule
named for the US chemist Donald James Cram (1919-2001)

-crase
derived from *krasis* (Greek: mixing, mixture), from *kerannynai* (Greek: to mix)

cream of tartar
$C_4H_5KO_6$, an obsolete name for potassium hydrogen tartrate, derived from tartar – referring to the fact that this crystalline salt is the most valuable part of the crude tartar deposited in wine casks

creatine
$C_4H_9N_3O_2$, derived from creat(o)- and -in(e)

creatinine
$C_4H_7N_3O$, derived from creat(o)- and -inin(e)

creat(o)-
derived from *kreas* (Greek: flesh)

cremor tartari
$C_4H_5KO_6$, New Latin name for cream of tartar, from *cremor* (Latin: slime) and *tartarum* (Medieval Latin: cream of tartar)

cre(o)-
derived from *kreas* (Greek: flesh)

creosol
$C_8H_{10}O_2$, derived from creosote and -ol

creosote
derived from cre(o)- and *soter* (Greek: preserver), from *sozein* (Greek: to preserve)

creosotic acid
$C_8H_8O_3$, derived from creosote

cresol (cresylic acid)
C_7H_8O, derived from creosote and -ol

cresolphthalein
$C_{22}H_{18}O_4$, derived from cresol and phthalein

cresol purple
$C_{21}H_{18}O_5S$, derived from cresol

cresol red
$C_{21}H_{18}O_3S$, derived from cresol

cresotic acid
$C_8H_8O_3$, derived from creosote

Criegee reaction
named for the German chemist Rudolf Criegee (1902-1975)

crinan
$C_{16}H_{19}NO_2$, derived from crin(o)- and -an(e)

crin(i)-
derived from *crinis* (Latin: hair)

[1]crinine
$C_{16}H_{17}NO_3$, derived from crin(o)- and -in(e)

[2]crinine (cherryline)
$C_{17}H_{19}NO_3$, derived from crin(o)- and -in(e)

crin(o)-
derived from the genus name *Crinum* (swamplilies), from *krinon* (Greek: lily)

cristobalite
SiO_2, named after this mineral's locality Cerro San Cristobal, Mexico

cRNA
an abbreviation for complementary ribonucleic acid

croceic acid (Bayer's acid)
$C_{10}H_8O_4S$, derived from *croceus* (Latin: of saffron, dark red yellow), from *crocus* (Latin: saffron), ultimately from *krokos* (Greek: saffron, crocus), of Semitic origin

crocetin
$C_{20}H_{24}O_4$, derived from croc(o)- and -etin(e)

crocidolite
$Na_2(Fe^{II},Mg)_3(Fe^{III})_2Si_8O_{22}(OH,F)_2$, derived from *krokys* (Greek: a nap of woollen cloth) and -lite – referring to this mineral's fibrous structure

croc(o)-
derived from crocus, from *krokos* (Greek: saffron, crocus), of Semitic origin

crocoite (crocoisite)
$PbCrO_4$, derived from *krokoeis* (Greek: saffron-colored) and -ite – referring to this mineral's saffron color

croconic acid
$C_5H_2O_5$, derived from croc(o)- and -one

cromoglycic acid
$C_{23}H_{16}O_{11}$, derived from chromonecarboxylic acid and glycerin

crookesite
$(Cu,Tl,Ag)_2Se$, named for the British chemist Sir William Crookes (1832-1919) who discovered thallium

crot(alo)-
ultimately derived from *krotalon* (Greek: rattle, castanet), from *krotein* (Greek: to clap)

crotamine
derived from the genus name *Crotalus* (rattlesnakes), from crot(alo)-, and amine

crot(on)-
derived from the genus name *Croton* (croton), ultimately from *kroton* (Greek: dog tick) – referring to the tick-shaped seeds of these plants

crotonaldehyde
C_4H_6O, derived from crot(on)-

crotonic acid
$C_4H_6O_2$, derived from crot(on)-

crotoxin
derived from crot(alo)- and toxin

crotyl alcohol
C_4H_8O, derived from crotonaldehyde

crucible
derived from *crucibulum* (Medieval Latin: crucible), probably generated by folk etymology from *crux* (Latin: cross) and *turibulum* (Latin: thurible, incense vessel), from *tus* (Latin: incense), from *thyos* (Greek: incense)

cry(o)-
derived from *kryos* (Greek: ice, frost)

cryolite (ice spar)
Na_3AlF_6, derived from cry(o)- and -lite – referring to this mineral's frosty appearance

cryolithionite
$Na_3Li_3Al_2F_{12}$, derived from cryolite, lithium, and -ite

cryoscopy
derived from cry(o)- and -scopy

cryptand
derived from crypt(o)- and -and

cryptate
derived from crypt(o)- and -ate

crypt(o)-
derived from *kryptos* (Greek: hidden), from *kryptein* (Greek: to hide)

cryptomelane
KMn_8O_{16}, derived from crypt(o)- and -melane – referring to the notion that the identity of this mineral is lost in the group of other manganese-bearing oxide minerals

cryptopine
$C_{21}H_{23}NO_5$, derived from crypto-, opium, and -in(e)

cryptopyrrole
$C_8H_{13}N$, derived from crypt(o)- and pyrrole

cryptoxanthin
$C_{40}H_{56}O$, derived from crypt(o)- and xanthin

crystal
derived from *krystallos* (Greek: ice, rock crystal), from *kryos* (Greek: ice, frost)

CS
$C_{10}H_5ClN_2$, a US military code name for (2-chlorobenzylidene)malononitrile, derived from the initials of the US chemists B. B. Corson (1896-1987) and R. W. Stoughton (born 1916)

CSF
an abbreviation for colony-stimulating factor

CT
an abbreviation for charge transfer

CTP
$C_9H_{16}N_3O_{14}P_3$, an abbreviation for cytidine 5'-triphosphate

CTX
$C_{176}H_{277}N_{57}O_{55}S_7$, an abbreviation for charybdotoxin

cubane (quadriprismane, tetraprismane)
C_8H_8, derived from *cubus* (Latin: dice), from *kybos* (Greek: dice), and -an(e) – referring to this hydrocarbon's cubic structure

cubanite (chalmersite)
$CuFe_2S_3$, named after this mineral's locality Barracanao, Cuba

cubebin
$C_{20}H_{20}O_6$, derived from the specific epithet of the species name *Piper cubeba* (cubeb tailed pepper), from *kubabah* (Arabic:

cubeb), and -in(e)

cucurbitacin
derived from the genus name *Cucurbita* (gourd), from *cucurbita* (Latin: gourd), and -icin(e)

cumene
C_9H_{12}, derived from *cumenum* (Latin: Roman caraway) and -ene

cumic acid (cuminic acid)
$C_{10}H_{12}O_2$, derived from cum(o)-

cumic alcohol
$C_{10}H_{14}O$, derived from cumic acid

cumidine
$C_9H_{13}N$, derived from cum(o)- and -idin(e)

cuminaldehyde
$C_{10}H_{12}O$, derived from cuminic acid

cummingtonite
$(Mg,Fe,Mn)_7(Si_8O_{22})(OH)_2$, named after this mineral's locality Cummington, MA, USA

cum(o)-
derived from cumin (*Cuminum cyminum*), ultimately from *kammon* (Hebrew: cumin)

cumulene
derived from cumulated, ultimately from *cumulus* (Latin: heap, mass), and -ene

cuoxam (Schweizer's reagent)
$CuH_{14}N_4O_2$, derived (with contraction) from cuprum, ox(o)-, and ammonia

cupel
derived from *cupella* (Latin: little tub), diminutive of *cupa* (Latin: cask)

cupferron
$C_6H_9N_3O_2$, derived from cuprum, ferrum, and [1]-on – referring to this compound's usefulness as a copper and iron reagent

cuprate
derived from cupr(o)- and -ate

cuprene (polyacetylene)
$(C_2H_2)_n$, derived from cuprum and -ene – referring to alleged copper catalysis of the polymerization of ethyne or to this polymer's copper-like luster

cupreine
$C_{19}H_{22}N_2O_2$, derived from cuprea bark (*Remijia*), ultimately from *cupreus* (Latin: of copper), and -in(e)

cupressene
$C_{20}H_{32}$, derived from *cupressus* (Latin: cypress) and -ene

cupric
Cu^{2+}, derived from cuprum and -ic

cuprite
Cu_2O, derived from cupr(o)- and -ite

cupr(o)-
derived from cuprum

cuproine
$C_{18}H_{12}N_2$, derived from cupr(o)- and -in(e) – referring to this compound's usefulness as a copper reagent

cuprone
$C_{14}H_{13}NO_2$, derived from cupr(o)- and -one – referring to the usefulness of this reagent for the determination of copper (as well as of molybdenum, tungsten, vanadium, and some of the platinum metals)

cuprosklodowskite
$Cu(UO_2)_2(SiO_3,OH)_2 \cdot 6H_2O$, derived from cupro- and, improperly, sklodowskite

cuprous
Cu^{+}, derived from cuprum and -ous

cuprum
New Latin name for copper, from *cuprum* (Late Latin: copper), ultimately from *aes cyprium* (Latin: metal of Cyprus)

curan
$C_{19}H_{26}N_2$, derived from cur(ar)- and -an(e)

cur(ar)-
derived from curare, ultimately from *kurari* (Tupi: he to whom it comes, falls)

curarine
derived from cur(ar)- and -in(e)

curcumin
$C_{21}H_{20}O_6$, derived from the genus name *Curcuma* (turmeric), ultimately from *kurkum* (Arabic: saffron, crocus)

curine
$C_{36}H_{38}N_2O_6$, derived from cur(ar)- and -in(e)

curium
Cm, named for the French physicist Pierre Curie (1859-1906) and the Polish-French physicist Marie Curie (1867-1934)

Curtius rearrangement
named for the German chemist Theodor Curtius (1857-1928)

curvularin
$C_{16}H_{20}O_5$, derived from the fungal genus name *Curvularia*, and -in(e)

cuscohygrine
$C_{13}H_{24}N_2O$, derived from cusco bark (*Cinchona pubescens*), named for the city of Cuzco, Peru, and hygrine

cutin
derived from *cutis* (Latin: skin) and -in(e)

cutinase
derived from cutin and -ase

CVD
an abbreviation for chemical vapor deposition

cyacetacide
$C_3H_5N_3O$, coined by contraction of cyanoacetic acid hydrazide

cyamelide
$C_3H_3N_3O_3$, derived (with contraction) from cyano-, melam, and -ide

cyanamide
CH_2N_2, derived from cyano- and amide

cyanate
NCO^{-}, derived from cyanic acid and -ate

cyanic acid
CHNO, derived from cyano-

cyanide
CN^{-}, derived from cyan(o)- – referring to *Blausäure* (German: hydrogen cyanide, literally blue acid), named after its preparation by acid treatment of Berlin blue) – and -ide

cyanidin chloride
$C_{15}H_{11}ClO_6$, derived from cyan(o)- and -idin(e)

cyanidine
$C_3H_3N_3$, derived from cyano- and -idin(e)

cyanin
$C_{27}H_{30}O_{16}$, derived from cyan(o)- and -in(e)

cyanine
derived from cyan(o)- and -in(e)

cyanite (kyanite)
$Al_2(O)SiO_4$, derived from cyan(o)- and -ite – referring to this mineral's blue color

cyano
NC–, derived from cyanide

cyan(o)-
derived from *kyanos* (Greek: dark blue)

cyanocobalamine
$C_{63}H_{88}CoN_{14}O_{14}P$, derived from cyano-, cobalt, and amine

[1]**cyanogen (cyanyl)**
CN, derived from cyano- and -gen; patterned after halogen

[2]**cyanogen**
C_2N_2, derived from cyano- and -gen; patterned after halogen

cyanogen bromide
CBrN, derived from cyanogen

cyanogen chloride
CClN, derived from cyanogen

cyanohydrin
derived from cyano-, hydr(o)-, and -in(e)

cyanopsin
derived from cyan(o)- and opsin

cyanuramide (melamine)
$C_3H_6N_6$, derived from cyanuric acid and amide

cyanuric acid
$C_3H_3N_3O_3$, derived from cyano- and urea

cyanuric chloride
$C_3Cl_3N_3$, derived from cyanuric acid and chloride

cyanyl (cyanogen)
CN, derived from cyano- and -yl

cyaphenine
$C_{21}H_{15}N_3$, derived from cyan(o)-, phene, and -in(e)

cybotactic
derived from cybotaxis, from *kybos* (Greek: cube) and *taxis* (Greek: order, position), from *tassein* (Greek: to arrange, to order)

cycasin
$C_8H_{16}N_2O_7$, derived from the genus name *Cycas* (cycad), ultimately from *koix* (Greek: doom palm), and -in(e)

cyclamate
$C_6H_{12}NO_3S^-$, derived from cyclamic acid and -ate

cyclamen aldehyde
$C_{13}H_{18}O$, derived from cyclamen – referring to the lily-of-the-valley odor of this compound

cyclamic acid
$C_6H_{13}NO_3S$, coined by contraction of cyclohexylsulfamic acid

cyclamin
derived from the genus name *Cyclamen* (Eurasian plants), from *kyklaminos*, *kyklamis* (Greek: cyclamen), from cycl(o)-, and -in(e)

cyclamine
coined by contraction of cyclic amine

cyclane
coined by contraction of cycloalkane

cyclanone
coined by contraction of cycloalkanone

cyclene
coined by contraction of cycloalkene

-cyclin
derived from tetracyclin

cyclitol
derived from cycl(o)- and -itol

cyclo-
derived from cycl(o)-

cycl(o)-
derived from *kyklos* (Greek: ring, circle, cycle, wheel)

cyclobuxine
derived (with contraction) from cyclopregnane and the genus name *Buxus* (boxwood), ultimately from *pyxos* (Greek: box), and -in(e)

cyclodextrin
derived from cycl(o)- and dextrin

cyclonite (hexogen, RDX)
$C_3H_6N_6O_6$, coined by contraction of *cyclo*-trimethylenetrinitramine

cyclonium
Pm, a name usuccessfully suggested for promethium, derived (with contraction) from cyclotron and -ium – referring to this element's preparation in a cyclotrone

cyclophane
derived from cycl(o)- and phane

cycloserine
$C_3H_6N_2O_2$, derived from cycl(o)- and serine

cyclosilicate
derived from cycl(o)- and silicate

cyclosporin
$C_{62}H_{111}N_{11}O_{12}$, derived from cycl(o)-, spor(o)-, and -in(e)

cyclotheonamide
derived from cycl(o)-, the genus name *Theonella* (marine sponges), possibly named for the Greek philosopher and mathematician Theon of Alexandria (335-405) or the Greek philosopher and mathematician Theon of Smyrna (70-135), and amide

cyclotron
derived from cycl(o)- and electron

cyclovalone
$C_{22}H_{22}O_5$, derived anagrammatically by contraction of 2,6-divanillylidene cyclo-hexanone

cyclovoltammetry
derived from cyclo- and voltammetry

cymarin
$C_{30}H_{44}O_9$, originally coined as a trademark, partially derived (with contraction) from the family name *Apocynaceae* (dogbanes), from *apocynum* (Latin: dogbane), from *kyon, kynos* (Greek: dog)

cymarose
$C_7H_{14}O_4$, derived from cymarin and -ose

cymene
$C_{10}H_{14}$, derived from the specific epithet of the species name *Cuminum cyminum* (cumin), from *kyminon* (Greek: cumin), ultimately from *kammon* (Hebrew: cumin), and -ene

cymophane (cat's eye)
$BeAl_2O_4$, derived from *kyma* (Greek: wave) and -phane – referring to this mineral's opalescence

cynarin(e)
$C_{25}H_{24}O_{12}$, derived from the genus name

Cynara (herbs), from *kynara* (Greek: a kind of artichoke), and -in(e)

cynth(i)-
derived from *Kynthia* (Greek: woman from Kynthos), i.e. Artemis, the Greek goddess of forests and hills

cystamine
$C_4H_{12}N_2S_2$, derived from cystine and amine

cystathionine
$C_7H_{14}N_2O_4S$, coined by contraction of cysteine and methionine

cysteamine
C_2H_7NS, derived from cysteine and amine

cysteic acid
$C_3H_7NO_5S$, derived from cysteine

cysteine
$C_3H_7NO_2S$, derived from cyst(o)- and -ein(e)

cystine
$C_6H_{12}N_2O_4S_2$, derived from cyst(o)- and -in(e)

cyst(o)-
derived from *kystis* (Greek: bladder)

cytidine
$C_9H_{13}N_3O_5$, derived from cyt(o)- and -idin(e)

cytidylic acid
$C_9H_{14}N_3O_8P$, derived from cytidine

cytisine
$C_{11}H_{14}N_2O$, derived from *kytisos* (Greek: moon trefoil, *Medicago arborea*) and -in(e)

cyt(o)-
derived from *kytos* (Greek: cell, hollow vessel)

cytochalasin (cytochalasan)
derived from cyt(o)-, *chalasis* (Greek: a slackening), from *chalan* (Greek: to slacken), and -in(e)

cytochrome
derived from cyt(o)- and -chrome

cytochrome P 450
derived (with contraction) from cytochrome, pigment, and (absorption at) 450 nm

cytocuprein (superoxide dismutase)
derived from cyt(o)-, cupr(o)-, and -ein(e)

cytohemin
$C_{49}H_{56}ClFeN_4O_6$, derived from cyt(o)-, heme, and -in(e)

cytokine
derived from cyt(o)- and kin(e)-

cytokinin
derived from cyt(o)- and kinin

cytosine
$C_4H_5N_3O$, derived from cyt(o)-, -ose, and -in(e)

Czochralski process
named for the Polish metallurgist Jan Czochralski (1885-1953)

D

d-
an abbreviation of dextr(o)-

D-
derived from *dexter* (Latin: right)

DABCO
$C_6H_{12}N_2$, an abbreviation for 1,4-diazabicyclo[2.2.2]octane

dacron
polyesters, a word invented as a trademark; patterned after nylon

dactinomycin
$C_{62}H_{86}N_{12}O_{16}$, coined by contraction of actinomycin D

daidz-
derived from *daidzu* (Japanese: soy bean)

daidzein
$C_{15}H_{10}O_4$, derived from daidz- and -ein(e)

daidzin
$C_{21}H_{20}O_9$, derived from daidz- and -in(e)

Dakin oxidation
named for the British chemist Henry Drysdale Dakin (1880-1952)

Dakin reaction
named for the British chemist Henry Drysdale Dakin (1880-1952)

Dakin's solution
named for the British chemist Henry Drysdale Dakin (1880-1952)

Dakin-West reaction
named for the British chemist Henry Drysdale Dakin (1880-1952) and the US chemist Randolf West (1890-1949)

dalapon
$C_3H_4Cl_2O_2$, coined by anagrammatical contraction of α,α-dichloropropionic acid

daltonide (proustide)
stoichiometric compound, named for the British chemist John Dalton (1766-1844)

Dalton's laws
named for the British chemist John Dalton (1766-1844)

damascenine (nigelline)
$C_{10}H_{13}NO$, derived from the specific epithet of the species name *Nigella damascena* L. (devil in the bush, love-in-a-mist), from *damascenus* (Latin: of Damascus, Syria), and -in(e)

dammarane
$C_{30}H_{54}$, derived from dammar (*Agathis*), from *damar* (Malay: dammar, resin), and -an(e)

damsyl
$C_8H_{10}NO_2S-$, coined by contraction of 4-(dimethylamino)benzenesulfonyl

danalite
$Be_3Fe_4(SiO_4)_3S$, named for the US mineralogist James Dwight Dana (1813-1895)

danburite
$CaB_2(SiO_4)_2$, named after this mineral's locality Danbury, CT, USA

Daniell element
named for the British chemist John Frederic Daniell (1790-1845)

Danishefsky's diene
$C_8H_{16}O_2Si$, named for the US chemist Samuel J. Danishefsky (born 1936)

Danishefsky reaction
named for the US chemist Samuel J. Danishefsky (born 1936)

danium
Hf, an alternative name suggested for hafnium, in honor of Denmark where hafnium was first isolated, derived from *Dania* (New Latin: Denmark)

dansyl
$C_{12}H_{12}NO_2S-$, coined by contraction of 5-(dimethylamino)naphthalene-1-sulfonyl

danubium
Tc, an unsuccessfully suggested name for technetium, after *Danubius fluvius* (Latin: Danube)

DAP
$C_{14}H_{14}O_4$, an abbreviation for diallyl phthalate

daphnane
$C_{22}H_{37}N$, derived from daphnin and -an(e)

daphnetin
$C_9H_6O_4$, coined by variation of daphnin

daphnin
$C_{15}H_{16}O_9$, derived from the genus name *Daphne* (European shrubs), from *daphne* (Greek: laurel), and -in(e)

daphnoline (trilobamine)
$C_{35}H_{36}N_2O_6$, derived from the genus name *Daphnandra* (plants), from *daphne* (Greek: laurel) and *aner* (Greek: man = male person, stamen), -ol, and -in(e)

D'Arcet's metal
(Bi,Pb,Sn), named for the French chemist Jean D'Arcet (1725-1801)

darmstadtium
Ds, named for the city of Darmstadt, Germany, where this element was prepared for the first time

Darzens condensation
named for the Russian-French chemist George Darzens (1867-1954)

DAST
$C_4H_{10}F_3NS$, an abbreviation for (diethylamino)sulfur trifluoride

datiscetin
$C_{15}H_{10}O_6$, derived from the genus name *Datisca* (herbs, trees), perhaps from *dateomai* (Greek: to divide, to cut in two), and -etin(e)

datolite
$CaBSiO_4(OH)$, derived from *datheisthai* (Greek: to divide) and -lite – referring to the fact that granular aggregates of this mineral crumble readily

daturine (duboisine, hyoscyamine)
$C_{17}H_{23}NO_3$, derived from the genus name *Datura* (jimsonweed), from *dhatura* (Hindi: jimsonweed), and -in(e)

daubreeite
BiO(OH,Cl), named for the French scientist Gabriel Auguste Daubrée (1814-1896)

daubreelite
Cr_2FeS_4, named for the French scientist Gabriel Auguste Daubrée (1814-1896)

daucol
$C_{15}H_{26}O_2$, derived from the genus name

Daucus (wild carrots), ultimately from *daukos*, *daukon* (Greek: a kind of parsnip or wild carrot), possibly akin to *daiein* (Greek: to ignite, to burn), and -ol

dauno-
derived from *daunius* (Latin: Apulian)

daunomycin (daunorubicin, rubido mycin)
$C_{27}H_{29}NO_{10}$, probably coined as trademark, derived from dauno- and -mycin

daunorubicin (daunomycin, rubido mycin)
$C_{27}H_{29}NO_{10}$, coined by contraction of daunomycin and rubidomycin

dauricine
$C_{38}H_{44}N_2O_6$, derived from the specific epithet of the species name *Menispermum dauricum* DC. (Asiatic moonseed), from Daur, a Manchu-Tungus people, and -in(e)

davidite
$(La,Ce,Ca)(Y,U)(Ti,Fe^{III})_{20}O_{38}$, named for the Australian geologist Tenatt William Edgeworth David (1858-1934)

Dawson anion (Wells-Dawson anion)
$E_2Mo_{18}O_{62}^{6-}$, $E_2W_{18}O_{62}^{6-}$ (E = main group element), named for the 20th century British chemist B. Dawson

DBN
$C_7H_{12}N_2$, an abbreviation for 1,5-diazabicyclo[4.3.0]non-5-ene

[1]DBP (BHT)
$C_{15}H_{24}O$, an abbreviation for di-*tert*-butylmethylphenol

[2]DBP
$C_{16}H_{22}O_4$, an abbreviation for dibutyl phthalate

DBU
$C_9H_{16}N_2$, an abbreviation for 1,8-diazabicyclo[5.4.0]undec-2-ene

DCC
$C_{13}H_{22}N_2$, an abbreviation for *N,N'*-dicyclohexylcarbodiimide

DCCC
an abbreviation for droplet counter-current chromatography

DDQ
$C_8Cl_2N_2O_2$, an abbreviation for 2,3-dichloro-5,6-dicyano-*p*-benzoquinone

DDT
$C_{14}H_9Cl_5$, an abbreviation for the obsolete name dichlorodiphenyltrichloroethane

DDVP
$C_4H_7Cl_2O_4P$, an abbreviation for *O,O*-dimethyl *O*-(2,2-dichlorovinyl) phosphate

DEA
$C_4H_{11}NO_2$, an abbreviation for diethanolamine

Deacon process
named for the British chemist Henry Deacon (1822-1876)

deanol
$C_4H_{11}NO$, coined by contraction of 2-(dimethylamino)ethanol

Dean-Stark apparatus
named for the US chemists Ernest Woodward Dean (1888-1959) and David D. Stark (born 1893)

Debye-Clausius- Mossotti equation
named for the Dutch-US physicist Peter Joseph Wilhelm Debye (1884-1966), the German physicist Rudolf Julius Emanuel Clausius (1822-1888), and the Italian

physicist O. F. Mossotti (1791-1863)

Debye-Hückel law
named for the Dutch-US physicist Peter Joseph Wilhelm Debye (1884-1966) and the German physicist Erich Hückel (1895-1973)

Debye-Scherrer method
named for the Dutch-US physicist Peter Joseph Wilhelm Debye (1884-1966) and the Swiss physicist P. Scherrer (1890-1969)

deca-
derived from *deka* (Greek: ten)

decakis-
derived from *dekakis* (Greek: ten times)

decalin
$C_{10}H_{18}$, originally a trademark, coined by contraction of *Decahydronaphthalin* (German: decahydronaphthalene)

decant
derived from de(s)- and *cantus* (Medieval Latin: side), from *canthus* (Latin: iron ring around a carriage wheel)

[1]deci-
derived from *decima pars* (Latin: one tenth), from *decimus* (Latin: tenth), from *decem* (Latin: ten)

[2]deci-
derived from *decies* (Latin: ten times), from *decem* (Latin: ten)

deet
$C_{12}H_{17}NO$, an acronym for *N,N*-diethyl-*m*-toluamide

defensin
derived from defense and -in(e)

deferoxamine (ferrioxamine)
$C_{25}H_{48}N_6O_8$, derived from de(s)-, ferrum, and hydroxylamine

deflagration
derived from de(s)- and *flagrare* (Latin: to burn)

deguelin
$C_{23}H_{22}O_6$, derived from the genus name *Deguelia* (derris), from *assa-ha-pagara undeguélé* (Galibi: derris), and -in(e)

Degussa process
named for the chemical company Degussa AG (an abbreviation for Deutsche Gold- und Silber-Scheide-Anstalt AG), Frankfurt/Main, Germany

DEHP
$C_{24}H_{38}O_4$, an abbreviation for the obsolete name di(2-ethylhexyl) phthalate

dehydroacetic acid
$C_8H_8O_4$, derived from de-, hydr(o)-, and acetic acid

d-electron
named after the 'diffuse' line groups in atomic spectroscopy

Delépine reaction
named for the French chemist Stephane Marcel Delépine (1871-1965)

deliquescent
derived from de(s)- and *liquescere* (Latin: to become liquid)

delphinidin
$C_{15}H_{11}ClO_7$, derived from delphin(o)- and -idin(e)

delphinine
$C_{33}H_{45}NO_9$, derived from delphin(o)- and -in(e)

delphin(o)-
derived from the genus name *Delphinium* (larkspur), from *delphinion* (Greek: larkspur), from *delphis* (Greek: dolphin)

delsoline
$C_{25}H_{41}NO_7$, coined by contraction of the species name *Delphinium consolida* L. (larkspur), from *delphinion* (Greek: larkspur), from *delphis* (Greek: dolphin), and *consolida* (Latin: black bryony, comfrey), from con- and *solidus* (Latin: solid), and -in(e)

deltanoid
derived from the letter D in vitamin D and -oid

deltic acid
$C_3H_2O_3$, named after its molecular shape resembling the Greek letter Δ

de Mayo reaction
named for the 20th century Canadian chemist Paul de Mayo

demeclocycline
$C_{21}H_{21}ClN_2O_8$, coined by contraction of the obsolete name demethylchlorotetracycline

demecolcine
$C_{21}H_{25}NO_5$, coined by contraction of *N*-deacetyl-*N*-methylcolchicine

Demyanov rearrangement
named for the Russian chemist Nikolai Jakovlevich Demyanov (1861-1938)

dendrimer
derived from dendr(o)- and -mer

dendritic
derived from dendr(o)-

dendr(o)-
derived from *dendron* (Greek: tree)

dendrobane
$C_{16}H_{27}NO$, derived from the genus name *Dendrobium* (orchids), from dendr(o)- and bio-, and -an(e)

dendrotoxin
derived from the genus name *Dendroaspis* (snakes), from dendr(o)-, and toxin

de novo synthesis
derived from *de* (Latin: from) and *novus* (Latin: new)

-dent
derived from *dens* (Latin: tooth)

-dentate
derived from *dentatus* (Latin: toothed), from -dent

dentin
derived from *dens* (Latin: tooth) and -in(e)

deoxy-
derived from de(s)- and oxygen

deoxyribonuclease
derived from deoxyribonucleic acid and -ase

deoxyribonucleic acid
derived from 2-deoxy-D-ribose and nucleic acid

DEP
$C_{10}H_{14}O_4$, an abbreviation for diethyl phthalate

dephlegmation
derived from *phlegma* (Greek: flame, inflammation, mucus)

dephlogisticated air (fire air, pure air, vital air)
O_2, an archaic name for oxygen, derived from phlogiston

depsi-
derived from *depsein* (Greek: to tan)

depside
derived from depsi- and -ide

depsipeptide
derived from depsi- and peptide

derm(o)-
derived from *derma* (Greek: skin), from *derein* (Greek: to skin)

dermostatin
derived from derm(o)- and -statin

derritol
$C_{21}H_{22}O_6$, derived from the genus name *Derris* (derris), from *derris* (Greek: leather covering, skin), from *derein* (Greek: to skin), and -ol

de(s)-
derived from *de* (Latin: away from)

desaspidin (rosapin)
coined by contraction of de(s)methylaspidin

desaurin
most probably coined by contraction of *Desoxybenzoin* (German: deoxybenzoin), a typical starting material for the synthesis of desaurins, aur(o)-, and -in(e) – referring to the yellow color of many desaurins

descloizite
$PbZnVO_4(OH)$, named for the French mineralogist Alfred Des Cloizeaux (1817-1897)

deserpidine
$C_{32}H_{38}N_2O_8$, coined by anagrammatical contraction of 11-de(s)methoxyreserpine

desiccator
derived from de(s)- and *siccare* (Latin: to dry), from *siccus* (Latin: dry)

deslanoside
$C_{47}H_{74}O_{19}$, coined by contraction of deacetyllanatoside C

desmine (stilbite)
$NaCa_2Al_5Si_{13}O_{36}\cdot 14H_2O$, derived from *desme* (Greek: bundle) and -in(e) – referring to this mineral's tufts of crystals

desm(o)-
derived from *desmos* (Greek: band, bond), from *dein* (Greek: to bind)

desmolase
derived from desmolysis and -ase

desmolysis
derived from desm(o)- and lysis

[1]**desmosine**
$C_{24}H_{40}N_5O_8$, derived from desm(o)- and -in(e) – referring to the role this compound plays in the cross-linking of the peptide chains of elastin

[2]**desmosine**
$C_{17}H_{15}NO_4$, derived from the genus name *Desmos* (shrubs), from desm(o)-, and -in(e)

desmosterol
$C_{27}H_{44}O$, derived from desm(o)- and sterol

desmotropism
derived from desm(o)-, -trope, and –ism

desorb, desorption
derived from de(s)- and sorption

desosamine
$C_8H_{17}NO_3$, derived from de(s)-, -ose, and amine

desoxy-
derived from de(s)- and oxygen

Dess-Martin oxidation
named for the US chemists Daniel B. Dess and James Cullen Martin (born 1928)

Dess-Martin periodinane
$C_{13}H_{13}IO_8$, named for the US chemists Daniel B. Dess and James Cullen Martin (born 1928)

detergent
derived from *detergere* (Latin: to cleanse, to wipe away), from de(s)- and *tergere* (Latin: to wipe off, to rub off), akin to *trogein* (Greek: to gnaw)

detoxin
derived from the subspecific epithet of the bacterial species name *Streptomyces caespitosus* var. *detoxicus* and -in(e)

deuterium
^2H, derived from deuter(o)- and -ium

deuter(o)-
derived from *deuteros* (Greek: the second)

deuteroheme
$C_{30}H_{28}FeN_2N_4O_4$, derived from deuter(o)- and heme

deuteron
$^2H^+$, derived (with contraction) from deuterium and [3]-on

deuteroporphyrin
$C_{30}H_{30}N_4O_4$, derived from deuter(o)- and porphyrin

Devarda's alloy
(Cu,Al,Zn), named for the Italian chemist Arturo De Varda (1859-1944)

Dewar benzene
C_6H_6, named for the Scottish chemist Sir James Dewar (1842-1923)

Dewar flask
named for the Scottish chemist Sir James Dewar (1842-1923)

Dexter-Silverton anion
$EMo_{12}O_{42}{}^{n-12}$ (n = valency of E), named for the 20th century US chemists David D. Dexter and J. V. Silverton

dextran
derived from dextrose and -an

dextrin
derived from dextrose and -in(e)

dextr(o)-
a prefix denoting dextrorotatory substances, derived from *dexter* (Latin: right)

dextronic acid
$C_6H_{12}O_7$, an obsolete name for D-gluconic acid, derived from dextrose

dextrose
$C_6H_{12}O_6$, an obsolete name for D-glucose, derived from dextr(o)- and -ose – referring to this sugar's dextrorotatory property

DHA
$C_{22}H_{32}O_2$, an abbreviation for (*all-Z*)-docosa-4,7,10,13,16,19-hexaenoic acid

dhurrin
$C_{14}H_{17}NO_7$, derived from dhurra (durra, *Sorghum bicolor*), from *dhurrah* (Arabic: durra, *Sorghum bicolor*), and -in(e)

di-
derived from *dyo* (Greek: two)

di(a)-
derived from *dia* (*di-*) (Greek: through, apart)

diaboline
$C_{21}H_{24}N_2O_3$, derived from the specific

epithet of the species name *Strychnos diaboli* Sandw. (a tropical tree), from *diabolos* (Greek: devil), from di(a)- and *ballein* (Greek: to throw), and -in(e)

diacetin
$C_7H_{12}O_5$, coined by contraction of *1,3-Diacetylglycerin* (German: 1,3-diacetyl-glycerol)

diacetone alcohol (diacetone)
$C_6H_{12}O_2$, derived from di- and acetone – referring to this compound being a formal dimer of acetone

diacetylene
C_4H_2, an obsolete name for buta-1,3-diyne, derived from di- and acetylene

diafiltration
coined by contraction of dialysis and ultrafiltration

diallage
$Ca(Mg,Al)[(Al,Si)SiO_6]$, derived from *diallage* (Greek: change), from *diallassein* (Greek: to interchange, to exchange, to change), from dia- and *allassein* (Greek: to change), from all(o)- – referring to this mineral's bronzy sheen

dialysis
derived from dia- and lysis

diamantane (congressane)
$C_{14}H_{20}$, derived from *diamas* (Latin: diamond) and -an(e) – referring to this hydrocarbon's structural similarity with adamantane

diamond
C_n, derived from *diamas* (Latin: diamond), coined by contraction of the words *adamas* (Greek: invincible) and *diaphanes* (Greek: transparent)

Diamond process
named for the chemical company Diamond Alkali Company, Pittsburgh, PA, USA

diaphragm
derived from dia- and *phragma* (Greek: fence, screen), from *phrassein*, *phrattein* (Greek: to enclose, to fence in)

diaspore
AlO(OH), derived from *diasporein* (Greek: to scatter, to spread), from *speirein* (Greek: to sow, to sprinkle) – referring to this mineral's easy disintegration in the blowpipe flame

diastase
derived from *diastasis* (Greek: cleavage, separation), from dia- and *histanai* (Greek: to set, to cause to stand)

diastereomer
derived from dia-, stereo-, and -mer

diastereotopic
derived from diastereomer and -topic

diatom
derived from *diatomos* (Greek: cut through), from *temnein* (Greek: to cut)

diatomaceous
derived from diatom

diatomite
derived from diatom and -ite

diatretyne
derived from the specific epithet of the species name *Clitocybe diatreta* (a mushroom), from *vas diatretum* (Latin: openwork vessel), ultimately from di(a)- and *rete* (Latin: net), and -yne

diatropic
coined by contraction of diamagnetism and

-tropic

diazo-
derived from di- and az(o)-

diazonium
derived from di-, az(o)-, and -onium

dicarbane
C_2H_6, systematic, but discouraged name for ethane, derived from di-, carb(o)-, and -an(e)

dicentrine
$C_{20}H_{21}NO_4$, derived from the genus name *Dicentra* (bleeding heart), from di- and *kentron* (Greek: sharp point), and -in(e)

dichroism
derived from *dichroos* (Greek: two-colored)

dichroite (cordierite, iolite)
$Mg_2Al_4Si_5O_{18}$, derived from dichroitic, from *dichroos* (Greek: two-colored), from di- and *chroma* (Greek: color), and -ite – referring to this mineral's dichroism

dickite
$Al_2Si_2O_5(OH)_4$, named for the Scottish mineralogist Allan Brugh Dick (1833-1926)

dicoumarol (dicumarol, melitoxin)
$C_{19}H_{12}O_6$, derived from di-, coumar(o)-, and -ol

dictamnine
$C_{12}H_9NO_2$, derived from the genus name *Dictamnus* (perennial herbs), probably from *diktamnon* (Greek: from Mount Dikte, Crete), and -in(e)

dicty(o)-
derived from *diktyon* (Greek: net), from *dikein* (Greek: to throw)

DIDA
$C_{26}H_{50}O_4$, an abbreviation for diisodecyl adipate

didanosine
$C_{10}H_{12}N_4O_3$, coined by contraction (and alteration) of dideoxyinosine

didemnin
derived from the genus name *Didemnum* (tunicates), from di- and *demnion* (Greek: bed, nest, home, den), and -in(e)

DIDP
$C_{28}H_{46}O_4$, an abbreviation for diisodecyl phthalate

didymium
(Nd,Pr), derived from *didymos* (Greek: twin) and -ium – referring to the presence of the two elements neodymium and praseodymium in this for a long time inseparable mixture

Dieckmann condensation
named for the German chemist Walter Dieckmann (1869-1925)

dieldrin
$C_{12}H_8Cl_6O$, named for the Diels-Alder reaction

Diels-Alder reaction
named for the German chemists Otto Paul Hermann Diels (1876-1954) and Kurt Alder (1902-1958)

Diels's hydrocarbon
$C_{18}H_{16}$, named for the German chemist Otto Paul Hermann Diels (1876-1954)

diene
derived from di- and -ene

dienone
derived from di-, -ene, and -on(e)

dienophile
derived from diene and -phile

diergolic
derived from di- and -ergolic

diesel oil (diesel fuel)
derived from diesel engine, named for the German mechanical engineer and inventor Rudolf Diesel (1858-1913)

digallic acid
$C_{14}H_{10}O_9$, derived from di- and gallic acid

digalogenin
$C_{27}H_{44}O_4$, derived from digalonin, -gen, and -in(e)

digalonin
$C_{56}H_{92}O_{28}$, derived by variation of digi(talo)- and -in(e)

digenic acid (kainic acid)
$C_{10}H_{15}NO_4$, derived from the genus name *Digenia* (red algae)

digenite
Cu_9S_5, derived from di- and *genus* (Latin: kind) – referring to this mineral's content of both cuprous and cupric ions

diginatigenin
$C_{23}H_{34}O_6$, derived from diginatin, -gen, and -in(e)

diginatin
$C_{41}H_{64}O_{15}$, derived (with contraction) from the species name *Digitalis lanata* Ehrk. (Grecian foxglove), from digi(talo)- and *lanatus* (Latin: woolly), from *lana* (Latin: wool), and -in(e)

diginin
$C_{28}H_{40}O_7$, derived from digi(talo)- and -in(e)

digitalin
$C_{36}H_{56}O_{14}$, derived digi(talo)- and -in(e)

digi(talo)-
derived from the genus name *Digitalis* (foxglove), from *digitalis* (Latin: belonging to a finger), from *digitus* (Latin: finger, toe)

digitalose
$C_7H_{14}O_5$, derived from digi(talo)- and -ose

digitogenin
$C_{27}H_{44}O_5$, derived from digi(talo)-, -gen, and -in(e)

digitonin
$C_{56}H_{92}O_{29}$, derived from digi(talo)-, -one, and -in(e)

digitoxigenin
$C_{23}H_{34}O_4$, derived from digitoxin, -gen, and -in(e)

digitoxin
$C_{41}H_{64}O_{13}$, derived from digi(talo)- and toxin

digitoxose
$C_6H_{12}O_4$, derived from digitoxin and -ose

diglyme
$C_6H_{14}O_3$, coined by contraction of diethylene glycol dimethyl ether

digol
$C_4H_{10}O_3$, coined by contraction of diethylene glycol

digoxigenin
$C_{23}H_{34}O_5$, derived from digoxin, -gen, and -in(e)

digoxin
$C_{41}H_{64}O_{14}$, derived (with contraction) from digi(talo)- and toxin

diguanide (biguanide)
$C_2H_7N_5$, derived from di- and guanidine

dihydrite (pseudomalachite)
$Cu_5(PO_4)_2(OH)_4$, derived from di-, hydr(o)-, and -ite – referring to this mineral's formal content of two water molecules

diketene
$C_4H_4O_2$, derived from di- and ketene – referring to this compound being an (unsymmetrical) dimer of ketene

diketopiperazine
$C_4H_6N_2O_2$, derived from di-, keto-, and piperazine

dilactide
$C_6H_8O_4$, derived from di- and lactide

dimedone
$C_8H_{12}O_2$, coined by contraction of the name 5,5-dimethylcyclohexane-1,3-dione

dimer
derived from di- and -mer

dimercaprol (British antilewisite, BAL)
$C_3H_8OS_2$, coined by contraction of (obsolete) 2,3-dimercaptopropan-1-ol

dimerization
derived from dimer

dimethylglyoxime (Chugaev's reagent)
$C_4H_8N_2O_2$, widely used, but improperly constructed, name derived from di-, methyl-, glyoxal, and oxime

dimorphism
derived from di- and *morphe* (Greek: shape, form)

dimorphite
As_4S_3, derived from dimorphism and -ite – referring to this mineral's perceived dimorphism

Dimroth condenser
named for the German chemist Otto Dimroth (1872-1940)

Dimroth rearrangement
named for the German chemist Otto Dimroth (1872-1940)

dimsyl
$C_2H_5OS^-$, coined by contraction of the name dimethyl sulfoxide anion and -yl

DINA
$C_{24}H_{46}O_4$, an abbreviation for diisononyl adipate

dinactin
$C_{42}H_{68}O_{12}$, coined by contraction of dihomononactin, from di-, homo-, and nonactin

dinosterol (Black Sea sterol)
$C_{30}H_{52}O$, derived from dinoflagellate, from *deinos* (Greek: terrible), and sterol – reflecting this sterol's isolation from the dinoflagellate *Gonyaulax tamarensis*

DIOP
$C_{24}H_{38}O_4$, an abbreviation for diisooctyl phthalate

diopside
$CaMgSi_2O_6$, derived from di-, *opsis* (Greek: appearance, sight), and -ide – referring to the antihemihedrism of this mineral's crystals

dioptase
$CuSiO_2(OH)_2$, derived from *dioptos* (Greek: visible), from *opsesthai* (Greek: to be going to see), and -ase – referring to the visibility of this mineral's internal cleavage planes

diorite
a rock species, derived from *diorizein* (Greek: to distinguish) and -ite

dioscin
$C_{45}H_{72}O_{16}$, derived from dios(cor)- and -in(e)

dios(cor)-
derived from the genus name *Dioscorea* (yams), named for the Greek physician Pedanius Dioscorides (40-90)

dioscorine
$C_{13}H_{19}NO_2$, derived from dios(cor)- and -in(e)

diose
$C_2H_4O_2$, derived from di- and -ose

diosgenin
$C_{27}H_{42}O_3$, derived from dioscin, -gen, and -ine).

dios(ma)-
derived from the genus name *Diosma* (shrubs), from *dios* (Greek: heavenly) and, ultimately, *-osme* (Greek: one having an odor)

diosmetin
$C_{16}H_{12}O_6$, derived from diosmin, -ete, and -in(e)

diosmin
$C_{28}H_{32}O_{15}$, derived from dios(ma)- and -in(e)

diosphenol
$C_{10}H_{16}O_2$, derived from dios(ma)- and phenol

dioxan
$C_4H_8O_2$, derived from di-, ox(a)-, and -an(e)

[1]dioxin
$C_4H_4O_2$, derived from di-, ox(a)-, and -in(e)

[2]dioxin
popular compound class name coined by contraction of polychlorodibenzo[1,4]dioxin

dioxygenase
derived from dioxygen and -ase

dipalmitin
$C_{35}H_{68}O_5$, coined by contraction of *Dipalmitoylglycerin* (German: dipalmitoyl glycerol)

dipentene (cajeputene)
$C_{10}H_{16}$, an obsolete name for 1-methyl-4-(prop-1-en-2-yl)cyclohex-1-ene, derived from di- and pentene

diphenic acid
$C_{14}H_{10}O_4$, derived from diphenyl

diphenolic acid
$C_{17}H_{18}O_4$, derived from di- and phenol

diphenyl
$C_{12}H_{10}$, an obsolete name for biphenyl, derived from di- and phenyl

diphos
$C_{26}H_{24}P_2$, an abbreviation for 1,2-bis(diphenylphosphino)ethane

diphosgene
$C_2Cl_4O_2$, derived from di- and phosgene – referring to the fact that this compound is a dimer of phosgene

dipicrylamine (hexyl)
$C_{12}H_5N_7O_{12}$, derived from di-, picryl, and amine

diploicin
$C_{16}H_{10}Cl_4O_5$, derived from the genus name *Diploicia* (lichens), ultimately from *diploos*

(Greek: double), and -in(e)

dipolarophile
derived from dipole and -phile

Dippel's oil
named for the German physician and alchemist Johann Conrad Dippel, a.k.a. Christianus Democritus (1673-1734)

diprenyl (geranyl)
$C_{10}H_{17}$–, derived from di- and prenyl

dipropargyl
C_6H_6, an obsolete name for hexa-1,5-diyne, derived from di- and propargyl

DIPT
$C_{10}H_{18}O_6$, an abbreviation for diisopropyl tartrate

dipyrrin
$C_9H_8N_2$, derived from di-, pyrrole, and -in(e)

dis-
derived from *dis-* (Latin: apart, asunder)

discodermolide
$C_{33}H_{53}NO_8$, derived from the genus name *Discodermia* (marine sponges), from *diskos* (Greek: quoit) and *derma* (Greek: skin), and -olide

discotic
derived from *diskos* (Greek: discus)

dismutase
derived from dismutation and -ase

dismutation
derived from dis- and *mutatio* (Latin: change)

disparlure
$C_{19}H_{38}O$, derived from the specific epithet of the species name *Lymantria dispar* (gypsy moth), from *dispar* (New Latin: unlike) – referring to these moths' sexual dimorphism – and lure

dispase
derived from *dispergere* (Latin: to disperse) and -ase

disrotatory
derived from dis- and rotation

dissymmetry
derived from dis- and symmetry

distearin
$C_{39}H_{76}O_5$, coined by contraction of *Distearoylglycerin* (German: distearoyl glycerol)

disthene (kyanite)
$Al_2(O)SiO_4$, derived from di- and -sthene – referring to this mineral's unequal hardness in different directions

distillation
derived from *destillare* (Latin: to drop down), from *stilla* (Latin: drop), diminutive of *stiria* (Latin: icicle)

distonic
derived from distant and ionic

disulphine blue
derived from di-, sulf(o)-, and -in(e)

disyndiotactic
derived from di- and syndiotactic

ditactic
derived from di- and tactic

diterpene
$(C_5H_8)_4$, derived from di- and terpene

dithionic acid
$H_2O_6S_2$, derived from di- and thi(o)-

dithionous acid
$H_2O_4S_2$, derived from di- and thi(o)-

dithizone
$C_{13}H_{12}N_4S$, coined by contraction of diphenylthiocarbazone

divicine
$C_4H_6N_4O_2$, derived from di- and vicine – referring to the formation of this compound by degradation of vicine

divinyl
C_4H_6, an obsolete name for buta-1,3-diene, derived from di- and vinyl

divinylenimine
C_4H_5N, an obsolete name for pyrrole, derived from di-, vinylene, and imine

djenkolic acid
$C_7H_{14}N_2O_4S_2$, derived from *djenkol* (Javanese: velvet bean, *Pithecolobium lobatum* Benth.)

DLVO theory
named for the Russian chemist B. V. Deryagin (1902-1994), the Russian physicist L. D. Landau (1908-1968), and the Dutch chemists E. J. W. Verwey (1905-1981) and J. T. G. Overbeek (born 1911)

DM ([1]adamsite)
$C_{12}H_9AsClN$, an arbitrary US military code for [1]adamsite

DMA
C_4H_9NO, an abbreviation for *N,N*-dimethylacetamide

DMF
C_3H_7NO, an abbreviation for *N,N*-dimethylformamide

DMP
$C_{10}H_{10}O_4$, an abbreviation for dimethyl phthalate

DMSO
C_2H_6OS, an abbreviation for dimethyl sulfoxide

DMT
$C_{10}H_{10}O_4$, an abbreviation for dimethyl terephthalate

DNA
an abbreviation for deoxyribonucleic acid

DNase (dornase)
coined by contraction of deoxyribonuclease

DNOC
$C_7H_6N_2O_5$, an abbreviation for 4,6-dinitro-*o*-cresol

DNOP
$C_{24}H_{38}O_4$, an abbreviation for di-*n*-octyl phthalate

DOA
$C_{32}H_{42}O_4$, an abbreviation for dioctyl adipate

Döbereiner lamp
named for the German chemist Johann Wolfgang Döbereiner (1780-1849)

Dobson unit (DU)
named for the British physicist G. M. B. Dobson (1889-1976)

DOC
an abbreviation for dissolved organic carbon

docimasy
derived from *dokimasia* (Greek: examination, scrutiny, test), from *dokimazein* (Greek: to assay, to test, to approve), from *dokimos* (Greek: approved,

assayed), from *dokein* (Greek: to seem good, to seem, to think)

doctor solution
Na_2PbO_2, derived from to doctor

doctor test
derived from doctor solution

dodeca-
derived from *dodeka* (Greek: twelve)

dodecahedrane
$C_{20}H_{20}$, derived from dodecahedron and -an(e)

dodecahedro-
derived from dodecahedron

dodecasil
SiO_2, derived from dodeca- and silicon

Doebner-Miller reaction
named for the German chemists Oscar Döbner (1850-1907) and Wilhelm von Miller (1848-1899)

Doebner reaction
named for the German chemist Oscar Döbner (1850-1907)

Doering-LaFlamme allene synthesis
named for the US chemists William von Eggers Doering (born 1917) and P. M. LaFlamme

doisynolic acid
$C_{18}H_{24}O_3$, derived from the name of the US chemist Edward A. Doisy (1893-1986) and -ol

dolabellane (palominol)
$C_{20}H_{32}O$, derived from the genus name *Dolabella* (sea hares) and -an(e)

dolichol
derived from *dolichos* (Greek: long) and -ol

dolomite
$CaMg(CO_3)_2$, named for the French mineralogist and geologist Déodat de Dolomieu (1750-1801); thus, contrary to popular belief, the Dolomite Mountains have been named after the mineral dolomite and not the other way round

dolostone
a rock species, derived from dolomite and stone

DOM (STP)
$C_{12}H_{19}NO_2$, an abbreviation for 2,5-dimethoxy-4-methylamphetamine

domesticine
$C_{19}H_{19}NO_4$, derived from the specific epithet of the species name *Nandina domestica* Thunb. (heavenly bamboo), ultimately from *domus* (Latin: house), and -in(e)

domoic acid
$C_{15}H_{21}NO_6$, derived from *domoi* (Japanese: the red alga *Chondria armata* Okamura)

Donnan equilibrium
named for the British chemist Frederick George Donnan (1870-1956)

DOP (DEHP)
$C_{24}H_{38}O_4$, an abbreviation for dioctyl phthalate

dopa
$C_9H_{11}NO_4$, an abbreviation for the obsolete name 3,4-dioxyphenylalanine

dopamine
$C_8H_{11}NO_2$, derived from dopa and amine

dopastin
$C_9H1_7N_3O_3$, coined by contraction of dopamine β-hydroxylase inhibitor

d-orbital
named after the spectroscopic 'diffuse' subseries

dornase (DNase)
coined by contraction of deoxyribonuclease

Dötz reaction
named for the 20th century German chemist Karl Heinz Dötz

Dowd-Beckwith ring expansion reaction
named for the US chemist Paul Dowd (1936-1996) and the Australian chemist Athelstan L. J. Beckwith (born 1930)

Dowex
a tradename, after The Dow Chemical Company, Midland, MI, USA

Dow metal (elektron)
(Mg,Al,Zn,Mn,Cu,Si,etc.), a trademark, after The Dow Chemical Company, Midland, MI, USA

doxorubicin
$C_{27}H_{29}NO_{11}$, coined by contraction of the name deoxydaunorubicin

doxycycline
$C_{22}H_{24}N_2O_8 \cdot H_2O$, coined by contraction of the name deoxytetracycline

DPM
$C_{11}H_{20}O_2$, an abbreviation for dipivaloyl-methane

Dragendorff's reagent
$KBiI_4$, named for the Estonian physician and pharmacist G. J. N. Dragendorff (1836-1898)

dravite
$NaMg_3Al_6(BO_3)_3Si_6O_{18}(OH)_4$, named after this mineral's locality Drau (Drava) River, Austria

Dreiding model
named for the Swiss chemist André S. Dreiding (born 1919)

drim-
derived from the genus name *Drimys* (woody evergreen flowering plants), from *drimys* (Greek: sharp, acrid)

drimane
$C_{15}H_{28}$, derived from drim- and -an(e)

drimenin
$C_{15}H_{22}O_2$, derived from drim-, -ene, and -in(e)

dry(o)-
derived from *drys* (Greek: tree, oak)

dryophantin
$C_{23}H_{28}O_{15}$, derived from the species name *Dryophanta* (oak galls), from dry(o)- and *-phantes* (Greek: showing), from *phainein* (Greek: to reveal, to show, to make known), and -in(e)

DSC
an abbreviation for differential scanning calorimetry

DSIP
$C_{35}H_{48}N_{10}O_{15}$, an abbreviation for delta sleep-inducing peptide

DTA
an abbreviation for differential thermal analysis

DTPA
$C_{14}H_{23}N_3O_{10}$, an abbreviation for diethylenetriamine- *N,N,N',N",N"*-pentaacetic

acid

DU
an abbreviation for Dobson unit

dubnium
Db, named for the town of Dubna, Russia, where this element was prepared for the first time

Duff reaction
named for the 20th century British chemist James C. Duff

Duhem-Margules equation
named for the French physicist Pierre Maurice Marie Duhem (1861-1916) and the Austrian physicist and meteorologist M. Margules (1856-1920)

dulcitol
$C_6H_{14}O_6$, derived from *dulcis* (Latin: sweet) and -itol

Dulong-Petit law
named for the French chemists Pierre Louis Dulong (1785-1838) and Alexis Thérèse Petit (1791-1820)

Dumas method
named for the French chemist Jean Baptiste André Dumas (1800-1884)

dumortierite
$(Al,Mg,Fe)_{27}B_4Si_{12}O_{69}(OH)_3$, named for the French paleontologist M. Eugène Dumortier (1803-1873)

dunite
a rock species, named after its locality Mount Dun, New Zealand

duralumin
coined as a trademark, derived from dur(o)- and aluminum

durapatite (hydroxylapatite)
$Ca_5(PO_4)_3(OH)$, derived from dur(o)- and apatite

durene
$C_{10}H_{14}$, derived from dur(o)- and -ene – referring to this hydrocarbon's crystallinity at ambient temperature

dur(o)-
derived from *durus* (Latin: hard)

duroquinone
$C_{10}H_{12}O_2$, derived from durene and quinone

Duthaler-Hafner reagent
$C_{36}H_{33}ClO_4Ti$, named for the Swiss chemists Rudolf O. Duthaler (born 1946) and Andreas Hafner (born 1956)

Dutt-Wormall reaction
named for the 20th century British chemists P. K. Dutt and A. Wormall

dvi-
derived from *dvi* (Sanskrit: two)

dvi-actinium
Usu, derived from dvi- and actinium

dvi-manganese
Re, derived from dvi- and manganese

dvi-rubidium
Fr, derived from dvi- and rubidium

dynamin
an enzyme, derived from dyn(amo)- and -in(e)

dynamite
derived from dyn(amo)- and -ite

dyn(amo)-
derived from *dynamis* (Greek: power)

dynorphin
derived from dyn(amo)- and endorphin

dy(o)-
derived from *dyo* (Greek: two)

dyotropic
derived from dy(o)- and -tropic

dypnone
$C_{16}H_{14}O$, derived from di- and hypnone – referring to this compound's formation by condensation of hypnone (i.e. acetophenone)

dys-
derived from *dys-* (Greek: difficult)

dyscrasite
Ag_3Sb, derived from *dyskrasis* (Greek: bad alloy), from dys- and -crase, and -ite

dysidiolide
$C_{25}H_{38}O_4$, derived from the genus name *Dysidea* (marine sponges), from *dysis* (Greek: immersion, diving), and -olide

dysprosia
Dy_2O_3, derived from dysprosium and -a

dysprosium
Dy, derived from *dysprositos* (Greek: hard to get at) and -ium – referring to this element's difficult isolation

dystectic
derived from dys- and -tectic

dystrophin
derived from dystrophy and -in(e)

E

E
derived from *entgegen* (German: opposite)

η (eta)
an affix in systematic inorganic nomenclature, derived from hapt(o)-

e- (ec-. ef-, ex)-
derived from *e(x)* (Latin: out of)

earth
archaic name for metal oxides, earlier often erroneously believed to be elements

earth alkali metals
derived from pre-1800 conceptions of calcium oxide and barium oxide as elements

Eastwood reaction
named for the 20th century Australian chemist F. W. Eastwood

eau de Javelle
KClO, named after the town of Javelle, France

eau de Labarraque
NaClO, named for the French pharmacist Antoine G. Labarraque (1777-1850)

Ebert and Merz acids
$C_{10}H_8O_6S_2$, named for the 19th century German chemists R. Ebert and Viktor Merz

ebonite
derived from ebony, from *ebenos* (Greek: ebony), from *hbnj* (Old Egyptian: ebony), and -ite

ebullioscopy
derived from *ebullire* (Latin: to boil up) and -scopy

eburna-
derived from the specific epithet of the species name *Hunteria eburnea* Pinchon (hunteria root), from *eburneus* (Latin: of ivory)

eburnamenine
$C_{19}H_{22}N_2$, derived from eburna-, amine, -ene, and -in(e)

eburnamonine
$C_{19}H_{22}N_2O$, derived from eburna-, amine, -one, and -in(e)

EC
an abbreviation for electron capture

E.C. (EC)
an abbreviation for Enzyme Commission

ec- (ex-)
derived from *ek, ex* (Greek: out of)

ECD
an abbreviation for electron capture detector

ecdysone
derived (with contraction) from *ekdysis* (Greek: act of getting out, escape), from *ek* (Greek: out of, out) and *dyein* (Greek: to dive in, to put on, to don), and hormone

ecdysteroid
coined by contraction of ecdysone and steroid

ecgonidine
$C_9H_{13}NO_2$, derived from ecgon(o)- and -idin(e) – referring to this compound's

formation from ecgonine

ecgonine
$C_9H_{15}NO_3$, derived from ecgon(o)- and -in(e) – referring to this compound's formation from cocaine

ecgon(o)-
derived from *ekgonos* (Greek: born of, sprung from)

echinenone
$C_{40}H_{54}O$, derived from echin(o)-, -ene, and -one

echin(o)-
derived from *echinos* (Greek: hedgehog, sea urchin)

echinochrome
derived from echin(o)- and -chrome

echinomycin
$C_{51}H_{64}N_{12}O_{12}S_2$, derived from the specific epithet of the bacterial species name *Streptomyces echinatus*, from *echinatus* (Latin: prickly like a hedgehog), from echin(o)-, and -mycin

echinopsin
$C_{10}H_9NO$, derived from the genus name *Echinops* (globe thistles), from echin(o)- and *ops* (Greek: eye), and -in(e)

echinuline
$C_{29}H_{39}N_3O_2$, derived from the specific epithet of the fungal species name *Aspergillus echinulatus*, from *echinulatus* (Latin: set with small prickles), from *echinulus* (Latin: small prickle), diminutive of *echinus* (Latin: hedgehog, sea urchin), and -in(e)

echitamine
$C_{22}H_{29}N_2O_4^+$, ultimately derived from *echis* (Greek: viper) and amine

eckermannite
$Na_3Mg_4Al(Si_8O_{22})(OH)_2$, named for the Swedish scientist Claes Walther Harry von Eckermann (1886-1969)

eclogite
a rock species, derived from *ekloge* (Greek: selection), from *eklegein* (Greek: to pick out, to select), and -ite

ECOIN
an EU abbreviation for European Core Inventory

ecstasy (Adam, E, MDMA, XTC)
$C_{11}H_{15}NO_2$ – the nickname of this hallucino-genic drug refers to its psychopharma-cological effects

ecteinascidin
derived from the genus name *Ecteinascidia* (marine tunicates), from *ekteinein* (Greek: to stretch out) and *askidion*, diminutive of *askos* (Greek: wineskin), and -in(e)

ect(o)-
derived from *ektos* (Greek: out of, out)

ectohormone
derived from ect(o)- and hormone

ectotoxin
derived from ect(o)- and toxin

EDAX
an abbreviation for energy dispersive analysis of X-rays

Edeleanu process
named for the Romanian chemist Lazar Edeleanu (1861-1941)

edenite
$NaCa_2Mg_5Si_7AlO_{22}(OH)_2$, named (with contraction) after this mineral's locality Edenville, Orange County, NY, USA

edestin
derived from *edestos* (Greek: edible), from *edein* (Greek: to eat), and -in(e)

edetate
$C_{10}H_{12}N_2O_8^{4-}$, derived from edetic acid

edetic acid
$C_{10}H_{16}N_2O_8$, derived from EDTA

edisyl
$-O_2SC_2H_4SO_2-$, coined by contraction of ethane-1,2-disulfonyl

Edman degradation
named for the Swedish chemist Pehr Victor Edman (1916-1977)

EDTA
$C_{10}H_{16}N_2O_8$, an abbreviation for ethylene-diamine-*N,N,N',N'*-tetraacetic acid

EE
an abbreviation for 2-ethoxyethyl

e.e.
an abbreviation for enantiomeric excess

EEDQ
$C_{14}H_{17}NO_3$, an abbreviation for *N*-ethoxycarbonyl-2-ethoxy-1,2-dihydroquinoline

EELS
an abbreviation for electron energy loss spectroscopy

effervescence
derived from *effervescere* (Latin: to boil over, to ferment), from *fervescere* (Latin: to begin to boil, to bubble), from *fervere* (Latin: to boil)

efflorescence
derived from *efflorescere* (Latin: to burst into flower), from *flos* (Latin: flower)

EGCG
$C_{22}H_{18}O_{11}$, an abbreviation for epigallo-catechol gallate

EGF
an abbreviation for epidermal growth factor

Eglinton reaction
named for the Welsh chemist Geoffrey Eglinton (born 1927)

EGTA
$C_{14}H_{24}N_2O_{10}$, an abbreviation for ethylene glycol bis(β-aminoethyl ether)-*N,N,N'N'*-tetraacetic acid

Ehrlich-Sachs reaction
named for the German physician and bacteriologist Paul Ehrlich (1854-1915) and the German chemist Franz Sachs

Ehrlich's reagent
C_7H_7NO, named for the German physician and bacteriologist Paul Ehrlich (1854-1915)

EHT
an abbreviation for extended Hückel theory

EI
an abbreviation for electron impact

EIA
an abbreviation for enzyme immunoassay

eicos(a)-
derived from *eikosi* (Greek: twenty)

eicosane
$C_{20}H_{42}$, derived from eicos(a)- and -an(e)

eicosanoid
derived from eicosane and -oid

eigenfunction
derived from *eigen* (German: proper)

eigenvalue
derived from *eigen* (German: proper)

-ein(e)
variant of -in(e)

EINECS
an EU abbreviation for European Inventory of Existing Commercial Chemical Substances

Einhorn-Brunner reaction
named for the German chemist Alfred Einhorn (1857-1917) and the Austrian chemist Karl Brunner (1855-1935)

einsteinium
Es, named in honor of the German-US physicist Albert Einstein (1879-1955)

eka-
derived from *eka* (Sanskrit: one)

eka-actinium
Ubu, derived from eka- and actinium

eka-aluminum
Ga, derived from eka- and aluminum

eka-astatine
Uus, derived from eka- and astatine

eka-bismuth
Uup, derived from eka- and bismuth

eka-boron
Sc, derived from eka- and boron

eka-cesium
Fr, derived from eka- and cesium

eka-gold
Rg, derived from eka- and gold

eka-hafnium
Rf, derived from eka- and hafnium

eka-iodine
At, derived from eka- and iodine

eka-iridium
Mt, derived from eka- and iridium

eka-lead
Uuq, derived from eka- and lead

eka-manganese
Tc, derived from eka- and manganese

eka-mercury
Uub, derived from eka- and mercury

eka-osmium
Hs, derived from eka- and osmium

eka-platinum
Ds, derived from eka- and platinum

eka-polonium
Uuh, derived from eka- and polonium

eka-radon
Uuo, derived from eka- and radon

eka-rhenium
Bh, derived from eka- and rhenium

eka-silicon
Ge, derived from eka- and silicon

eka-tantalum
Db, derived from eka- and tantalum

eka-tellurium
Po, derived from eka- and tellurium

eka-thallium
Uut, derived from eka- and thallium

eka-tungsten
Sg, derived from eka- and tungsten

elaidic acid
$C_{18}H_{34}O_2$, derived from elai(o)- and -ide

elai(o)-
derived from *elaion* (Greek: olive oil)

elaiomycin
$C_{13}H_{26}N_2O_3$, derived from elai(o)- and -mycin

elastase
derived from elastin and -ase

elastin
derived from elast(o)- and -in(e)

elast(o)-
derived from *elastos* (Late Greek: ductile), from *elaunein* (Greek: to drive, to beat out)

elastodiene
coined by contraction of elastomer and diene

elastomer
coined by contraction of elastic polymer

elatol
$C_{15}H_{22}BrClO_4$, derived from the specific epithet of the species name *Laurencia elata* (an alga), from *elatus* (Latin: high, lofty, proud), from *efferre* (Latin: to make proud), and -ol

elbaite
$Na(Li,Al)_3Al_6(BO_3)_3Si_6O_{18}(OH)_4$, named after this mineral's locality Island of Elba, Italy

Elbs anthracene synthesis
named for the German chemist Karl Elbs (1858-1933)

Elbs persulfate oxidation
named for the German chemist Karl Elbs (1858-1933)

elcatonin
$C_{148}H_{244}N_{42}O_{47}$, coined by alteration of calcitonin

electret
coined by contraction of electr(o)- and magnet

electride
derived from electron and -ide

electr(o)-
derived from *elektron* (Greek: amber) – referring to the discovery of static electricity by wiping an amber staff with a piece of cloth

electrocyclic
derived from electron and cyclic

electrode
derived from electr(o)- and *hodos* (Greek: road)

electrofuge
derived from electron and -fuge

electrolysis
derived from electr(o)- and lysis

electrolyte
derived from electrolysis

electron
derived from electr(o)- and ^{3}-on

electronegativity
derived from electron and negative

electrophile
derived from electron and -phile

electrophoresis
derived from electr(o)- and *pherein* (Greek: to carry, to bear)

electrum
(Au,Ag), ultimately derived from *elektor* (Greek: beaming sun), ultimately from *ulka* (Sanskrit: fiery phenomenon in the sky, meteor)

eledoisin
derived from the genus name *Eledone* (octopi), from *eledone* (Greek: a kind of octopus), and -in(e)

elektron (Dow metal)
(Mg,Al,Zn,Mn,Cu,Si,etc.), coined as a trademark

element
derived from *elementum* (Latin: basic substance, originally letter of the alphabet) of obscure etymology

elemol
$C_{15}H_{26}O$, derived from Manila elemi (oil of *Canarium* trees), from *al-lami* (Arabic: elemi), and -ol

elenolide
$C_{11}H_{12}O_5$, derived from ele(o)- and -olide

ele(o)-
derived from *elaion* (Greek: olive oil)

eleoptene
the liquid part of a partially solidified natural oil, derived from ele(o)- and -pten(e)

eleostearic acid
$C_{18}H_{30}O_2$, derived from ele(o)- and stearic acid – referring to the fact that this acid is the C_{18} analog of oleic acid

eleutherobin
$C_{35}H_{48}N_2O_{10}$, derived from the genus name *Eleutherobia* (marine soft corals), from *eleutheros* (Greek: free) and bio-, and -in(e)

elfamycin
derived (with contraction) from the arbitrary code name LL-E19020 alpha and -mycin

ELINCS
an EU abbreviation for European List of Notified Chemical Substances

ELISA
an abbreviation for enzyme-linked immuno-sorbent assay

elixir
derived from *al-iksir* (Arabic: elixir), from *xeron* (Greek: desiccating powder), from xer(o)-

ellagic acid
$C_{14}H_6O_8$, derived anagramatically from gallic acid

ellipticine
$C_{17}H_{14}N_2$, derived from the specific epithet of the species name *Ochrosia elliptica* Labill. (elliptic yellowwood), from *ellipticus* (Latin: elliptic), from *ellipsis* (Greek: omission, deficiency), and -icin(e)

elliptone
$C_{20}H_{16}O_6$, derived from the specific epithet of the species name *Derris elliptica* (Wall.) Benth. (tuba root), from *ellipticus* (Latin: elliptic), from *ellipsis* (Greek: omission, deficiency), and -one

Ellis carbonylate
named for the US chemist John Emmett Ellis (born 1943)

Ellman's reagent
$C_{14}H_8N_2O_8S_2$, named for the US chemist George Leon Ellman (born 1923)

elpidite
$Na_2ZrSi_6O_{15}\cdot 3H_2O$, derived from *elpis* (Greek: hope) and -ite – referring to the

discoverer's wish to find another mineral and -in(e)

ELS
an abbreviation for electron loss spectroscopy

El-Sayed's rule
named for the Egyptian-US chemist Mostafa Amr El-Sayed (born 1933)

eluate, eluend, elute, elution
derived from elu(o)-

elu(o)-
derived from *eluere* (Latin: to wash out)

eluotropic
derived from elu(o)- and -tropic

elutriation
derived from *elutriatus* (Latin: washed out), from *elutriare* (Latin: to wash out), from elu(o)-

elymoclavine
$C_{16}H_{18}N_2O$, derived from the genus name *Elymus* (dunegrass), from *elymos* (Greek: millet), and -clavine

EMA
an abbreviation for electron microprobe analysis

emamectin
derived from the name of the chemical E. Merck & Co., Rahway, NJ, USA and -mectin

emanation
derived from *emanatio* (Latin: outflow), from *manare* (Latin: to flow)

embelin (embelic acid)
$C_{17}H_{26}O_4$, derived from the genus name *Embelia* (Old World tropical woody vines), from a Ceylonese indigenous plant name, and -in(e)

Embden-Meyerhof-Parnas path (EMP path)
named for the German chemists Gustav Embden (1874-1933) and Otto F. Meyerhof (1884-1951), and the Polish-Russian chemist Jakub Karol Parnas (1884-1949)

embonic acid (pamoic acid)
$C_{23}H_{16}O_6$, coined as an approximate abbreviation for 4,4'-methylenebis(3-hydroxynaphthalene-2-carboxylic acid)

Emde reaction
named for the German chemist Hermann Karl Christian Maximilian Emde (1880-1935)

emerald
$Al_3Be_2Si_6O_{18}$, ultimately derived from *smaragdos* (Greek: emerald), akin to *bareqet* (Hebrew: emerald)

Emerson reaction
named for the US chemist E. Emerson

emery
a heterogeneous mineral, derived from *smyris* (Greek: powdered emery)

emetan
$C_{25}H_{32}N_2$, derived from emetine and -an(e)

emetine
$C_{29}H_{40}N_2O_4$, ultimately derived from *emetos* (Greek: vomiting) and -in(e)

EMIT
an abbreviation for enzyme-multiplied immunoassay technique

Emmert reaction
named for the German chemist Bruno Emmert (1880-1951)

emodin (frangulic acid)
$C_{15}H_{10}O_5$, derived from the specific epithet of the species name *Rheum emodi* (Gilgiti rhubarb), from *Emodos* (Greek: Himalaya), and -in(e)

EMP
an abbreviation for Embden-Meyerhof-Parnas

emplectite
$CuBiS_2$, derived from *emplektos* (Greek: interwoven) and -ite – referring to this mineral's intimate association with quartz

empyreuma
the peculiar odor of organic substances burnt in closed vessels, derived from *empyreumata* (Greek: live coal covered with ashes), from *empyreuein* (Greek: to light a fire), from en- and pyr(o)-

emulsin
derived from emulsion and -in(e)

emulsion
derived from *emulsus* (Latin: milked out), from *mulgere* (Latin: to milk)

en- (el-, em-)
derived from *en* (Greek: in, on)

enamide
derived from -ene and amide

enamine
derived from -ene and amine

enanthic acid
$C_7H_{14}O_2$, derived from en(o)- and anth(o)-

enanthotoxin
$C_{17}H_{22}O_2$, derived from the genus name *Oenanthe* (Old World herbs), from en(o)- and anth(o)-, and toxin

enantio-
derived from *enantios* (Greek: opposite)

enantiomer
derived from enantio- and -mer

enantiotopic
derived from enantio- and -topic

enantiotropy
derived from enantio- and -tropy

enargite
Cu_3AsS_4, derived from *enarges* (Greek: distinct, visible), from *argos* (Greek: white), and -ite – referring to this mineral's distinct cleavage

endergonic
derived from end(o)- and erg(o)-

endiandric acid
$C_{21}H_{22}O_2$, derived from the genus name *Endiandra* (plants), from end(o)- and *aner* (Greek: man = male person, stamen) – referring to these plants' fertile stamens

endo-
derived from end(o)-

end(o)-
derived from *endon* (Greek: within)

endocyclic
derived from end(o)- and cyclic

endohedral
derived from end(o)- and -hedral

ENDOR
an abbreviation for electron nuclear double resonance

endorphin
coined by contraction of endogenous morphine

endosmosis
derived from end(o)- and osmosis

endostatin
coined by contraction of endogenous and -statin

endothelin
derived from endothelium, ultimately from *thele* (Greek: nipple), and -in(e)

endothermic (endothermal)
derived from end(o)- and thermic

endotoxin
derived from end(o)- and toxin

endotropic
derived from end(o)- and entropy

endrin
$C_{12}H_8Cl_6O$, coined by contraction of *endo*-dieldrin

-ene
derived from *-ene* (Greek suffix for female descendant)

enediol
derived from -ene, di-, and -ol

ene reaction
derived from alkene

energy
derived from *energeia* (Greek: activity, operation), from *energos* (Greek: active, effective), from *en* (Greek: in) and *ergon* (Greek: work)

Engel's sulfur (Aten's sulfur)
S_6, named for the 19th century French chemist C. R. Engel

enigmatite (aenigmatite)
$Na_2Fe_5TiSi_6O_{20}$, derived from enigma, from *ainigma* (Greek: riddle), from *ainissesthai* (GreeK. to speak in riddles), from *ainos* (Greek: tale, fable), and -ite – referring to a perceived imperfect knowledge of this mineral

enkephalin
derived from *enkephalos* (Greek: brain) and -in(e)

ennea-
derived from *ennea* (Greek: nine); in contemporary IUPAC nomenclature replaced by nona-

enneadeca-
derived from *enneadeka* (Greek: nineteen); in contemporary IUPAC nomenclature replaced by nonadeca-

enneaconta-
derived from *enneakonta* (Greek: ninety); in contemporary IUPAC nomenclature replaced by nonaconta-

enniatin
derived from the subspecific epithet of the fungal species name *Fusarium orthoceras* var. *enniatum*, and -in(e)

en(o)-
derived from *oinos* (Greek: wine)

enol
derived from -ene and -ol

enone
derived from -ene and -one

enophile
derived from ene reaction and -phile

enose
derived from -ene and -ose

enstatite
$(Mg,Fe)SiO_3$, derived from *enstates* (Greek: adversary) and -ite – referring to this mineral's refractory quality

enter(o)-
derived from *enteron* (Greek: intestine)

enterobactin
derived from the bacterial family name *Enterobacteriaceae*, from enter(o)- and bacterium, and -in(e)

enterogastrone
derived (with contraction) from enter(o)-, gastr(o)-, and hormone

enterolactone
$C_{18}H_{18}O_4$, derived from enter(o)- and lactone

enterotoxin
derived from enter(o)- and toxin

enthalpy
derived from *en* (Greek: in) and *thalpos* (Greek: heat)

Entner-Doudoroff path
named for the US biochemists N. Entner (born 1920) and Michael Doudoroff (1911-1975)

entropy
derived from *entrepein* (Greek: to turn around), from *en* (Greek: in) and *trepein* (Greek: to turn)

E number
derived from European Community

enviomycin (viomycin)
$C_{25}H_{43}N_{13}O_{10}$, derived by variation of viomycin

enzyme
derived from *enzymos* (Middle Greek: leavened), from *en* (Greek: in) and *zyme* (Greek: leaven)

eosin(e)
$C_{20}H_6Br_3Na_2O_5$, derived from *eos* (Greek: dawn) and -in(e) – referring to this dye's pink color

EPA (IPA)
$C_{20}H_{30}O_2$, an abbreviation for (*all-Z*)-eicosa-5,8,11,14,17-pentaenoic acid

-epane
derived from hepta- and -an(e)

ephedrine
$C_{10}H_{15}NO$, derived from the genus name *Ephedra* (shrubs), derived from epi- and *hedra* (Greek: chair), and -in(e)

ep(i)-
derived from *epi* (Greek: beside, on, upon, after)

epiandrosterone
$C_{19}H_{30}O_2$, derived from ep(i)- and androsterone

epibatidine
$C_{11}H_{13}ClN_2$, derived (with contraction) from the genus name *Epipedobates* (frogs), from *epipedos* (Greek: on the bottom), from ep(i)- and *pous* (Greek: foot), and *-bates* (Greek: walker), from *bainein* (Greek: to walk), and -idin(e)

epichlorohydrin
C_3H_5ClO, derived from ep(i)- and chlorohydrin

epididymite
$NaBeSi_3O_7(OH)$, derived from ep(i)- and didymite – referring to this mineral's dimorphous relationship with eudidymite

epidote (pistacite)
$Ca_2(Fe,Al)Al_2(SiO_4)(Si_2O_7)O(OH)$, derived from *epidotos* (Greek: added), from *epididonai* (Greek: to give besides, to increase) – referring to the greater length of this mineral's crystals compared with those of related minerals

epihydric acid
$C_3H_4O_3$, derived from ep(i)- and hydr(o)-

epilachnene
$C_{16}H_{29}NO_2$, derived from the genus name *Epilachnis* (ladybirds), from ep(i)- and *lachne* (Greek: soft woolly hair), and -ene

epimer
coined by contraction of ep(i)- and isomer

epimerase
derived from epimer and -ase

epimino-
derived from ep(i)- and imino-

-epine
derived from hepta- and -in(e)

epinephrine (adrenaline)
$C_9H_{13}NO_3$, derived from *epinephron* (Greek: adrenal gland), from ep(i)- and *nephron* (Greek: kidney), and -in(e)

epiquinidine
$C_{20}H_{24}N_2O_2$, derived from ep(i)- and quinidine

epiquinine
$C_{20}H_{24}N_2O_2$, derived from ep(i)- and quinine

epirubicin
$C_{27}H_{29}NO_{11}$, derived (with contraction) from ep(i)- and doxorubicin

epistolite
$Na_4Nb_2Ti(Si_2O_7)_2O_2(OH)_2 \cdot 4H_2O$, derived from *epistole* (Greek: letter) and -lite – referring to this mineral's envelope-shaped crystals

epitaxy
derived from epi- and -taxy

EPMA
an abbreviation for electron probe microanalysis

EPO
an abbreviation for erythropoietin

epothilone
derived from epoxy-, thi(o)-, and –one

epoxide
derived from ep(i)- and oxide

epoxomicin
$C_{28}H_{50}N_4O_7$, derived from epoxy- and -mycin

epoxy-
derived from ep(i)- and oxygen

EPR
an abbreviation for electron paramagnetic resonance

epsomite (Epsom salt)
$MgSO_4 \cdot 7H_2O$, named after this mineral's locality, the town of Epsom, England, UK

Epsom salt (epsomite)
$MgSO_4 \cdot 7H_2O$, named after this mineral's locality, the town of Epsom, England, UK

Epton titration
named for the 20th century British chemist Sidney Robert Epton

equ(i)-
derived from *equs* (Latin: horse)

equilenine
$C_{18}H_{18}O_2$, derived from equ(i)-, -ene, and -in(e)

equilin
$C_{18}H_{20}O_2$, derived from equ(i)- and -in(e)

equol
$C_{15}H_{14}O_3$, derived from equ(i)- and -ol

erabutoxin
derived (with contraction) from *erabu-umihebi* (Japanese: the sea snake *Laticauda semifasciata*) and toxin

erbia
Er_2O_3, derived from erbium and -a

erbium
Er, derived (with contraction) from Ytterby, Sweden where the first erbium ores were found, and -ium

erem(o)-
derived from *eremos* (Greek: lonely, solitary) or *eremos* (Greek: desert)

eremophilane
$C_{15}H_{28}$, derived from the genus name *Eremophila* (shrubs, trees), from erem(o)- and -philic, and -an(e)

erepsin
derived from *eripere* (Latin: to take away), from *rapere* (Latin: to seize, to rob), and -in(e); patterned after pepsin

erg(o)-
derived from *ergon* (Greek: work)

ergocalciferol (calciferol, vitamin D$_2$)
$C_{28}H_{44}O$, derived from ergo(t)- and calciferol

ergochrome
derived from ergo(t)- and -chrome

ergocornine
$C_{31}H_{39}N_5O_5$, derived from ergo(t)-, either *Korn* (German: grain) or *cornu* (Latin: horn), and -in(e)

ergocorninine
$C_{31}H_{39}N_5O_5$, derived from ergocornine and -in(e) – referring to the isomerism of ergocornine and ergocorninine

ergocristine
$C_{35}H_{39}N_5O_5$, derived from ergo(t)-, crystalline, and -in(e)

ergocristinine
$C_{35}H_{39}N_5O_5$, derived from ergocristine and -in(e) – referring to the isomerism of ergocristine and ergocristinine

ergocryptine
$C_{32}H_{41}N_5O_5$, derived from ergo(t)-, crypt(o)-, and -in(e)

ergocryptinine
$C_{32}H_{41}N_5O_5$, derived ergocryptine and -in(e) – referring to the isomerism of ergocryptine and ergocryptinine

ergoflavin
$C_{30}H_{26}O_{14}$, derived from ergo(t)- and flavin

-ergolic
derived from erg(o)- and -ol(e)

ergoline
$C_{14}H_{16}N_2$, derived (with contraction) from ergo(t)-, indole, and -in(e)

ergometrine (ergonovine)
$C_{19}H_{23}N_3O_2$, derived (with contraction) from ergo(t)-, endometrium, and -in(e)

ergometrinine (ergonovinine)
$C_{19}H_{23}N_3O_2$, derived from ergometrine and -in(e) – referring to the isomerism of ergometrine and ergometrinine

ergone
coined by contraction of erg(o)- and hormone

ergonovine (ergometrine)
$C_{19}H_{23}N_3O_2$, derived from ergo(t)-, nov(o)-, and -in(e)

ergosine
$C_{30}H_{37}N_5O_5$, derived from ergo(t)- and -in(e)

ergosinine
$C_{30}H_{37}N_5O_5$, derived from ergosine and -in(e) – referring to the isomerism of ergosine and ergosinine

ergostane
$C_{28}H_{50}$, derived from ergo(t)-, sterol, and -an(e)

ergosterol
$C_{28}H_{44}O$, derived from ergo(t)- and sterol

ergo(t)-
derived from ergot

ergot
derived from *ergot* (French: cock's spur)

ergotaman
$C_{25}H_{33}N_5O$, derived from ergo(t)-, amine, and -an(e)

ergotamine
$C_{33}H_{35}N_5O_5$, derived from ergo(t)- and amine

ergotaminine
$C_{33}H_{35}N_5O_5$, derived from ergotamine and -in(e) – referring to the isomerism of ergotamine and ergotaminine

ergothioneine (thioneine)
$C_9H_{15}N_3O_2S$, derived from ergo(t)-, -thione, and -ein(e)

ergotoxin(e)
$C_{35}H_{39}N_5O_5$, derived from ergo(t)- and toxin(e)

erio-
derived from *erion* (Greek: wool)

eriochalcite (antofagastite)
$CuCl_2 \cdot 2H_2O$, derived from erio- and chalcite – referring to this mineral's woolly appearance

eriochrome
derived from erio- and -chrome

eriodictyol
$C_{15}H_{12}O_6$, derived from the genus name *Eriodictyon* (North American shrubs), from erio- and *diktyon* (Greek: net), and -ol

eritadenine
$C_9H_{11}N_5O_4$, derived (with contraction) from erythronic acid and adenine

Erlenmeyer flask
named for the German chemist Richard August Karl Emil Erlenmeyer (1825-1909)

Erlenmeyer-Plöchl azlactone and amino acid synthesis
named for the German chemists Richard August Karl Emil Erlenmeyer (1825-1909) and J. Plöchl

Erlenmeyer rule
named for the German chemist Richard August Karl Emil Erlenmeyer (1825-1909)

Ernst angle
named for the Swiss chemist Richard

Robert Ernst (born 1933)

erucic acid
$C_{22}H_{42}O_2$, named after the genus name *Eruca* (cruciferous plants), from *eruca* (Latin: garden rocket, *Eruca sativa*)

erysimin (helveticoside)
$C_{29}H_{42}O_9$, derived from the genus name *Erysimum* (wallflower), from *erysimon* (Greek: a kind of mustard), from *erysthai* (Greek: to protect, to defend, to save), and -in(e)

erythorbic acid
$C_6H_8O_6$, coined by contraction of erythrose and ascorbic acid

erythraric acid
$C_4H_6O_6$, derived from erythrose and -aric acid

erythrin
$C_{20}H_{22}O_{11}$, derived from erythr(o)- and -in(e) – referring to the red color observed upon this compound's oxidation

erythrinan
$C_{16}H_{21}N$, derived from the genus name *Erythrina* (tropical shrubs, trees), from erythr(o)-, and -an(e)

erythrite
$Co_3(AsO_4)_2 \cdot 8H_2O$, derived from erythr(o)- and -ite – referring to this mineral's rose red color

erythritol
$C_4H_{10}O_4$, derived from erythrin which is an erythrityl ester

erythr(o)-
derived from *erythros* (Greek: red)

erythro-
derived from erythrose

erythrocentaurin
$C_{10}H_8O_3$, derived (with contraction) from the species name *Erythraea centaurium* Pers. (European centaury), from erythr(o)- and, ultimately, from *kentauros* (Greek: centaur) – referring to the ancient belief that this herb was discovered by the centaur Chiron – and -in(e)

erythrocuprein (superoxide dismutase)
derived from erythr(o)-, cupr(o)-, and -ein(e)

erythrogenic acid (bolecic acid, isanic acid)
$C_{18}H_{26}O_2$, derived from erythr(o)- and -gen

erythroidine
$C_{16}H_{19}NO_3$, derived from the genus name *Erythrina* (tropical shrubs, trees), from erythr(o)-, and -idin(e)

erythromycin
$C_{37}H_{67}NO_{13}$, derived from the specific epithet of the bacterial species name *Streptomyces erythreus*, from erythr(o)-, and -mycin

erythronic acid
$C_4H_8O_5$, derived from erythrose and -onic acid

erythronium
$PbCrO_4$, an obsolete name for lead chromate – coined in the erroneous belief that this compound was a previously unknown element

erythrophleine
$C_{24}H_{39}NO_5$, derived from the genus name *Erythrophleum* (sasswood), from erythr(o)- and *phloios* (Greek: bark), and -in(e)

erythropoietin
derived from erythropoietic, from erythr(o)- and *poietikos* (Greek: productive,

formative), from *poiein* (Greek: to make, to do, to create, to compose)

erythropterin
$C_9H_7N_5O_5$, derived from erythr(o)- and pterin

erythrose
$C_4H_8O_4$, derived from erythritol and -ose

erythrosine
$C_{20}H_6I_4Na_2O_5$, derived from erythr(o)- and eosine

erythrulose
$C_4H_8O_4$, derived from erythritol and -ulose

Esbach reagent
named for the French physician G. H. Esbach (1843-1890)

ESCA
an abbreviation for electron spectroscopy for chemical analysis

-escent
derived from *-escens* (Latin: beginning, beginning to be, becoming)

Eschenmoser coupling reaction
named for the Swiss chemist Albert Eschenmoser (born 1925)

Eschenmoser's salt
C_3H_8IN, named for the Swiss chemist Albert Eschenmoser (born 1925)

Eschenmoser-Tanabe fragmentation
named for the Swiss chemist Albert Eschenmoser (born 1925) and the Japanese chemist M. Tanabe

Eschweiler-Clarke reaction
named for the German chemist Wilhelm Eschweiler (1860-1936) and the British chemist Hans Thatcher Clarke (1888-1972)

eschynite (aeschynite, blomstrandine, priorite)
$(Ce,Ca,Fe)(Ti,Nb)_2(O,OH)_6$, derived from *aischyne* (Greek: shame) and -ite – referring to the inability of chemists, at the time of its discovery, to separate the constituents of this mineral

escin
derived from the genus name *Aesculus* (trees, shrubs), from *aesculus* (Latin: evergreen winter oak), of obscure etymology, and -in(e)

esculetin
$C_9H_6O_4$, coined by variation of esculin

esculin
$C_{15}H_{16}O_9$, derived from the genus name *Aesculus* (trees, shrubs), from *aesculus* (Latin: evergreen winter oak), of obscure etymology, and -in(e)

ESD
an abbreviation for electron-stimulated desorption

eser-
derived from *ésére* (French: Calabar bean, *Physostigma venenosum*)

eseridine
$C_{15}H_{21}N_3O_3$, derived from eser- and -idin(e)

eserine (physostigmine)
$C_{15}H_{21}N_3O_2$, derived from eser- and -in(e)

eseroline
$C_{13}H_{18}N_2O$, derived from eser-, -ol, and -in(e)

esperite
$PbCa_3Zn_4(SiO_4)_4$, named for the US petrologist Esper Signius Larsen, Jr. (1878-1961)

ESR
an abbreviation for electron spin resonance

essonite (hessonite, cinnamon stone)
$Ca_3Al_2(SiO_4)_3$, from *hesson* (Greek: inferior, less), comparative of *heka* (Greek: slightly) – referring to the fact that this mineral is less hard than true hyacinth

ester
derived from *Ester* (German: ester), coined by contraction of the obsolete name *Essigäther* (German: ester), from *Essig* (German: vinegar) and *Äther* (German: ³ether)

esterase
derived from ester and -ase

estradiol
$C_{18}H_{24}O_2$, derived from estrane, di-, and -ol

estragole
$C_{10}H_{12}O$, derived from estragon, from tarragon, from the city of Tarragona, Spain, and -ol(e)

estrane
$C_{18}H_{30}$, derived from estr(o)- and -an(e)

estriol
$C_{18}H_{24}O_3$, coined by contraction of (16α,17β)-estra-1,3,5(10)-triene-3,6,17-triol

estr(o)-
ultimately derived from *oistros* (Greek: gadfly, frenzy)

estrogen
derived from estr(o)- and -gen

estrone
$C_{18}H_{22}O_2$, coined by contraction of 3-hydroxyestra-1,3,5(10)-trien-17-one

esyl
$C_2H_5O_2S-$, derived by contraction of ethanesulfonyl

ET
an abbreviation for electron transfer

-etane
derived from tetra- and -an(e)

Etard reaction
named for the French chemist Alexandre Léon Etard (1852-1910)

-etene
derived from tetra- and -en(e)

ethacrynic acid
$C_{13}H_{12}Cl_2O_4$, coined by contraction of ethyl, acetic, butyryl, and phenol

ethamphetamine
$C_{11}H_{17}N$, coined by contraction of *N*-ethylamphetamine

ethane
C_2H_6, derived from ethyl and -an(e)

¹ether
R^1OR^2, derived from ²ether

²ether
$C_4H_{10}O$, diethyl ether, derived from ³ether – referring to this compound's high volatility

³ether
an archaic name for volatile substances, derived from ⁴ether

⁴ether
hypothetical, essentially massless or nearly massless, medium everywhere in space, derived from *aither* (Greek: ether), from *aithein* (Greek: to kindle, to blaze)

ethionine
$C_6H_{13}NO_2S$, coined by contraction of *S*-ethyl-L-homocysteine

Ethiops iron (ferrosoferric oxide)
Fe_3O_4, derived from *Aethiops* (Latin: Ethiopia) and iron

-ethrin
derived from pyrethrin

ethyl
C_2H_5-, derived from ether and -yl

ethylene
C_2H_4, derived from ether, -yl, and -ene

ethylene blue
$C_{20}H_{26}N_3S^+$, derived from ethylene and blue – referring to the formal insertion of four ethylene groups into the four N–H bonds of the parent diamine

ethylene glycol
$C_2H_6O_2$, derived from ethylene and glycol

ethylene oxide (oxirane)
C_2H_4O, derived from ethylene and oxide

-etidine
derived from tetra- and -idin(e)

etidronic acid
$C_2H_8O_7P_2$, coined by contraction of the unsystematic name ethane-1-hydroxy-1,1-diphosphonic acid

-etine
derived from tetra- and -in(e)

-etin(e)
variant of -in(e)

etio-
derived from *aitia* (Greek: cause) – this obsolete prefix has been used for degradation products of undetermined structure, derived from known natural products

etiocholanic acid
$C_{20}H_{32}O_2$, derived from etio- and cholanic acid

etiocobalamine
$C_{50}H_{72}CoN_{13}O_8$, derived from etio- and cobalamine

etioporphyrin
$C_{32}H_{38}N_4$, derived from etio- and porphyrin

ettringite
$Ca_6Al_2(SO_4)_3(OH)_{13}\cdot 26H_2O$, named after this mineral's locality Ettringen, Germany

etryptamine
$C_{12}H_{16}N_2$, coined by contraction of α-ethyltryptamine

eu-
derived from *eu* (Greek: easily, fine)

eucalyptol (cineole, cajeputol, cajuputol)
$C_{10}H_{18}O$, derived from the genus name *Eucalyptus* (eucalyptus), from eu- and *kalyptos* (Greek: covered), from *kalyptein* (Greek: to cover, to conceal), and -ol

euchlorine
$(K,Na)_8Cu_9(SO_4)_{10}(OH)_6$, derived from eu-, [1]chlor(o)-, and -in(e) – referring to this mineral's olive green color

euclase
$BeAl(SiO_4)(OH)$, derived from eu- and -clase – referring to this mineral's brittleness

eucryptite
$LiAlSiO_4$, derived from *eukryptos* (Greek: easily hidden), from eu-, crypt(o)-, and -ite – referring to this mineral's occurrence

embedded in albite

eudialyte
$Na_4(Ca,Fe)ZrSi_6O_{17}(OH,Cl)_2$, derived from *eudialytos* (Greek: easy to dissolve), from eu- and dialysis – referring to this mineral's ready dissolution in acids

eudesm-
derived from the genus name *Eudesmia* (eucalyptus), from eu- and desm(o)-

eudesmane
$C_{15}H_{28}$, derived from eudesm- and -an(e)

eudesmol
$C_{15}H_{26}O$, derived from eudesm- and -ol

eudidymite
$NaBeSi_3O_7(OH)$, derived from eu-, *didymos* (Greek: twin), and -ite – referring to this mineral's occurrence as twin crystals

eudiometer
derived from *eudios* (Greek: quiet), from *eudia* (Greek: fair weather) and -meter

eugenite
Ag_9Hg_2, named for the Austrian mineralogist Eugen Friedrich Stumpfl (born 1931)

eugenol
$C_{10}H_{12}O_2$, derived from the genus name *Eugenia* (tropical trees, shrubs), named for the Austrian general Prince Eugen of Savoy (1663-1736), and -ol

eulytite (agricolite, eulytine)
$Bi_4(SiO_4)_3$, derived from *eulytos* (Greek: easily dissolved) and -ite – referring to this mineral's relatively low melting point

eumelanin
derived from eu- and melanin

euosmophore
derived from eu-, osm(e)-, and -phore

euparin
$C_{13}H_{12}O_3$, derived from eupa(tor)- and -in(e)

eupa(tor)-
derived from the genus name *Eupatorium* (herbs), from *eupatorion* (Greek: hemp agrimony, *Eupatorium cannabinum* L.), named for Mithridates VI Eupator, king of Pontus (132 BC-63 BC), from *eupator* (Greek: born of a noble father)

eupatorin
$C_{18}H_{16}O_7$, derived from eupa(tor)- and -in(e)

europium
Eu, named for Europe

eutectic
derived from eu- and -tectic

eutrophication
derived from eu- and *trephein* (Greek: to nourish)

euxenite
$Y(Nb,Ta,Ti)_2O_6$, derived from *euxenos* (Greek: hospitable), from eu- and xen(o)-, and -ite – referring to the small amounts of rare elements (Ce, Th, U, Y) contained in this mineral

Evans aldol reaction
named for the US chemist David A. Evans (born 1941)

Evans blue
$C_{34}H_{24}N_6Na_4O_{14}S_4$, named for the US anatomist Herbert M. Evans (1882-1971)

Evans-Polanyi relation
named for the British chemist Meredith Gwynne Evans (1904-1952) and the Hungarian-British physicist Michael

Polanyi (1891-1976)

Evans principle
named for the British chemist Meredith Gwynne Evans (1904-1952)

evaporation
derived from e(x)- and vapor

evodiamine
$C_{19}H_{17}N_3O$, derived from the genus name *Evodia* (shrubs, trees), from *euodia* (Greek: fragrance), from eu- and *ozein* (Greek: to smell), and amine

evonimine
$C_{36}H_{43}NO_{17}$, derived from evon(o)- and -imine

evonine
$C_{36}H_{43}NO_{17}$, derived from evon(o)- and -in(e)

evon(o)-
derived from the genus name *Euonymus* (shrubs, trees), from *euonymos* (Greek: having an auspicious name), from eu- and *onyma*, *onoma* (Greek: name)

Ewens-Bassett system
named for the British chemists V. G. Ewens (1913-1948) and H. Bassett (1881-1965)

ex- (ef-)
derived from *ex* (Latin: out of, from)

EXAFS
an abbreviation for extended X-ray absorption fine structure

exaltolide
$C_{15}H_{28}O_2$, originally coined as a trademark, derived from exaltation and -olide

exaltone
$C_{15}H_{28}O$, originally coined as a trademark, derived from exaltation and -one

excimer
coined by contraction of excited dimmer

exciplex
coined by contraction of excited complex

exciton
derived from excited and [3]-on

exendin
derived from ex(o)-, end(o)-, and -in(e)

exergonic
derived from ex(o)- and erg(o)-

exinite (liptinite)
derived from exine and -ite

ex(o)-
derived from *exo* (Greek: outside)

exocyclic
derived from ex(o)- and cyclic

exothermic (exothermal)
derived from ex(o)- and thermic

exotoxin
derived from ex(o)- and toxin

exotropic
derived from ex(o)- and entropy

extra-
derived from *extra* (Latin: outside)

extract
derived from *extractus* (Latin: drawn out), from ex- and *trahere* (Latin: to draw)

Eyring equation
named for the Mexican-US chemist Henry Eyring (1901-1981)

F

FAAS
an abbreviation for flame atom absorption spectroscopy

FAB
an abbreviation for fast atom bombardment

fac-
derived from *facies* (Latin: face)

FAD
$C_{27}H_{33}N_9O_{15}P_2$, an abbreviation for flavine adenine dinucleotide

FADH₂

$FADH_2$
$C_{27}H_{35}N_9O_{15}P_2$, an abbreviation for reduced flavine adenine dinucleotide

fagarine
$C_{13}H_{11}NO_3$, derived from the genus name *Fagara* (zanthoxylum), from *fagara* (Arabic: zanthoxylum), and -in(e)

Fahrenheit temperature scale
named for the German physicist Gabriel Daniel Fahrenheit (1686-1736)

Fajans' rules
named for the Polish-German chemist Kasimir Fajans (1887-1975)

Faraday constant
named for the British chemist Michael Faraday (1791-1867)

farnes-
derived from the specific epithet of the species name *Acacia farnesiana* (sweet acacia), named for the Italian Cardinal Odoardo Farnese (1573-1626)

farnesene
$C_{15}H_{24}$, derived from farnes- and -ene

farnesol
$C_{15}H_{26}O$, derived from farnes- and -ol

farnesyl (triprenyl)
$C_{15}H_{25}-$, derived from farnesol and -yl

fasciculin (FAS)
derived from *fasciculus*, diminutive of *fascis* (Latin: bundle), and -in(e)

fassaite (augite)
$(Ca,Na)(Al,Fe,Mg)(Si,Al)_2O_6$, named after this mineral's locality Fassa Valley, Italy

fat
an indigenous English word

faujasite
$(Na_2,Ca)Al_2Si_4O_{12})\cdot 6H_2O$, named for the French geologist Barthélemy Faujas de Saint-Fond (1741-1819)

Fauser process
named for the Italian engineer Giacomo Fauser (1892-1971)

Favorsky-Babayan synthesis
named for the Russian chemist Aleksei Yefgrafovich Favorsky (1860-1945) and the Armenian chemist A. Babayan

Favorsky rearrangement
named for the Russian chemist Aleksei Yefgrafovich Favorsky (1860-1945)

fayalite
Fe_2SiO_4, named after this mineral's locality Fayal Island, Azores, Portugal

FCC
an abbreviation for fluid catalytic cracking

FD
an abbreviation for field desorption

FDP
an abbreviation for fructose 1,6-diphosphate

febrifugine
$C_{16}H_{19}N_3O_3$, derived from the specific epithet of the species name *Dichroa febrifuga* Lour. (antifeverile dichroa), from *febris* (Latin: fever) and *fugare* (Latin: to put to flight), and -in(e)

fecapentaene
derived from feces, penta-, and -ene

Fehling's reagent
named for the German chemist Hermann Christian von Fehling (1812-1885)

Feist-Benary synthesis
named for the Swiss-German chemist Franz Feist (1864-1941) and the 20th century German chemist Erich Benary

Feist's acid
$C_6H_6O_4$, named for the Swiss-German chemist Franz Feist (1864-1941)

feldspar
derived from *Feldspat* (German: feldspar), from *Feld* (German: field) and *Spat* (German: spar, crystalline material)

f-electron
derived from 'fundamental' line groups in atomic spectroscopy

felinine
$C_8H_{17}NO_3S$, derived from *felis* (Latin: cat) and -in(e)

fench-
derived from *Fenchel* (German: fennel), from *foeniculum* (Latin: fennel), diminutive of *foenum* (Latin: hay)

fenchane
$C_{10}H_{18}$, derived from fench- and -an(e)

fenchene
$C_{10}H_{16}$, derived from fench- and -ene

fenchol (fenchyl alcohol)
$C_{10}H_{18}O$, derived from fench- and -ol

fenchone
$C_{10}H_{16}O$, derived from fench- and -one

fenchyl
$C_{10}H_{17}-$, derived from fenchyl alcohol

fenchyl alcohol (fenchol)
$C_{10}H_{18}O$, derived from fench- and -yl

fenestrane
derived from *fenestra* (Latin: window), probably of Etruscan origin, and -an(e) – referring to these hydrocarbons' window-like structure

Fenske rings
named for the US chemist Merrell Robert Fenske (1904-1971)

Fenton reaction
named for the British chemist Henry John Horstman Fenton (1854-1929)

-fer
derived from *ferre* (Latin: to bear, to carry)

ferberite
$FeWO_4$, named for the German mineralogist Moritz Rudolph Ferber (1805-1875)

fergusonite
$YNbO_4$, named for the Scottish mineralogist Robert Ferguson (1767-1840)

fermentation
derived from *fermentare* (Latin: to cause to

rise), from *fervere* (Latin: to boil)

fermium
Fm, named for the Italian-US physicist Enrico Fermi (1901-1954)

ferrate
derived from ferrum and -ate

ferredoxin
coined by contraction of ferrum, redox, and -in(e)

ferri-
derived from ferrum

ferrichrome
derived from ferri- and -chrome

Ferrier rearrangement
named for the 20th century Scottish-New Zealand chemist Robert J. Ferrier

ferrimycin
derived from ferri- and -mycin – referring to this antibiotic's effect on the cellular uptake of ferrioxamine

ferrioxamine
$C_{25}H_{45}FeN_6O_8$, derived from ferri-, oxamic acid, and -in(e)

[1]**ferrite**
(Fe,Fe_3C), derived from ferrum and -ite

[2]**ferrite (ferrospinel)**
$M^{II}Fe_2O_4$, derived from ferrum and -ite

ferritin
derived from ferr(o)-, -ite, and -in(e)

ferr(o)-
derived from ferrum

ferroactinolite
$Ca_2(Fe,Mg)_5Si_8O_{22}(OH)_2$, derived from ferr(o)- and actinolite

ferroaxinite
$Ca_2FeAl_2(BO_3)Si_4O_{12}(OH)$, derived from ferr(o)- and axinite

ferrocene
$C_{10}H_{10}Fe$, coined by contraction of ferr(o)- and cyclopentadiene or derived from ferr(o)- and -acene

ferrocolumbite
$(Fe,Mn)(Nb,Ta)_2O_6$, derived from ferr(o)- and columbite

ferroin
$C_{36}H_{24}FeN_6$, derived from ferr(o)- and -in(e)

ferron
$C_9H_6INO_4S$, derived from ferr(o)- and [1]-on – referring to this compound's usefulness as a reagent for iron (and aluminum)

ferrosilite
$(Fe,Mg)SiO_3$, derived from ferr(o)-, silicon, and -ite

ferrosoferric oxide (Ethiops iron)
Fe_3O_4, derived from *ferrosus* (New Latin: ferrous), ferric, and oxide

ferrospinel (ferrite)
$M^{II}Fe_2O_4$, derived from ferr(o)- and spinel

ferrotantalite
$Fe(Ta,Nb)_2O_6$, derived from ferr(o)- and tantalite

ferrotapiolite (tapiolite)
$(Fe,Mn)(Ta,Nb)_2O_6$, derived from ferr(o)- and tapiolite

ferrum
Fe, New Latin name of iron, derived from *ferrum* (Latin: iron), of Asiatic origin

ferulane (aristolane)
$C_{15}H_{26}$, derived from ferul(o)- and -an(e)

ferulic acid
$C_{10}H_{10}O_4$, derived from ferul(o)-

ferul(o)-
derived from the genus name *Ferula* (fennel), from *ferula* (Latin: fennel)

fervenulin
$C_7H_7N_5O_2$, derived from the specific epithet of the bacterial species name *Streptomyces fervens*, from *fervens* (Latin: boiling, glowing) from *fervere* (Latin: to boil, to glow), and -ulin(e)

FES
an abbreviation for flame emission spectroscopy

festucine (loline)
$C_8H_{14}N_2O$, derived from the genus name *Festuca* (perennial grasses), from *festuca* (Latin: stalk, straw, rod), and -in(e)

Fétizon's reagent
Ag_2CO_3/celite, named for the French chemist Marcel Ernest Paul Fétizon (born 1926)

fetoprotein
derived from fetus and protein

FFF
an abbreviation for field flow fractionation

FFS
an abbreviation for flame fluorescence spectroscopy

FH_2
$C_{19}H_{21}N_7O_6$, an abbreviation for dihydrofolic acid

FH_4 (coenzyme F)
$C_{19}H_{23}N_7O_6$, an abbreviation for tetrahydrofolic acid

FI
an abbreviation for field ionization

^{1}FIA
an abbreviation for flow injection analysis

^{2}FIA
an abbreviation for fluorescence immunoassay

fibrin
derived from fibr(o)- and -in(e)

fibrinogen
derived from fibrin and -gen

fibr(o)-
derived from *fibra* (Latin: fiber)

fibroin
derived from fibr(o)- and -in(e)

fibrolite (sillimanite)
$Al(AlSiO_5)$, derived from fibr(o)- and -lite – referring to this mineral's fibrous crystal habit

fibronectin
derived from fibr(o)-, *nectere* (Latin: to tie), and -in(e)

ficin
derived from the genus name *Ficus* (fig tree), from *ficus* (Latin: fig tree), and -in(e)

FID
an abbreviation for flame ionization detector

Fieser's reagent
$CrO_3{\cdot}C_2H_4O_2$, named for the US chemist Louis Frederick Fieser (1899-1977)

Fieser's solution
named for the US chemist Louis Frederick Fieser (1899-1977)

Fieser-Woodward rules
named for the US chemists Louis Frederick Fieser (1899-1977), Mary Peters Fieser (1909-1997), and Robert Burns Woodward (1917-1979)

filicinic acid
$C_8H_{10}O_3$, derived from the genus name *Filix* (fern), from *filix* (Latin: fern)

filipin
derived from the specific epithet of the bacterial species name *Streptomyces filipinensis*, from *filipinensis* (New Latin: Filipino), and -in(e)

filixic acid
$C_8H_{10}O_3$, derived from the specific epithet of the species name *Dryopteris filixmas* (male fern), from *filix mas* (Latin: male fern)

filter
derived from *feltrum* (Medieval Latin: felt), from *filta* (Gothic: felt)

FIM
an abbreviation for field ion microscopy

FIMS
an abbreviation for field ion mass spectroscopy

Finkelstein reaction
named for the German chemist Hans Finkelstein (1885-1938)

FIR
an abbreviation for far infrared

Fischer carbene
named for the German chemist Ernst Otto Fischer (born 1918)

Fischer esterification
named for the German chemist Emil Hermann Fischer (1852-1919)

Fischer-Hafner synthesis
named for the German chemists Ernst Otto Fischer (born 1918) and Walter Hafner

Fischer-Hepp rearrangement
named for the German chemists Philipp Otto Fischer (1852-1932) and Eduard Hepp (1851-1917)

Fischer indole synthesis
named for the German chemist Emil Hermann Fischer (1852-1919)

Fischer oxazole synthesis
named for the German chemist Emil Hermann Fischer (1852-1919)

Fischer peptide synthesis
named for the German chemist Emil Hermann Fischer (1852-1919)

Fischer phenylhydrazine synthesis
named for the German chemist Emil Hermann Fischer (1852-1919)

Fischer phenylhydrazone and osazone reaction
named for the German chemist Emil Hermann Fischer (1852-1919)

Fischer projection
named for the German chemist Emil Hermann Fischer (1852-1919)

Fischer's salt
$K_3[Co(NO_2)_6]$, named for the German chemist Nikolaus Wolfgang Fischer (1782-1850)

Fischer-Speier esterification
named for the German chemists Emil Hermann Fischer (1852-1919) and A. Speier

Fischer-Tropsch synthesis
named for the German chemists Franz Joseph Emil Fischer (1877-1947) and Hans Tropsch (1889-1935)

fisetin
$C_{15}H_{10}O_6$, derived from *Fisetholz* (German: wood from a species of fustic) and -in(e)

fissium
derived from fission and -ium

Fittig reaction
named for the German chemist Rudolph Fittig (1835-1910)

fixed air
CO_2, an archaic name for carbon dioxide

flagellin
derived from flagellum, from *flagellum*, diminutive of *flagrum* (Latin: whip), and -in(e)

flavan
$C_{15}H_{14}O$, derived from flav(o)- and -an(e)

flavanone
$C_{15}H_{12}O_2$, derived from flavan and -one

flavanthrone
$C_{28}H_{12}N_2O_2$, derived from flav(o)-, anthr(o)-, and -one

flavaspidic acid
$C_{24}H_{30}O_8$, derived from flav(o)- and aspidin

flavianic acid
$C_{10}H_6N_2O_8S$, derived from flav(o)-

[1]**flavin**
$C_{12}H_{10}N_4O_2$, derived from flav(o)- and -in(e)

[2]**flavin (riboflavin 5'-phosphate)**
$C_{17}H_{21}N_4O_9P$, derived from flav(o)- and -in(e)

flav(o)-
derived from *flavus* (Latin: light yellow)

flavodoxin
coined by contraction of flavin mononucleotide and redoxin

flavoenzyme
derived from flav(o)- and enzyme

flavone
$C_{15}H_{10}O_2$, derived from flav(o)- and -one

flavonoid
derived from flavone and -oid

flavonol
$C_{15}H_{10}O_3$, derived from flavone and -ol

flavopereirine (melinonine G)
$C_{17}H_{14}N_2$, derived from flav(o)-, pereira tree (*Geissospermum vellosii* Allem.), named for the British pharmacologist Jonathan Pereira (1804-1853), and -in(e)

flavoprotein
derived from flav(o)- and protein

flavopurpurin
$C_{14}H_8O_5$, derived from flav(o)- and purpurin

flavoxanthin
$C_{40}H_{56}O_3$, derived from flav(o)- and xanthin

flindersine
$C_{14}H_{13}NO_2$, derived from the genus name *Flindersia* (Australian trees), named for the British mariner Matthew Flinders (1774-

1814), and -in(e)

flint
SiO_2, a native Germanic word

flocculation
derived from *flocculus*, diminutive of *floccus* (Latin: flock of wool)

Flood reaction
named for the 20th century US chemist E. A. Flood

florentinum
Pm, a name unsuccessfully suggested for promethium, after the city of Florence, Italy

florigen
(a hypothetical plant hormone), derived from *florere* (Latin: to blossom), from *flos* (Latin: flower), and -gen

Flory equation
named for the US chemist Paul John Flory (1910-1985)

Flory temperature
named for the US chemist Paul John Flory (1910-1985)

flos ferri
$CaCO_3$, derived from *flos* (Latin: flower) and ferrum – referring to the occurrence of this mineral in beds of iron ore

fluate
coined by contraction of fluorosilicate

fluoranthene
$C_{16}H_{10}$, derived (with contraction) from fluorescence and anthracene

[1]**fluorane**
HF, derived from fluorine and -an(e)

[2]**fluorane**
$C_{20}H_{12}O_3$, derived from fluorescence and -an(e)

fluorapatite (apatite)
$Ca_5(PO_4)_3F$, derived from fluoro- and apatite

fluorapophyllite
$KCa_4(Si_4O_{10})_2F{\cdot}8H_2O$, derived from fluoro- and apophyllite

fluorene
$C_{13}H_{10}$, derived from fluorescence and -ene

fluorescamine
$C_{17}H_{10}O_4$, derived from fluorescence and amine – referring to this compound's reaction with primary amines to yield fluorescent products

fluorescein
$C_{20}H_{12}O_5$, derived from fluorescence and -ein(e)

fluorescence
derived from fluorspar and -escence

fluorescin
$C_{20}H_{14}O_5$, derived from fluorescence and -in(e)

fluoride
F^-, derived from fluorine and –ide

fluorine
F, derived from fluorspar, ultimately from *fluor lapis* (New Latin name for fluorite); from *fluor* (Latin: flow), from *fluere* (Latin: to flow) and *lapis* (Latin: stone) – referring to the usefulness of fluorite as a flux

fluorite (fluorspar)
CaF_2, derived from fluorspar and -ite – referring to this mineral's usefulness as a flux

fluoro acid air
SiF_4, an archaic name for silicon tetrafluoride, derived from fluorine, acid, and air

fluorochrome
derived from fluorescence and -chrome

fluoroform
CHF_3, derived from fluor(o)- and -form

fluorspar (fluorite)
CaF_2, derived from *fluor* (Latin: flow), from *fluere* (Latin: to flow), and *Spat* (German: spar, crystalline mineral) – referring to this mineral's usefulness as a flux

fluorum
F, New Latin name for fluorine, from *Fluor* (German: fluorine) and -um

fluosilicic acid
H_2SiF_6, derived from fluorine and silicon

flustramine
derived from the genus name *Flustra* (bryozoans), from *flustra* (Latin: quiet sea), and amine

flux
derived from *fluxus* (Latin: flowing) from *fluere* (Latin: to flow)

FMN (riboflavin 5'-phosphate)
$C_{17}H_{21}N_4O_9P$, an abbreviation for flavin mononucleotide

FMNH$_2$
$C_{17}H_{23}N_4O_9P$, an abbreviation for dihydroflavin mononucleotide

fol-
derived from *folium* (Latin: leaf)

foliamenthin
$C_{26}H_{36}O_{12}$, derived (with contraction) from the species name *Menyanthes trifoliata* (buckbean), from *Menyanthes* (bog plants), from *meny-,* of unknown origin and meaning, and anth(o)-, and fol-, and -in(e)

folic acid
$C_{19}H_{19}N_7O_6$, derived from fol-

Folin-Ciocâlteu reagent
named for the Swedish-US chemist Otto Knut Olof Folin (1867-1934) and the Romanian chemist Vintila Ciocâlteu (1890-1947)

folinic acid (citrovorum factor)
$C_{20}H_{23}N_7O_7$, derived from fol- and -in(e)

Folin's reagents
named for the Swedish-US chemist Otto Knut Olof Folin (1867-1934)

follitropin (follicle-stimulating hormone)
derived from follicle, from *folliculus*, diminutive of *follis* (Latin: leather bag), and -tropin

fomannosin
$C_{15}H_{18}O_4$, derived (with contraction) from the fungal species name *Fomes annosus*, from *fomes* (Latin: touchwood, tinder), from *fovere* (Latin: to warm), and *annosus* (Latin: long lived), and -in(e)

fomecin
derived from the fungal genus name *Fomes*, from *fomes* (Latin: touchwood, tinder), from *fovere* (Latin: to warm), and -in(e)

fomentariol
$C_{17}H_{16}O_7$, derived from the specific epithet of the fungal species name *Fomes fomentarius* (white spongy trunk rot), from *fomentarius* (Latin: soothing), ultimately from *fovere* (Latin: to warm), and -ol

fommanoxin
$C_{12}H_{12}O_2$, coined by variation of fomannosin

footballene (buckminsterfullerene)
C_{60}, derived from football (soccer ball) and -ene

f-orbital
derived from f-electron

-form
HC≡, derived from [2]formyl

formal
coined by contraction of formaldehyde acetal

formaldehyde
CH_2O, derived from form(o)-

formalin
derived (with contraction) from formaldehyde and -in(e)

formate
CHO_2^-, derived from formic acid

formazan
CH_4N_4, derived from form(o)-, az(o)-, and -an(e)

formic acid
CH_2O_2, derived from *formica* (Latin: ant)

formicin
$C_3H_7NO_2$, derived from form(o)- and -icin(e)

form(o)-
derived from formic acid

formosanan
$C_{18}H_{22}N_2O$, derived from formosanine and and -an(e)

formosanine
$C_{21}H_{24}N_2O_4$, derived from the specific epithet of the species name *Uncaria formosana* Matsuma and Hayata (a tropical woody vine), from *formosanus* (New Latin: from the Island of Formosa, Republic of China), and -in(e)

formose
derived from formaldehyde and -ose

formycin
$C_{10}H_{13}N_5O_4$, derived (with contraction) from the specific epithet of the bacterial species name *Nocardia interformis*, from inter- and *forma* (Latin: shape, form), and -mycin

[1]formyl
CHO−, derived from formic acid and -yl

[2]formyl
CH≡, derived from formic acid and -yl

forskolin (colforsin)
$C_{22}H_{34}O_7$, derived (with contraction) from the specific epithet of the species name *Coleus forskohlii* Briq. (makandi), after the Swedish-Danish botanist Peter Forskål (1736-1763), and -in(e)

Forster diazoketone synthesis
named for the British chemist Sir Martin Onslow Forster (1872-1945)

Forster reaction
named for the British chemist Sir Martin Onslow Forster (1872-1945)

forsterite
Mg_2SiO_4, named for the German naturalist Johann Reinhold Forster (1729-1798)

Fourier transform
named for the French mathematician and physicist Jean Baptiste Joseph Baron de

Fourier (1768-1830)

Fowler's solution
KH_2AsO_4, named for the British physician Thomas Fowler (1736-1801)

fragranol
$C_{10}H_{18}O$, derived from the specific epithet of the species name *Artemisia fragrans* (a wormwood), from *fragrans* (Latin: to give off an odor, fragrant), and -ol

Franchimont reaction
named for the Dutch chemist Antoine Paul Nicolas Franchimont (1844-1919)

francium
Fr, named for France. derived from *Francia* (New Latin: France)

Franck-Condon principle
named for the German-US physicist James Franck (1882-1964) and the US physicist Edward Uhler Condon (1902-1974)

Franck-Rabinovitch cage effect
named for the German-US physicist James Franck (1882-1964) and the Russian-US chemist Eugene Rabinovitch (1901-1973)

frangula emodin (emodin, frangulic acid)
$C_{15}H_{10}O_5$, derived from the specific epithets of the species names *Rhamnus frangula* L. (alder buckthorn), from *frangulus* (Latin: brittle, fragile), from *frangere* (Latin: to break), and *Rheum emodi* (Gilgiti rhubarb), from *rheum* (Greek: rhubarb) and *Emodos* (Greek: Himalaya), respectively, and -in(e)

frangulin
derived from the specific epithet of the species name *Rhamnus frangula* L. (alder buckthorn), from *frangulus* (Latin: brittle, fragile), from *frangere* (Latin: to break), and -in(e)

Frank-Caro process
named for the German chemists Adolf Frank and Nikodem Caro (1871-1935)

Frankland-Duppa reaction
named for the British chemists Sir Edward Frankland (1825-1899) and B. F. Duppa

Frankland synthesis
named for the British chemist Sir Edward Frankland (1825-1899)

franklinite
$(Zn,Mn^{II},Fe^{II})(Fe^{III},Mn^{III})_2O_4$, named after this mineral's locality Franklin and Sterling Hill, NJ, USA

Frasch process
named for the German-US chemist Herman Frasch (1851-1914)

fraxetin
$C_{10}H_8O_5$, coined by variation of fraxin

fraxin
$C_{16}H_{18}O_{10}$, derived from the genus name *Fraxinus* (ash tree), from *fraxinus* (Latin: ash tree), and -in(e)

fredericamycin
named after the Frederick Cancer Research Center, Frederick, MD, USA, where this antibiotic was isolated and characterized

Free-Wilson analysis
named for the US chemists S. M. Free, Jr. (born 1923) and J. W. Wilson (born 1919)

freibergite
$(Ag,Cu,Fe)_{12}(Sb,As)_4S_{13}$, named after this mineral's locality, the town of Freiberg, Germany

Frémy's salt
$K_2NO_7S_2$, named for the French chemist Edmond Frémy (1814-1894)

Frenkel defect
named for the Russian physicist Yakov Ilyich Frenkel (1894-1952)

frenolicin
$C_{18}H_{18}O_7$, a name coined without any stated reason

freon
coined by contraction of freezing and (halogenated) hydrocarbon

frequentin
$C_{14}H_{20}O_4$, derived from the specific epithet of the species name *Penicillium frequentans* Westling (a fungus), *frequentans* (Latin: crowding), from *frequentare* (Latin: to crowd), from *frequens* (Latin: crowded, frequent), and -in(e)

Freund reaction
named for the Austrian chemist August Freund (1835-1892)

Freund's acid
$C_{10}H_9NO_6S_2$, named for the German chemist Martin Freund (1863-1920)

Freund's adjuvant
named for the US pathologist J. T. Freund (1890-1960)

Fricke dosimeter
named for the German chemist R. Fricke (1895-1950)

friedelane
$C_{30}H_{52}$, derived from friedelin and -an(e)

Friedel-Crafts reaction
named for the French chemist Charles Friedel (1832-1899) and the US chemist James Mason Crafts (1839-1917)

friedelin
$C_{30}H_{50}O$, named for the French chemist Charles Friedel (1832-1899)

Friedländer synthesis
named for the German chemist Paul Friedländer (1857-1923)

Fries rearrangement
named for the German chemist Karl Theophil Fries (1875-1962)

Fritsch-Buttenberg-Wiechell rearrangement
named for the German chemists Paul Ernst Moritz Fritsch (1859-1913), Wilhelm Paul Buttenberg (1864-1946), and Heinrich G. Wiechell (born 1865)

Fröhde's reagent
named for the German chemist K. F. A. Fröhde (born 1830)

frondelite
$Mn(Fe^{III})_4(PO_4)_3(OH)_5$, named for the US mineralogist Clifford Frondel (1907-2002)

frontalin
$C_8H_{14}O_2$, derived from the specific epithet of the species name *Dendroctonus frontalis* (southern pine beetle), from *frontalis* (Latin: frontal), from *frons* (Latin: forehead), and -in(e)

fructan (fructosan)
derived (with contraction) from fructose and -an

fructosan (fructan)
derived from fructose and -an

fructose
$C_6H_{12}O_6$, derived from *fructus* (Latin: fruit) and –ose

Frumkin effect
named for the Russian chemist Aleksandr Naumovich Frumkin (1895-1976)

FSGO
an abbreviation for floating spherical Gauss orbital

FSH
an abbreviation for follicle stimulating hormone

FT
an abbreviation for Fourier transform

fucan (fucosan)
derived (with contraction) from fucose and -an

fuchsin (magenta)
$C_{20}H_{20}ClN_3$, derived from the genus name *Fuchsia* (shrubs), named for the German botanist Leonhard Fuchs (1501-1566), and -in(e)

fuc(o)-
derived from the genus name *Fucus* (algae), from *fucus* (Latin: archil, red dye, rouge, deceit), ultimately from *phykos* (Greek: seaweed, rouge), of Semitic origin

fucosamine
$C_6H_{12}O_5$, derived from fucose and amine

fucosan (fucan)
derived from fucose and -an

fucose
$C_6H_{12}O_5$, derived from fuc(o)- and -ose

fucosterol
$C_{29}H_{48}O$, derived from fuc(o)- and sterol

fucoxanthin
$C_{42}H_{58}O_6$, derived from fuc(o)- and xanthin

fugacity
derived from *fugax* (Latin: apt to flee), from *fugere* (Latin: to flee)

-fuge
derived from from *fugere* (Latin: to flee)

Fujimoto-Belleau reaction
named for the US chemist George Iwao Fujimoto (born 1920) and the 20th century Canadian chemist B. Belleau

fullerene
named for the US architect Richard Buckminster Fuller (1895-1983) – referring to the similarity of these compounds' structural symmetry to characteristic elements of Buckminster Fuller's architecture

fulleride
derived from fullerene and -ide

fuller's earth (floridin)
derived from fuller, from *fuler* (Old French: to cleanse and thicken cloth)

fulmin-
derived from *fulmen* (Latin: lightning)

fulminic acid
CHNO, derived from fulmin-

fulvalene
$C_{10}H_8$, derived from fulvene and -alene

fulvene
C_6H_6, derived from fulv(a)- and -ene

fulvinic acid
derived from fulv(o)-

fulv(o)-
derived from *fulvus* (Latin: tawny)

fulvoplumierin
$C_{14}H_{12}O_4$, derived from fulv(o)-, the genus name *Plumeria* (shrubs, trees), after the French botanist Charles Plumier (1646-1704), and -in(e)

fumagillin
$C_{26}H_{34}O_7$, derived (with anagrammatical contraction) from the fungal species name *Aspergillus fumigatus*, from *fumigatus* (Latin: smoked), from *fumigare* (Latin: to smoke), from *fumus* (Latin: smoke), and -in(e)

fumarase
derived from fumaric acid and -ase

fumaric acid
$C_4H_4O_4$, derived from the genus name *Fumaria* (fumitory), from *fumus* (Latin: smoke)

fumigatin
$C_8H_8O_4$, derived from the specific epithet of the fungal species name *Aspergillus fumigatus*, from *fumigatus* (Latin: smoked), from *fumigare* (Latin: to smoke), from *fumus* (Latin: smoke), and -in(e)

fumonisin
derived (with contraction) from the fungal species name *Fusarium moniliforme*, from *fusus* (Latin: spindle) and *monile* (Latin: necklace), and -in(e)

fungichromin
$C_{35}H_{58}O_{12}$, derived from fungus, chrom(o)-, and -in(e)

fungicide
derived from fungus and -cide

fungisterol
$C_{28}H_{44}O$, derived from fungus and sterol

funtumine
$C_{21}H_{35}NO$, derived from the genus name *Funtumia* (tropical African trees), from *funtum* (Ewe dialect: funtumia), and -in(e)

fural
$C_5H_4O_2$, derived from *furfur* (Latin: bran) and -al

furan
C_4H_4O, coined by contraction of furfuran

furanose
derived from furan and -ose – referring to a five-membered cyclic hemiacetal

furcellaran
derived from the genus name *Furcellaria* (algae), from *furcella*, diminutive of *furca* (Latin: fork), and -an

furfural
$C_5H_4O_2$, derived from *furfur* (Latin: bran) and -al

furfuran
C_4H_4O, an obsolete name for furan, derived from furfural

furfurol
$C_5H_4O_2$, an obsolete name for furfural, derived from *furfur* (Latin: bran) and *oleum* (Latin: oil)

furfuryl
C_5H_5O-, derived from furfural and -yl

furil
$C_{10}H_6O_2$, derived from furan and -il(e); patterned after benzil

furo-
derived from furan

furoic acid
$C_5H_4O_3$, derived from furan; patterned after benzoic acid

furoin
$C_{10}H_8O_2$, derived from furan and -oin; patterned after [1]benzoin

furostan
$C_{27}H_{46}O$, derived from furo-, sterol, and -an(e)

furyl
C_4H_3O-, derived from furan and -yl; patterned after phenyl

fus(a)-
derived from the fungal genus name *Fusarium*, from fus(i)-

fusaric acid
$C_{10}H_{13}NO_2$, derived from fus(a)-

fusarubin
$C_{15}H_{14}O_7$, derived from fus(a)-, rubi-, and -in(e)

fuscin
$C_{15}H_{16}O_5$, derived from the specific epithet of the fungal species name *Oidiodendron fuscum* Robak, from fusc(o)-, and -in(e)

fusc(o)-
derived from *fuscus* (Latin: dark, tawny, brownish orange)

fusel oil
derived from *Fusel* (German: poor-quality liquor)

fus(i)-
derived from *fusus* (Latin: spindle)

fusicoccin
derived from the fungal genus name *Fusicoccum*, ultimately from fus(i)- and *kokkos* (Greek: grain, kernel, kermes berry)

fusidic acid
$C_{31}H_{48}O_6$, derived from the fungal genus name *Fusidium*, from fus(i)-

fusidin
$C_{31}H_{47}NaO_6$, derived from fusidic acid and -in(e)

fustin
$C_{15}H_{12}O_6$, derived from fustet (*Rhus cotinus* L.), ultimately from *fustaq*, *fustuq* (Arabic: fustet), and -in(e)

G

GA (tabun)
$C_5H_{11}N_2O_2P$, a US military code for tabun derived from German and the letter A

GABA
$C_4H_9NO_3$, an abbreviation for γ-aminobutyric acid

gabbro
a rock species, named after its locality Gabbro, near Livorno, Italy, ultimately probably derived from *glaber* (Latin: bare, smooth)

Gabriel-Colman rearrangement
named for the German chemists Siegmund Gabriel (1851-1924) and J. Colman

Gabriel-Marckwald synthesis
named for the German chemists Siegmund Gabriel (1851-1924) and W. Marckwald

Gabriel synthesis
named for the German chemist Siegmund Gabriel (1851-1924)

gadolinia
Gd_2O_3, derived from gadolinium and -a

gadolinite
$Be_2FeY_2Si_2O_{10}$, named for the Finnish chemist Johan Gadolin (1760-1852)

gadolinium
Gd, named for the Finnish chemist Johan Gadolin (1760-1852); the name of this element has also been related to the source mineral gadolinite, likewise named for Johan Gadolin

gagate (jet)
C_n, named after the river and ancient town of Gagas, Lycia, Asia Minor

gahnite
$ZnAl_2O_4$, named for the Swedish chemist and mineralogist Johann Gottlieb Gahn (1745-1818)

galactan
derived (with contraction) from galactose and -an

galactaric acid
$C_6H_{10}O_8$, derived from galact(o)- and -aric acid

galactitol
$C_6H_{14}O_6$, derived from galact(o)- and -itol

galact(o)-
derived from galactose

galactomannan
derived from galactose and mannan

galactosamine
$C_6H_{13}NO_5$, derived from galactose and amine

galactose
$C_6H_{12}O_6$, derived from *gala* (Greek: milk) and -ose

galactoside
derived from galactose and ^{3}-ide

galactosidase
derived from galactoside and -ase

galacturonic acid
$C_6H_{10}O_7$, derived from galact(o)- and -uronic acid

galangin
$C_{15}H_{10}O_5$, derived from galanga root (*Alpinia officinarum* Hance), ultimately from *khalanjan* (Arabic: galanga root) and -in(e)

galanthamine
$C_{17}H_{21}NO_3$, derived from galanth(o)- and amine

galanthan
$C_{15}H_{19}N$, derived from galanth(o)- and -an(e)

galanth(o)-
derived from the genus name *Galanthus* (snowdrops), from *gala* (Greek: milk) and anth(o)-

galaxite
$MnAl_2O_4$, named after this mineral's locality, the town of Galax, NC, USA

galegine
$C_6H_{13}N_3$, derived from the genus name *Galega* (goat's rues), ultimately from *Gallica herba* (Medieval Latin: Gallic herb), and -in(e)

galena
PbS, derived from *galena* (Latin: galena), by some, probably improperly, related to *galene* (Greek: serenity)

galenical
named for the Greek-Roman physician Galen (Galenos) (129-199)

galenite
PbS, derived from galena and -ite

galenobismutite
$PbBi_2S_4$, derived from galena and bismutite

gallacetophenone
$C_8H_8O_4$, coined by contraction of gallic acid and acetophenone

gallane
GaH_3, derived from gallium and -an(e)

gallein
$C_{20}H_{12}O_7$, coined by contraction of pyrogallolphthalein

gallic acid
$C_7H_6O_5$, derived from *galle* (French: gall), from *galla* (Latin: gall nut)

gallium
Ga, named for France, derived from *Gallia* (Latin: Gaul, New Latin: France); sometimes the name gallium is also seen as derived from *Gallus*, from *gallus* (Latin: cock), the latinized form of the name of gallium's discoverer Poul Emile Lecoq de Boisbaudran (1838-1912)

gallo-
derived from gallic acid

gallocyanine
$C_{15}H_{13}ClN_2O_5$, derived from gallo- and cyanine

gallotannin
derived from gallo- and tannin

galloyl
$C_7H_5O_4-$, derived from gallo- and -oyl

galvanization
named for the Italian physiologist Luigi Galvani (1737-1798)

gambogic acid
$C_{38}H_{44}O_8$, derived from gamboge (*Garcinia hanburyi* Hook. f.), from Cambodia

gamet(o)-
derived from *gametes* (Greek: spouse)

gammacerane
$C_{30}H_{52}$, derived from the Greek letter γ, cer(o)-, and -an(e)

gam(o)-
derived from *gamos* (Greek: marriage)

gamone
derived from gam(o)- and hormone

gangli(o)-
derived from ganglion, from *ganglion* (Greek: cystic tumor, mass of nerve tissue)

ganglioside
derived from gangli(o)-, -ose, and ^{3}-ide

gardenin
$C_{21}H_{22}O_9$, derived from the genus name *Gardenia* (gardenia), named for the Scottish-US physician and naturalist Alexander Garden (1730-1791), and -in(e)

Garner aldehyde
$C_{11}H_{19}NO_4$, named for the US chemist Philip Paul Garner (born 1955)

garnet
$A_3B_2(SiO_4)_3$ (A = Mg, Fe, Mn, Ca; B = Al, Fe, Cr), derived from *granatum* (Latin: pomegranate), from *granum* (Latin: grain) – referring to these minerals' pomegranate red color

garryine
$C_{22}H_{33}NO_2$, derived from the genus name *Garrya* (shrubs, small trees), named for the Hudson's Bay Company official Nicholas Garry (1782-1856), and -in(e)

gas
an artificial word derived from chaos, coined by the Flemish alchemist Jan Batista van Helmont (1577-1644)

gasohol
coined by contraction of gasoline and alcohol

gasoline
derived from gas, -ol(e), and -in(e)

gastricsin
derived (with contraction) from gastric and pepsin

gastrin
derived from gastr(o)- and -in(e)

gastr(o)-
derived from *gaster* (Greek: stomach)

Gattermann aldehyde synthesis
named for the German chemist Ludwig Gattermann (1860-1920)

Gattermann-Hopff synthesis
named for the German chemists Ludwig Gattermann (1860-1920) and H. Hopff (1896-1920)

Gattermann-Koch synthesis
named for the German chemist Ludwig Gattermann (1860-1920) and the US chemist Julius Arnold Koch (1864-1932)

Gattermann reaction
named for the German chemist Ludwig Gattermann (1860-1920)

Gattermann-Skita synthesis
named for the German chemist Ludwig Gattermann (1860-1920) and the Austrian-German chemist Aladar Skita (1876-1953)

gauche
derived from *gauche* (French: left, on the left)

Gauss orbital
named for the German mathematician and

astronomer Carl Friedrich Gauß (1777-1855)

Gay-Lussac's law
named for the French chemist Joseph Louis Gay-Lussac (1778-1850)

gaylussite (natrocalcite)
$Na_2Ca(CO_3)_2{\cdot}5H_2O$, named for the French chemist Joseph Louis Gay-Lussac (1778-1850)

GB (sarin, trilon 46)
$C_4H_{10}FO_2P$, a US military code name derived from German and the letter B

GC
an abbreviation for gas chromatography

GD (soman, trilon 300)
$C_7H_{16}FO_2P$, a US military code name derived from German and the letter D

GDP
$C_{10}H_{15}N_5O_{11}P_2$, an abbreviation for guanosine 5'-diphosphate

gedrite
$(Mg,Fe)_5Al_2(Si_6Al_2)O_{22}(OH)_2$, named after this mineral's locality Heas Valley, near Gedres, France

gehlenite
$Ca_2Al(Si,Al)_2O_7$, named for the German chemist Adolph Ferdinand Gehlen (1775-1815)

geisso-
derived from *geis(s)on* (Greek: cornice, tile, eaves of a roof)

geissoschizoline (pereirine)
$C_{19}H_{26}N_2O$, derived (with contraction) from geissospermine, schiz(o), and indoline – referring to the fact that this compound is formed by hydrolytic cleavage of geissospermine and contains an indoline fragment

geissospermine
$C_{40}H_{48}N_4O_3$, derived from the genus name *Geissospermum* (tropical trees), from geisso- and sperm(o)-, and -in(e)

gel
derived from gelatin

gelatin
ultimately derived from *gelatus* (Latin: frozen), from *gelare* (Latin: to freeze, to congeal) and -in(e)

gelation
derived from *gelare* (Latin: to freeze, to congeal)

gelsemine
$C_{20}H_{22}N_2O_2$, derived from the genus name *Gelsemium* (jessamine), ultimately from *yasamin* (Arabic: jasmine), and -in(e)

gem
derived from *gemma* (Latin: precious stone)

gem-
coined by contraction of geminal

geminal
derived from *gemini* (Latin: twins)

-gen
derived from *-genes* (Greek: of a certain kind), from *gennaein* (Greek: to generate)

Geneva nomenclature
named for the 1892 IUPAC congress in Geneva, Switzerland

-genin
derived from -gen and -in(e)

genist-
derived from the genus name *Genista* (shrubs), from *genista* (Latin: the plant broom)

genistein
$C_{15}H_{10}O_5$, derived from genist- and -ein(e)

genistin
$C_{21}H_{20}O_9$, derived from genist- and -in(e)

genotoxic
derived from gene and toxic

gentamycin
derived from gentian violet – referring to the color of the producing microorganism (*Micromonospora purpurea*) – and -mycin

genthelvite
$Be_3Zn_4(SiO_4)_3S$, coined by contraction of the name of the German-US mineralogist Frederick August Ludwig Genth (1820-1893) and helvite

gentian
derived from the genus namc *Gentiana* (gentian), probably named for its discoverer Gentius, 2nd century BC king of Illyria

gentianine
$C_{10}H_9NO_2$, derived from gentian and -in(e)

gentianose
$C_{18}H_{32}O_{16}$, derived from gentian and -ose

gentian violet (crystal violet)
$C_{25}H_{30}ClN_3$, derived from gentian

gentio-
derived from gentian

gentiobiase
derived from gentiobiose and -ase

gentiobiose
$C_{12}H_{22}O_{11}$, derived from gentio-, bi-, and -ose

gentiopicrin
$C_{16}H_{20}O_9$, derived from gentio-, picr(o)-, and -in(e)

gentisic acid
$C_7H_6O_4$, derived from gentian

gentisin
$C_{14}H_{10}O_5$, derived from gentian and -in(e)

gentisyl
$C_7H_7O_2-$, derived from gentisic acid and -yl

ge(o)-
derived from *ge* (Greek: earth, land)

geosmin
$C_{12}H_{22}O$, derived from ge(o)-, osm(e)- and -in(e) – referring to this terpenoid's earthy odor

gephyrotoxin
$C_{19}H_{29}NO$, derived from *gephyra* (Greek: bridge) and toxin – referring to this compound's bridged structure

geran(i)-
derived from the genus name *Geranium* (geranium), ultimately trom *geranıon* (Greek: geranium), diminutive of *geranos* (Greek: crane)

geranial (citral)
$C_{10}H_{16}O$, derived from geran(i)- and -al

geraniol
$C_{10}H_{18}O$, derived from geran(i)- and -ol

geranyl (diprenyl)
$C_{10}H_{17}-$, derived from geraniol and -yl

germacrane
$C_{15}H_{30}$, derived (with contraction) from the species name *Geranium macrorhizum* (bigroot perennial), from geran(i)- and macro-, and -an(e)

germane
GeH_4, derived from germanium and -an(e)

germanite
$Cu_3(Ge,Fe)S_4$, named for this mineral's germanium content

germanium
Ge, named for Germany, derived from *Germania* (New Latin: Germany)

germicide
derived from *germen* (Latin: germ) and -cide

geronic acid
$C_9H_{16}O_3$, coined by variation of geran(i)-

gersdorffite
NiAsS, named for the 19th century Austrian mine owner Johann von Gersdorff

gestagen
derived from gestation and -gen

getter
derived from to get

geyserite
SiO_2, derived from geyser, from *geysir* (Icelandic: geyser)

GFC
an abbreviation for gel filtration chromatography

GGG
$Ga_5Gd_3O_{12}$, an abbreviation for gadolinium-gallium garnet

GH (somatotropin)
an abbreviation for growth hormone

gibbane
$C_{15}H_{24}$, derived from gibb(erell)- and -an(e)

gibb(erell)-
derived from the fungal genus name *Gibberella*, from *gibberella*, diminutive of *gibber* (Latin: hump on the back)

gibberellane
$C_{20}H_{34}$, derived from gibb(erell)- and -an(e)

gibberellic acid
derived from gibb(erell)-

gibberellin
derived from gibb(erell)- and -in(e)

Gibbs-Duhem equation
named for the US mathematician and physicist Josiah Willard Gibbs (1839-1903) and the French physicist Pierre Maurice Marie Duhem (1861-1916)

Gibbs energy
named for the US mathematician and physicist Josiah Willard Gibbs (1839-1903)

Gibbs-Helmholtz equation
named for the US physicist Josiah Willard Gibbs (1839-1903) and the German physicist and physiologist Hermann Ludwig Ferdinand von Helmholtz (1821-1894)

gibbsite (hydrargillite)
$Al(OH)_3$, named for the US mineral collector George Gibbs (1776-1833)

Gibbs phase rule
named for the US mathematician and physicist Josiah Willard Gibbs (1839-1903)

Gibbs reaction
$C_6H_2Cl_3NO$, named for the US chemist

Harry Drake Gibbs (1872-1934)

Gibbs reagent
$C_6H_2Cl_3NO$, named for the US chemist Harry Drake Gibbs (1872-1934)

Giemsa's solution
named for the German pharmacist and chemist Gustav Giemsa (1867-1948)

Giese salt
$(NH_4)Fe^{III}[Fe^{II}(CN)_6]$, named for the German physicist and veterinarian W. Giese (born 1936)

Gif oxidation
named for the town of Gif-sur-Yvette, France, the home of the CNRS institute where this reaction was discovered

gigantine
$C_{13}H_{19}NO_3$, derived from the specific epithet of the species name *Carnegia gigantea* Engelm. (saguaro), from *gigas* (Greek: giant), and -in(e)

Gillespie-Nyholm model (VSEPR model)
named for the British chemists Ronald James Gillespie (born 1924) and Sir Ronald Sydney Nyholm (1917-1971)

gillespite
$BaFe^{II}Si_4O_{10}$, named for the 20th century US mineral collector Frank Gillespie

gilsonite
named for the US industrialist Samuel H. Gilson (1836-1913)

gingerol
$C_{17}H_{26}O_4$, derived from ginger and -ol

ginkgolide
derived from ginkgo, from *ginkyo* (Japanese: gingko, *Ginkgo biloba*), from *gin* (Japanese: silver) and *kyo* (Japanese: apricot), and -olide

ginsenoside (panaxoside)
derived from ginseng (*Panax shinseng*), from *jen-shen* (Chinese: ginseng), -ose, and ^{3}-ide

GIP
an abbreviation for gastric inhibitory polypeptide

Girard reagents
named for the French chemist André Charles Girard (1837-1918)

girasol (opal)
$SiO_2{\cdot}nH_2O$, derived from *girare* (Italian: to turn) and *sole* (Italian: sun)

girbotol process
a name coined by contraction of the company name Girdler Corporation, Louisville, KY, USA and the name of the inventor, the US chemist Robert R. Bottoms (born 1890)

gitogenin
$C_{27}H_{44}O_4$, derived from gitonin, -gen, and -in(e)

gitonin
$C_{50}H_{82}O_{23}$ derived (with contraction) from the genus name *Digitalis* (foxglove), from *digitalis* (Latin: belonging to a finger), from *digitus* (Latin: finger, toe), and -in(e)

gitoxigenin
$C_{23}H_{34}O_5$, derived from gitoxin, -gen, and -in(e)

gitoxin
$C_{41}H_{64}O_{14}$, derived (by anagrammatical contraction) from the genus name *Digitalis* (foxglove), from *digitalis* (Latin: belonging to a finger), from *digitus* (Latin: finger, toe), and toxin

glacial acetic acid
$C_2H_4O_2$, a term based on this acid's melting point of 16.6 °C

Glaser coupling
named for the German chemist Carl Andreas Glaser (1841-1935)

glass
derived from *glas* (Old High German: amber), ultimately from *geolu* (Old English: yellow) – referring to the common discoloration of ancient glass by iron impurities

Glauber salt (mirabilite)
$Na_2SO_4 \cdot 10H_2O$, named for the German-Dutch chemist Johann Rudolf Glauber (1604-1670)

glauberite
$Na_2Ca(SO_4)_2$, named for the German-Dutch chemist Johann Rudolf Glauber (1604-1670)

glaucine
$C_{21}H_{25}NO_4$, derived from the genus name *Glaucium* (herbs), from *glaukion* (Greek: juice of a papaveraceous plant), from glauc(o)-, and -in(e)

glauc(o)-
derived from *glaukos* (Greek: blue green)

glaucodot
(Co,Fe)AsS, derived from glauc(o)- and *doter* (Greek: giver), from *didonai* (Greek: to give) – referring to the use of this mineral in the manufacture of dark blue glass

glauconite
$(K,Na)(Fe,Al,Mg)_2(Si,Al)_4O_{10}(OH)_2$, derived from glauc(o)- and -ite – referring to this mineral's blue green color

glaucophane
$Na_2(Mg,Fe)_3Al_2Si_8O_{22}(OH)_2$, derived from glauc(o)- and -phan(e) – referring to this mineral's blue color

GLC
an abbreviation for gas liquid chromatography

gliadin
derived from gli(o)- and -in(e)

gli(o)-
ultimately derived from *glia* (Middle Greek: glue)

gliotoxin
$C_{13}H_{14}N_2O_4S_2$, derived from the fungal genus name *Gliocladium*, from gli(o)-, and toxin

globin
derived from *globus* (Latin: globe) and -in(e)

-globulin
derived from *globulus*, diminutive of *globus* (Latin: globe), and -in(e)

Glover acid
H_2SO_4, named for the British chemist John Glover (1817-1902)

GLP
an abbreviation for good laboratory practice

glucagon
derived from glucose and *agon* (Greek: leading, driving), from *agein* (Greek: to lead, to drive)

glucamine
$C_6H_{15}NO_5$, derived from ^{2}gluc(o)- and amine

glucan (glucosan)
derived (with contraction) from glucose and -an

glucanase
derived from glucan and -ase

glucaric acid
$C_6H_{10}O_8$, derived from glucose and -aric acid

glucina
BeO, an obsolete name for berylla, derived from gluc(o)-, -in(e), and -a – referring to this compound's sweet taste

glucinium
Be, an obsolete name for beryllium, derived from glucina and -ium

glucitol (sorbitol)
$C_6H_{14}O_6$, derived from glucose and -itol

[1]**gluc(o)-**
derived from *gleukos* (Greek: must), akin to *glykys* (Greek: sweet)

[2]**gluc(o)-**
derived from glucose

glucofrangulin
$C_{27}H_{30}O_{14}$, derived from [2]gluc(o)-, the specific epithet of the species name *Rhamnus frangula* L. (alder buckthorn), from *frangulus* (Latin: brittle, fragile), from *frangere* (Latin: to break), and -in(e)

glucofuranose
$C_6H_{12}O_6$, derived from [2]gluc(o)- and furanose

glucogallin
$C_{13}H_{16}O_{10}$, derived from [2]gluc(o)-, gallo-, and -in(e)

gluconic acid
$C_6H_{12}O_7$, derived from glucose and -one

gluconolactone
$C_6H_{10}O_6$, derived from gluconic acid and lactone

glucopyranose
$C_6H_{12}O_6$, derived from [2]gluc(o)- and pyranose

glucosamine
$C_6H_{13}NO_5$, derived from glucose and amine

glucosan (glucan)
derived from glucose and -an

glucose
$C_6H_{12}O_6$, derived from *gleukos* (Greek: must), akin to *glykys* (Greek: sweet) – referring to this sugar's sweet taste

glucoside
derived from glucose and [3]-ide

glucosinolate
derived from glucose, *sinapis* (Latin: mustard), -ol(e), and –ate

glucovanillin
$C_{14}H_{18}O_8$, derived from [2]gluc(o)- and vanillin

glucuronic acid
$C_6H_{10}O_7$, derived from glucose and *ouron* (Greek: urine)

glucuronide
derived from glucuronic acid and [3]-ide

glucuronidase
derived from glucuronide and -ase

glucuronolactone
$C_6H_8O_6$, derived from glucuronic acid and lactone

glut-
derived from *gluten* (Latin: glue)

glutaconic acid
$C_5H_6O_4$, derived from glut- and aconic acid

glutamic acid (glutaminic acid)
$C_5H_9NO_4$, derived from glut- and amine

glutamine
$C_5H_{10}N_2O_3$, derived from glut- and amine

glutaminic acid (glutamic acid)
$C_5H_9NO_4$, derived from glut- and amine

glutaraldehyde
$C_5H_8O_2$, derived from glutaric acid and aldehyde

glutaric acid
$C_5H_8O_4$, derived from glut- and -aric acid

glutathione
$C_{10}H_{17}N_3O_6S$, derived from glutamic acid, thi(o)-, and -one

glutelin
derived from glut- and -in(e)

gluten
derived from *gluten* (Latin: glue) – referring to this protein's elastic properties

glutenin
derived from gluten and -in(e)

glutin
derived from *glutinum* (Latin: glue) and -in(e)

glycan (polysaccharide)
derived from glyc(o)- and -an

glyceraldehyde
$C_3H_6O_3$, derived from glycer(o)- and aldehyde

glyceride
derived from glycerol and [3]-ide

glyceric acid
$C_3H_6O_4$, derived from glycer(o)-

glyceride
derived from glycerol and -ide

glycerin
$C_3H_8O_3$, an obsolete name for glycerol, derived from *glykeros* (Greek: sweet) and -in(e) – referring to this compound's sweet taste

glycer(o)-
derived from glycerin

glycero-
derived from glyceraldehyde

glycerol
$C_3H_8O_3$, derived from glycerin and –ol

glycerol formal
$C_4H_8O_3$, derived from glycerol and formal

[1]**glyceryl**
$C_3H_7O_2-$, derived from glycerol and -yl

[2]**glyceryl**
$-CH_2CH(-)CH_2-$, derived from glycerol and -yl

glycide
$C_3H_6O_2$, an obsolete name for glycidol, derived from glyc(o)- and [3]-ide

glycidic acid
$C_3H_4O_3$, derived from glycide

glycidol
$C_3H_6O_2$, derived from glycide and -ol

glycine
$C_2H_5NO_2$, derived from *glykys* (Greek:

sweet) and -in(e) – referring to this amino acid's sweet taste

glycinin
derived from glycine and -in(e)

glyc(o)-
derived from *glykys* (Greek: sweet)

glycocholic acid
$C_{26}H_{43}NO_6$, derived from glyc(o)- and cholic acid

glycocoll
$C_2H_5NO_2$, an obsolete name for glycine, derived from glyc(o)- and *kolla* (Greek: glue)

glycogen
derived from glyc(o)- and -gen

[1]**glycol**
coined by contraction of glycerol and alcohol

[2]**glycol**
$C_2II_6O_2$, coincd by contraction of ethylene glycol

glycolaldehyde
$C_2H_4O_2$, derived from [2]glycol and aldehyde

glycol(l)ic acid
$C_2H_4O_3$, derived from [2]glycol

glycol(l)ide
$C_4H_4O_4$, derived from glycol(l)ic acid and [3]-ide

glycoluril
$C_4H_6N_4O_2$, derived from [2]glycol, urea, and -il(e)

glycolysis
derived from glyc(o)- and lysis

glycosamine
derived from glycose and amine

glycosaminoglycan (mucopolysaccharide)
derived from glycosamine and glycan

glycose (monosaccharide)
$C_nH_{2n}O_n$, derived from glyc(o)- and -ose

glycosidase (carbohydrase)
derived from glycoside and -ase

glycoside
derived from glycose and [3]-ide

glycosine (arborine)
$C_{16}H_{14}N_2O$, derived from the genus name *Glycosmis* (trees, shrubs), from glyc(o)- and *osme* (Greek: smell), and -in(e)

glycosphingolipid
derived from glyc(o)-, sphing(o)-, and lipid

glycosyl
derived from glycose and -yl

glycuronan
derived from glycuronic acid and -an

glycuronic acid (uronic acid)
derived from glyc(o)- and -uronic acid

glycyl
C_2H_4NO-, derived from glycine and -yl

glycyrrhizic acid
$C_{42}H_{62}O_{16}$, derived from the genus name *Glycyrrhiza* (licorice), from glyc(o)- and rhiz(o)-

glyme
$C_4H_{10}O_2$, originally coined as a trademark, derived (with contraction) from glycol dimethyl ether

glyoxal
$C_2H_2O_2$, coined by contraction of [2]glycol and oxalic acid

glyoxalin (imidazole)
$C_3H_4N_2$, derived (with contraction) from glyoxal and amine

glyoxime
$C_2H_4N_2O_2$, coined by contraction of glyoxal dioxime

glyoxylic acid
$C_2H_2O_3$, derived from glyoxal

gmelinite
$Na_4(Al_4Si_8)O_{24}\cdot 11H_2O$, named for the German mineralogist and chemist Christian Gottlob Gmelin (1792-1860)

Gmelin test
named for the German mineralogist and chemist Leopold Gmelin (1788-1853)

[1]**GMP**
an abbreviation for good manufacturing practice

[2]**GMP**
$C_{10}H_{14}N_5O_8P$, an abbreviation for guanosine 5'-monophosphate

gneiss
a rock species, derived from *Gneis* (German: gneiss), from *gneista* (Old High German: spark) – probably referring to this rock's sparkling appearance

GO
an abbreviation for Gauss orbital

goethite
Fe(OH)O, named for the German poet and writer Johann Wolfgang von Goethe (1749-1832)

goitrin
C_5H_7NOS, derived from goiter, ultimately from *guttur* (Latin: throat), and -in(e)

Golay column
named for the Swiss-US physicist M. J. E. Golay (1902-1989)

gold
Au, ultimately derived from *geolu* (Old English: yellow)

golden beryl (heliodor)
$Be_3Al_2Si_6O_{18}$, named after its color

Goldschmidt process
named for the German chemist Hans Goldschmidt (1861-1923)

Goldschmidt rule
named for the German chemist V. M. Goldschmidt (1888-1947)

Goldschmidt's radical
$C_{18}H_{12}N_5O_6$, named for the German chemist Stefan Goldschmidt (1889-1971)

Gold's reagent
$C_7H_{14}ClN_3$, named for the German chemist Heinrich Gold (born 1912)

Gomberg-Bachmann reaction
named for the Ukrainian-US chemist Moses Gomberg (1866-1847) and the US chemist Werner Emmanuel Bachmann (1901-1951)

Gomberg's radical
$C_{19}H_{15}$, named for the Ukrainian-US chemist Moses Gomberg (1866-1847)

gomphidic acid
$C_{18}H_{12}O_9$, derived from the genus name *Gomphidius* (spike mushrooms), from *gomphos* (Greek: tooth, peg, bolt, bond)

-gon
derived from *gonia* (Greek: angle), from *gony* (Greek: knee)

gonad(o)-
derived from gonad

gonadoliberin (gonadorelin)
derived from gonadotropin and liberin

gonadotropin
derived from gonad(o)- and -tropin

gonane
$C_{17}H_{28}$, derived from gonad and -an(e)

gonnardite
$(Na,Ca)_2(Si,Al)_5O_{10}\cdot 3H_2O$, named for the French mineralogist Ferdinand Gonnard (1833-1923)

Gooch crucible
named for the US chemist Frank Austin Gooch (1852-1929)

Good's buffer
named for thc US botanist and plant pathologist N. E. Good (born 1917)

gorgostane
$C_{30}H_{52}$, derived from the order name *Gorgonacea* (gorgonians), from *gorgonia* (Latin: coral), ultimately from *gorgos* (Greek: terrible), sterol, and -an(e)

gossyp-
derived from *gossypion* (Latin: cotton plant)

gossyplure
$C_{18}H_{32}O_2$, derived from the specific epithet of the species name *Pectinophora gossypiella* (Saunders) (pink bollworm), from gossyp-, and lure

gossypol
$C_{30}H_{30}O_8$, derived from the genus name *Gossypium* (cotton plant), from gossyp-, and -ol

GOT
an abbreviation for glutamate oxalacetate transaminase

gougerotin (asteromycin)
$C_{16}H_{25}N_7O_8$, derived from the specific epithet of the bacterial species name *Streptomyces gougerotii*, probably after the French dermatologist Henri Gougerot (1881-1955), and -in(e)

Gould-Jacobs reaction
named for the 20th century US chemists R. Gordon Gould, Jr. and Walter A. Jacobs

Gouy-Chapman double layer
named for the French physicist L. G. Gouy (1854-1926) and the British chemist David Leonard Chapman (1869-1958)

GPC
an abbreviation for gel permeation chromatography

GPD
an abbreviation for glucose 6-phosphate dehydrogenase

G-protein
coined by contraction of GTP-binding protein

GPT
an abbreviation for glutamate pyruvate transaminase

Graebe-Ullmann synthesis
named for the German chemists Carl Graebe (1841-1927) and Fritz Ullmann

Graham reaction
named for the US chemist William Hardin Graham (born 1932)

Graham's salt
$NaPO_3$, named for the British chemist Thomas Graham (1805-1867)

gramicidin
derived from Gram-positive, named for the Danish physician and bacteriologist Hans Christian Joachim Gram (1853-1938), -cide, and -in(e)

gramine
$C_{11}H_{14}N_2$, derived from *gramen* (Latin: grass) and -in(e)

grammatite (tremolite)
$Ca_2(Mg,Fe)_5Si_8O_{22}(OH)_2$, derived from *gramme* (Greek: line, stroke) and -ite

granaticin
$C_{22}H_{20}O_{10}$, derived from *Granat* (German: garnet) and -icin(e) – referring to the garnet-like color of this compound

grandisol
$C_{10}H_{18}O$, derived from the specific epithet of the species name *Anthonomus grandis* Boheman (boll weevil), from *grandis* (Latin: large, great, grand), and -ol

granite
a rock species, ultimately derived from *granum* (Latin: grain) and -ite

graphene
derived from graphite and -ene

graphite
C_n, derived from *graphein* (Greek: to write) and -ite – referring to the use of graphite in pencils etc.

graphitic acid (graphitic oxide)
$(C_8H_2O_4)_n$, derived from graphite – referring to this compound's formation by oxidation of graphite

GRAS
an abbreviation for generally recognized as safe

gravimetry
derived from *gravis* (Latin: heavy) and -metry

grayanotoxane
$C_{20}H_{34}$, derived from grayanotoxin and -an(e)

grayanotoxin
derived from the specific epithet of the species name *Leucothoe grayana* Max. (hana-hirinoki, a Japanese shrub), after the US botanist Asa Gray (1810-1888), and toxin

greenockite
CdS, named for the British soldier Charles Murray Cathcart, Lord Greenock (1783-1859)

Griess-Ilosvay reaction
named for the German-British chemist Johann Peter Griess (1829-1888) and the Hungarian chemist Lajos Ilosvay (1851-1936)

Griess reaction
named for the German-British chemist Johann Peter Griess (1829-1888)

Grignard reaction
named for the French chemist Victor François Auguste Grignard (1871-1935)

Grimm's hydride rule
named for the German chemist Hans Georg Grimm (1887-1958)

Grimm-Sommerfeld rule
named for the German chemist Hans Georg Grimm (1887-1958) and the German physicist Arnold Sommerfeld (1868-1951)

grindelic acid
$C_{20}H_{32}O_3$, derived from the genus name *Grindelia* (gumweed), after the Russian chemist and physician David Hieronymus Grindel (1777-1836)

grisein
$C_{40}H_{61}FeN_{10}O_{20}S$, derived from the specific epithet of the fungal species name *Penicillium griseofulvum*, from *griseus* (Medieval Latin: greenish gray), and -in(e)

griseofulvin
$C_{17}H_{17}ClO_6$, derived from the specific epithet of the fungal species name *Penicillium griseofulvum*, from *griseus* (Medieval Latin: greenish gray) and *fulvus* (Latin: dark, tawny)

Grob fragmentation
named for the British-Swiss chemist Cyril A. Grob (1917-2003)

Grosheintz-Fischer-Reissert aldehyde synthesis
named for the German chemists J. M. Grosheintz, Herrmann Otto Laurenz Fischer (1888-1960), and C. Arnold Reissert (1860-1945)

grossular (grossularite, hessonite, tsavorite)
$Ca_3Al_2(SiO_4)_3$, derived from the specific epithet of the species name *Ribes grossularia* (gooseberry), from *groseille à maquereau* (French: gooseberry), probably from *Kräuselbeere* (obsolete German name for gooseberry), or from *grossulus* (Latin: small unripe fig) – referring to this mineral's pale green color

Grotthuss-Draper law
named for the German-Russian naturalist Theodor Christian Johann Dietrich Freiherr von Grotthuß (1785-1822) and the British-US chemist and naturalist John William Draper (1811-1882)

GRP
an abbreviation for gastrin-releasing peptide

Grubbs' first generation catalyst
$C_{43}H_{72}Cl_2P_2Ru$, named for the US chemist Robert H. Grubbs (born 1942)

Grubbs' second generation catalyst
$C_{46}H_{65}Cl_2N_2PRu$, named for the US chemist Robert H. Grubbs (born 1942)

Grundmann aldehyde synthesis
named for the 20th century German-US chemist Christoph Grundmann

grünerite (amosite)
$(Fe,Mg)_7Si_8O_{22}(OH)_2$, named for the Swiss-French chemist Louis Emanuel Grüner (1809-1883)

Grunwald-Winstein equation
named for the German-US chemist Ernest Max Grunwald (born 1923) and the Canadian-US chemist Saul Winstein (1912-1969)

GSC
an abbreviation for gas solid chromatography

GSH
$C_{10}H_{17}N_3O_6S$, an abbreviation for glutathione, derived from glutathione and −SH (thiol group)

GSSG
$C_{20}H_{32}N_6O_{12}S_2$, an abbreviation for oxidized glutathione, derived from glutathione and −SS− (disulfide group)

GT
an abbreviation for glutamyl transferase

GTO
an abbreviation for Gaussian-type orbital

GTP
$C_{10}H_{16}N_5O_{14}P_3$, an abbreviation for guanosine 5'-triphosphate

guai(ac)-
derived from the genus name *Guaiacum* (guaiac tree), ultimately from *guayacan* (Taino: guaiac)

guaiacol
$C_7H_8O_2$, derived from guai(ac)- and -ol

guaiane
$C_{15}H_{28}$, derived from guai(ac)- and -an(e)

guaiazulene
$C_{15}H_{18}$, derived from guai(ac)- and azulene

guaiol
$C_{15}H_{26}O$, derived from guai(ac)- and -ol

guan-
derived guano, from *huano* (Spanish: guano), ultimately from *huanu* (Quechua: dung)

guanidine
CH_5N_3, derived from guan- and -idin(e)

guanine
$C_5H_5N_5O$, derived from guan- and -in(e)

guanosine
$C_{10}H_{13}N_5O_5$, derived from guan-, -ose, and -in(e)

guanylic acid (GMP)
$C_{10}H_{14}N_5O_8P$, derived from guanosine

guaran (guar gum)
derived from guar (*Cyamopsis tetragonaloba* (L.) Taub.), from *guar* (Hindi: guar), and -an

Guareschi-Thorpe condensation
named for the Italian chemist Icilio Guareschi (1847-1918) and the British chemist Sir Jocelyn Field Thorpe (1872-1940)

Gudden-Pohl effect
named for the German physicists Bernhard Friedrich Adolf Gudden (1892-1945) and Robert Wiechert Pohl (1884-1976)

Guerbet reaction
named for the French chemist Marcel Guerbet (1861-1938)

Guignet's green
named for the French industrialist E. Guignet (born 1829)

Guldberg-Waage law
named for the Norwegian mathematician and chemist Cato Maximilian Guldberg (1836-1902) and the Norwegian chemist Peter Waage (1833-1900)

gulitol (glucitol, sorbitol)
$C_6H_{14}O_6$, derived from gulose and -itol

gulonic acid
$C_6H_{12}O_7$, derived from gulose

gulose
$C_6H_{12}O_6$, anagrammatically derived from glucose

gum
ultimately derived from *qmy* (Old Egyptian: resin)

Günzburg reagent
named for the German physician A. Günzburg (1861-1945)

Gustavson reaction
named for the Russian chemist Gavriil Gavriilovich Gustavson (1842-1908)

Guthrie's alloy
(Bi,Sn,Pb,Cd), named for the US physician and chemist Samuel Guthrie, Jr. (1782-1848)

Gutknecht pyrazine synthesis
named for the 19th century German chemist H. Gutknecht

gutta-percha
derived from *getah* (Malay: sap, latex) and percha, the name of the trees (of the genera *Payena* and *Palaquium*) producing gutta-percha

Gutzeit test (for arsenic)
named for the German chemist Heinrich Wilhelm Theodor Gutzeit (1845-1888)

guvacine
$C_6H_9NO_2$, derived from *guvaca* (Old Indian: betel palm tree, *Areca catechu*) and -in(e)

gymnemic acid (gymnemin)
derived from the genus name *Gymnema* (periploca), from gymn(o)- and *nema* (Greek: thread)

gymn(o)-
derived from *gymnos* (Greek: naked)

gyn(o)-
derived from *gyne* (Greek: woman)

gynogamone
derived from gyn(o)- and gamone

gynotermone
derived from gyn(o)- and termone

gypsum
$CaSO_4 \cdot 2H_2O$, ultimately derived from *gypsos* (Greek: chalk, gypsum), of Semitic origin, akin to *jibs* (Arabic: plaster, mortar)

gyrase
derived from gyr(o)- and -ase

gyr(o)-
derived from *gyros* (Greek: ring)

gyrolite
$NaCa_{16}AlSi_{24}O_{60}(OH)_8 \cdot 14H_2O$, derived from gyr(o)- and -lite – referring to the round form of the crystalline groupings of this mineral

H

Haber-Bosch process
named for the German chemists Fritz Haber (1868-1934) and Carl Bosch (1874-1940)

Haber-Weiss mechanism
named for the German chemists Fritz Haber (1868-1934) and L. Weiss

hachimycin (trichomycin)
derived from the specific epithet of the bacterial species name *Streptomyces hachijoensis,* after the island of Hachijo Jima, Japan, and -mycin

hadacidin
$C_3H_5NO_4$, a name coined without any stated reasons

haem-
derived from *haima* (Greek: blood)

hafnia
HfO_2, derived from hafnium and -a

hafnium
Hf, named for Copenhagen, Denmark, where this element was isolated for the first time, derived from *Hafnia* (New Latin: Copenhagen, Denmark) and -ium

hafnocene
$C_{10}H_{10}Hf$, derived from hafnium and -ocene

[1]hahnium
Db, unsuccessfully suggested name for dubnium, in honor of the German physicist Otto Hahn (1879-1968)

[2]hahnium
Hs, unsuccessfully suggested name for hassium, in honor of the German physicist Otto Hahn (1879-1968)

Hahn's rule
named for the German physicist Otto Hahn (1879-1968)

Haine's test
named for the US chemist W. S. Haine (1850-1923)

halite
NaCl, derived from hal(o)- and -lite

Haller-Bauer reaction
named for the French chemists Albin Haller (1849-1925) and E. Bauer

halloysite
$Al_2Si_2O_5(OH)_4$, named for the Belgian geologist Jean-Baptiste Julien Baron Omalius d'Halloy (1783-1875)

halo-
derived from halogen

hal(o)-
derived from *hals* (Greek: salt)

halochromism (acichromism)
derived from hal(o)- and -chromism

haloform
CHX_3, derived from halo- and -form

halogen
derived from and hal(o)- and -gen

halon
coined by contraction of halohydrocarbon

haloperoxidase
derived from halo- and peroxidase

halostachine
$C_9H_{13}NO$, derived from the genus name *Halostachys* (shrubs), from hal(o)- and and *stachys* (Greek: ear of grain), and -in(e)

halothane
$C_2HBrClF_3$, coined by contraction of halogen and ethane

halotrichite (feather alum)
$FeAl_2(SO_4)_4{\cdot}22H_2O$, derived from hal(o)-, trich(o)-, and -ite – referring to this alum's feathery appearance

hamamel(i)-
derived from the genus name *Hamamelis* (shrubs, small trees), from *hamamelis* (Greek: medlar), ultimately from *hama* (Greek: together with) and *melon* (Greek: apple, fruit)

hamamelitannin
$C_{20}H_{20}O_{14}$, derived from hamamel(i)- and tannin

hamamelose
$C_6H_{12}O_6$, derived from hamamel(i)- and -ose

hambergite
$Be_2BO_3(OH)$, named for the Swedish mineralogist Axel Hamberg (1863-1933)

Hammett equation
named for the US chemist Louis Plack Hammett (1894-1987)

Hammick reaction
named for the British chemist Dalziel Llewellyn Hammick (1887-1966)

hamycin (primamycin)
derived from the initials of the company name Hindustan Antibiotics Limited, Pune, India and -mycin

Hanes plot
named for the Canadian biochemist Charles Samuel Hanes (1903-1990)

hanksite
$KNa_{22}(SO_4)_9(CO_3)_2Cl$, named for the US geologist Henry G. Hanks (1826-1907)

Hansch analysis
named for the US chemist Corwin H. Hansch (born 1918)

Hantzsch pyridine synthesis
named for the German chemist Arthur Rudolf Hantzsch (1857-1935)

Hantzsch pyrrole synthesis
named for the German chemist Arthur Rudolf Hantzsch (1857-1935)

Hantzsch-Widman name
named for the German chemist Arthur Rudolf Hantzsch (1857-1935) and the Swedish chemist K. O. Widman (1852-1930)

Hanuš reagent
named for the Czech chemist Josef Hanuš (1872-1955)

hapl(o)-
derived from *haploos* (Greek: single, simple)

haplophytine
$C_{37}H_{40}N_4O_7$, derived from the genus name *Haplophyton* (herbs), from hapl(o)- and phyt(o)-, and -in(e)

hapticity
derived from hapt(o)-

hapt(o)-
derived from *haptein* (Greek: to touch, to adhere)

haptoglobin
derived from hapt(o)- and globin

Harden-Young ester
$C_6H_{14}O_{12}P_2$, named for the British biochemist Sir Arthur Harden (1865-1940) and the Australian biochemist William John Young (1878-1942)

Harkins rule
named for the US chemist William Draper Harkins (1873-1951)

harm(al)-
derived from harmal (*Peganum harmala*), from *harmalah* (Arabic: harmal)

harmaline
$C_{13}H_{14}N_2O$, derived from harm(al)- and -in(e)

harmalol
$C_{12}H_{12}N_2O$, derived from harm(al)- and -in(e)

harmane
$C_{12}H_{10}N_2$, derived from harmine and -an(e)

harmine
$C_{13}H_{12}N_2O$, derived from harm(al-) and -in(e)

harmotome
$BaAl_2Si_6O_{16}\cdot 6H_2O$, derived from *harmos* (Greek: joint) and *tome* (Greek: section), from *temnein* (Greek: to cut) – referring to details of this mineral's crystal structure

Harper's alloy
(Bi,Pb,Sn), named for the US chemist H. W. Harper (1859-1943)

Harries ozonolysis reaction
named for the German chemist Carl Dietrich Harries (1866-1923)

Hartree
H, a unit named for the British physicist Douglas Rayner Hartree (1897-1958)

Hartree-Fock method (Hartree-Fock-Roothaan method)
named for the British physicist Douglas Rayner Hartree (1897-1958) and the Russian physicist Vladimir Alexandrovich Fock (1898-1974)

Hartree-Fock-Roothaan method (Hartree-Fock method)
named for the British physicist Douglas Rayner Hartree (1897-1958), the Russian physicist Vladimir Alexandrovich Fock (1898-1974), and the Dutch-US physicist Clemens C. J. Roothaan (born 1918)

hashish
derived from *hashish* (Arabic: dried grass)

Hass cyclopropane process
named for the 20th century US chemist Henry B. Hass

hassium
Hs, named for Hesse, Germany, where this element was prepared for the first time, derived from *Hassia* (New Latin: Hesse, Germany), and -ium

hastingsite
$NaCa_2(Fe^{II})_4Fe^{III}Si_6Al_2O_{22}(OH)_2$, named after this mineral's locality Dungannon, Hastings County, Ontario, Canada

hasubanan
$C_{16}H_{21}N$, derived from hasubanonine and -an(e)

hasubanonine
$C_{21}H_{27}NO_5$, derived from *hasu-no-ha-kazura* (Japanese: snake vine, *Stephania japonica*), -one, and -in(e)

Hatch-Slack cycle
named for the British physiologist T. F. Hatch (1901-1987) and the US plant physiologist D. A. Slack (born 1924)

hausmannite
Mn_3O_4, named for the German mineralogist Johann Friedrich Ludwig Hausmann (1782-1859)

haüyne
$(Na,Ca)_{4-8}Al_6Si_6(O,S)_{24}(SO_4,Cl)_{1-2}$, named for the French crystallographer René-Just Haüy (1743-1822)

Haworth methylation
named for the British chemist Sir Walter Norman Haworth (1883-1950)

Haworth phenanthrene synthesis
named for the British chemist Robert Downs Haworth (1898-1990)

Haworth projection
named for the British chemist Sir Walter Norman Haworth (1883-1950)

Hayashi rearrangement
named for the 20th century Japanese chemist Mosuke Hayashi

Hayem's solution
named for the French physician Georges Hayem (1841-1933)

HCB
C_6Cl_6, an abbreviation for hexachlorobenzene

HCH (BHC, lindane)
$C_6H_6Cl_6$, an abbreviation for hexachlorocyclohexane

HDI
$C_8H_{12}N_2O_2$, an abbreviation for hexamethylene diisocyanate

HDPE
an abbreviation for high-density polyethylene

HEC
an abbreviation for hydroxyethyl cellulose

Heck reaction
named for the US chemist Richard Fred Heck (born 1931)

Heck-Stille reaction
named for the US chemists Richard Fred Heck (born 1931) and J. K. Stille (born 1930)

hecogenin
$C_{27}H_{42}O_4$, derived (with contraction) from the genus name *Hechtia* (bromeliads), after the counselor to the Prussian king Julius Gottfried Conrad Hecht (1771-1837), and sapogenin

hect(o)-
derived from *hekaton* (Greek: hundred)

hectorite
$(Mg,Li)_3Si_4O_{10}(OH)_2$, named after this mineral's locality Hector, CA, USA

hedenbergite
$CaFeSi_2O_6$, named for the Swedish mineralogist M. A. Ludwig Hedenberg (1781-1809)

hederagenin (caulosapogenin, melanthigenin)
$C_{30}H_{48}O_4$, derived from hederin, -gen, and -in(e)

hederin (helixin)
$C_{41}H_{64}O_{11}$, derived from the genus name *Hedera* (ivy), from *hedera* (Latin: ivy), and -in(e)

HEDTA
$C_{10}H_{18}N_2O_7$, an abbreviation for *N*-(2-hydroxyethyl)ethylenediaminetriacetic acid

Hedvall effect
named for the Swedish chemist J. Arvid Hedvall (1888-1974)

HEED
an abbreviation for high-energy electron diffraction

HEIS
an abbreviation for hig-energy ion scattering

Heisenberg's principle
named for the German physicist Werner Karl Heisenberg (1901-1976)

Heitler-London-Slater-Pauling method
named for the German physicist Walter Heinrich Heitler (1904-1981), the German-US physicist Fritz Wolfgang London (1900-1954), the US physicist John Clarke Slater (1900-1978), and the US chemist Linus Carl Pauling (1901-1994)

helenalin
$C_{15}H_{18}O_4$, derived (with contraction) from the genus name *Helenium* (herbs), ultimately perhaps from *helene* (Greek: wicker basket), from *eilein* (Greek: to wind, to roll), and -in(e)

helenynolic acid
$C_{18}H_{30}O_3$, derived from the genus name *Helichrysum* (strawflower), from heli(co)-, -ene, -yne, and -ol

Helferich method
named for the German chemist Burckhardt Helferich (1887-1982)

helianthine (methyl orange)
$C_{14}H_{14}N_3NaO_3S$, named for the genus name *Helianthus* (sunflower), from heli(o)- and anth(o)-, and -in(e) – referring to this dye's orange color

helicene
derived from heli(co)- and -ene

helicin (spirein)
$C_{13}H_{16}O_7$, derived from heli(co)- and -in(e) – referring to this compound's derivation from spiraeic acid

helicity
derived from helix

heli(co)-
derived from *helix* (Greek: spiral)

heli(o)-
derived from *helios* (Greek: sun)

heliodor (golden beryl) $Be_3Al_2Si_6O_{18}$, derived from heli(o)- and *dosis* (Greek: gift) – referring to this mineral's golden-yellow color

helion
$^3He^{2+}$, derived from helium and ^{3}-on

heliosupine
$C_{20}H_{31}NO_7$, derived (with contraction) from the species name *Heliotropium supinum* L. (dwarf heliotrope), from heli(o)-, -trope, and *supinus* (Latin: lying on the back), and -in(e)

heliotrope
SiO_2, derived from heli(o)- and -trope – referring to this mineral's red color

heliotropin
$C_8H_6O_3$, derived from the genus name *Heliotropium* (herbs, shrubs), from heli(o)- and -trope, and -in(e)

helium
He, derived from heli(o)- and -ium –

referring to the discovery of this element in the sun's chromosphere

helix
derived from *helix* (Greek: spiral)

helixin (hederin)
$C_{41}H_{64}O_{11}$, derived from the specific epithet of the species name *Hedera helix* L. (English ivy), from heli(co)-, and -in(e)

Hell-Volhard-Zelinsky reaction
named for the German chemists Carl Magnus von Hell (1849-1926) and Jacob Volhard (1834-1910), and the Russian chemist Nikolay Dmitrievich Zelinsky (1861-1953)

Helmholtz equation
named for the German physicist Hermann Ludwig Ferdinand von Helmholtz (1821-1894)

helminth(o)-
derived from *helmis* (Greek: intestinal worm, parasitic worm)

helminthosporal
$C_{15}H_{22}O_2$, derived from helminthospor(o)- and -al

helminthosporin
$C_{15}H_{10}O_5$, derived from helminthospor(o)- and -in(e)

helminthospor(o)-
derived from the fungal genus name *Helminthosporium*, from helminth(o)- and spor(o)-

helveticoside (erysimin)
$C_{29}H_{42}O_9$, derived from the specific epithet of the species name *Erysimum helveticum* (Jacq.) DC. (Swiss wallflower), from *helveticus* (New Latin: Swiss), -ose, and -[3]ide

helo-
derived from *helos* (Greek: stake, palisade)

helodermin (helospectin I)
$C_{176}H_{285}N_{47}O_{49}$, derived from the genus name *Heloderma* (lizards), from helo- and derm(o)-, and -in(e)

helospectin
derived (with contraction) from the species name *Heloderma suspectum* (Gila monster), from helo-, derm(o)-, and *suspectus* (Latin: awesome), and -in(e)

helvetium
At, a name unsuccessfully suggested for astatine, derived from *Helvetia* (New Latin: Switzerland) and -ium

helvite
$Be_3Mn_4(SiO_4)_3S$, derived from *helvus* (Latin: light-bay-colored) and -ite – referring to this mineral's color

helvolic acid
$C_{33}H_{44}O_8$, derived from the subspecific epithet of the fungal species name *Aspergillus fumigatus* mut. *helvola*, from *helvolus* (Latin: pale yellow), diminutive of *helvus* (Latin: light-bay-colored)

hematein
$C_{16}H_{12}O_6$, derived from hemat(o)- and -ein(e)

hematin
$C_{34}H_{33}FeN_4O_5$, derived from hemat(o)- and -in(e)

hematite (haemat ite)
Fe_2O_3, derived from hemat(o)- and -ite – referring to this mineral's blood red color

hemat(o)-
derived from *haimatites* (Greek: bloodlike), from *haima* (Greek: blood)

hematoxylin
$C_{16}H_{14}O_6$, derived from the genus name *Haematoxylon* (trees), from hemat(o)- and xyl(o)-, and -in(e)

heme
$C_{34}H_{32}FeN_4O_4$, derived from hem(o)-

hementin
derived from the genus name *Haementeria* (leeches), from hem(o)- and enter(o)-, and -in(e)

hemerythrin
derived from hem(o)-, erythr(o)-, and -in(e)

hemi-
derived from *hemisys* (Greek: half)

hemicellulase
derived from hemicellulose and -ase

hemicellulose
derived from hemi- and cellulose – referring to the obsolete belief that this polysaccharide is a biosynthetic precursor of cellulose

hemi-Dewar biphenyl
$C_{12}H_{10}$, derived from hemi-, Dewar benzene, and biphenyl

hemimellitene
C_9H_{12}, derived from hemi-, mellite, and -ene – referring to this hydrocarbon's formation by formal reduction of hemimellitic acid

hemimellitic acid
$C_9H_6O_6$, derived from hemi- and mellitic acid – referring to the fact that this acid contains three carboxyl groups, i.e. half as many as mellitic acid

hemimorphism
derived from hemi- and -morphy

hemimorphite (calamine)
$Zn_4Si_2O_7(OH)_2{\cdot}H_2O$, derived from hemimorphic and -ite – referring to the hemimorphic crystals of this mineral

hemin
$C_{34}H_{32}ClFeN_4O_4$, derived from hem(o)- and -in(e)

hemipyocyanin
$C_{12}H_8N_2O$, derived from hemi- and pyocyanin

hemiterpene
C_5H_8, derived from hemi- and terpene – referring to the presence of just one isoprene (C_5) unit

hem(o)-
derived from *haima* (Greek: blood)

hemocyanin
derived from hem(o)- and cyanin

hemoglobin
derived from hem(o)- and globin

Hempel pipet
named for the German chemist Walther Matthias Hempel (1851-1916)

hen-
derived from *heis* (Greek: one)

hendeca-
derived from *hendeka* (Greek: eleven); in contemporary IUPAC nomenclature replaced by undeca-

Henderson-Hasselbalch equation
named for US chemist Lawrence Joseph Henderson (1878-1942) and the Danish physician Karl Albert Hasselbalch (1874-1962)

henicosa-
derived from *heneikosi* (Greek: twenty-one)

Henkel process (Raecke process)
named for the chemical company Henkel & Cie., Düsseldorf, Germany

Henry reaction
named for the Belgian chemist Louis Henry (1834-1913)

Henry's law
named for the British chemist William Henry (1775-1836)

hepar
an archaic name for any (liver-colored) metal sulfide, derived from hepa(r)-

hepa(r)-
derived from *hepar* (Greek: liver)

heparamine
derived from heparin and amine

heparin
derived from hepa(r)- and -in(e)

hepar sulfuris
K_2S_n, an archaic New Latin name for potassium oligosulfides, derived from hepar and sulfur

hepar test
derived from hepar sulfuris

hepatic air
H_2S, an archaic name for hydrogen sulfide, derived from hepar

hepaxanthin
$C_{20}H_{30}O_2$, derived from hepa(r)- and xanthin

HEPES
$C_8H_{18}N_2O_4S$, an abbreviation for 4-(2-hydroxyethyl)piperazine-1-(ethanesulfonic acid)

hept(a)-
derived from *hepta* (Greek: seven)

heptaconta-
derived from *heptakonta* (Greek: seventy)

heptadeca-
derived from *heptakaideka* (Greek: seventeen)

heptakis-
derived from *heptakis* (Greek: seven times)

heptalene
$C_{12}H_{10}$, derived from hepta- and -alene

heptane
C_7H_{16}, derived from hept(a)- and -an(e)

heptose
$C_7H_{14}O_7$, derived from hept(a)- and -ose

heptulose
$C_7H_{14}O_7$, derived from hept(a)- and -ulose

herb(i)-
derived from *herba* (Latin: grass, herb)

herbicide
derived from herb(i)- and -cide

herbimycin
derived from herb(i)- and –mycin

hercyn-
derived from the ancient *Hercynia Silva* (Hercynian Forest), the Romans' name for the wooded regions east of the Rhine

hercynine (herzynine)
$C_9H_{15}N_3O_2$, ultimately derived from hercyn- and -in(e)

hercynite
$FeAl_2O_4$, named after its locality Poběžovice, Czech Republic, derived from hercyn-

herderite
$CaBePO_4(F,OH)$, named for the German mining official Siegmund August Wolfgang von Herder (1776-1838)

heroin
$C_{21}H_{23}NO_5$, originally coined as a trademark, from *heros* (Greek: hero) and -in(e) – referring to this compound's former use in a 'heroic' therapy of tuberculosis

HERON rearrangement
coined as an abbreviation for heteroatom rearrangements on nitrogen

herqueinone
$C_{20}H_{20}O_7$, derived from the specific epithet of the fungal species name *Penicillium herquei*, -in(e), and -one

Hershberg stirrer
named for the US chemist Emanuel Benjamin Hershberg (born 1908)

Herz reaction
named for the German chemist Richard Herz (1867-1936)

Herzig-Meyer method
named for the Austrian chemists J. Herzig (1853-1924) and H. L. Meyer (1871-1942)

hesper-
derived from the genus name *Hesperidium* (citrus fruit trees), from *Hesperides* (Greek: mythical paradisiacal garden with golden apples guarded by nymphs, the Hesperides)

hesperetin
$C_{16}H_{14}O_6$, derived from hesper- and -etin(e)

hesperidin
$C_{28}H_{34}O_{15}$, derived from hesper- and -idin(e)

Hess' law
named for the Swiss-Russian chemist Germain Henri Hess (1802-1850)

hessonite (essonite, cinnamon stone)
$Ca_3Al_2(SiO_4)_3$, from *hesson* (Greek: inferior, less), comparative of *heka* (Greek: slightly) – referring to the fact that this mineral is not as hard as true hyacinth

5-HETE
$C_{20}H_{32}O_3$, an abbreviation for 5-hydroxyeicosa-8,11,14-tetraenoic acid

heteratisine
$C_{22}H_{33}NO_5$, derived from heter(o)- and atisine

heter(o)-
derived from *heteros* (Greek: other, different)

heterocyclic
derived from heter(o)- and cyclic

heterodetic
derived from hetero- and *detos* (Greek: bound, linked)

heterogeneous
derived from *heterogenes* (Greek: of different kind)

heteroleptic
derived from heter(o)- and -leptic

heterolysis
derived from heter(o)- and lysis

heteropoly acid
derived from heter(o)-, poly-, and acid

heteroside
coined by contraction of heter(o)- and glycoside

heterosite
$FePO_4$, derived from heter(o)- and -ite – referring to this being the second manganese-containing mineral discovered in the same layer

hetisan
$C_{20}H_{27}N$, derived from hetisine and -an(e)

hetisine (delatine)
$C_{20}H_{27}NO_3$, coined by variation of heteratisine

HETP
an abbreviation for height equivalent to a theoretical plate

HETS
an abbreviation for height equivalent to a theoretical stage

heulandite
$CaAl_2Si_7O_{18}\cdot 6H_2O$, named for the British mineral collector John Henry Heuland (1777-1856)

Heusler's alloy
[Cu,Mn,(Al,Sn)], named for the German metallurgist Friedrich Heusler (1866-1947)

Hevesy-Paneth analysis
named for the Hungarian chemist George de Hevesy (1885-1966) and the Austrian chemist Friedrich Adolf Paneth (1887-1958)

hex(a)-
derived from *hex* (Greek: six)

hexacelsian
$Ba(AlSiO_4)_2$, coined by contraction of hexagonal celsian

hexaconta-
derived from *hexakonta* (Greek: sixty)

hexadeca-
derived from *hexakaideka* (Greek: sixteen)

hexahedro-
derived from hexahedron

hexakis-
derived from *hexakis* (Greek: six times)

hexaose
hexasaccharides, derived from hexa- and -ose

hexaprismo-
derived (with contraction) from hexagonal prism

hexitol
$C_6H_{14}O_6$, derived from hexa- and -itol

hexogen (cyclonite, RDX)
$C_3H_6N_6O_6$, coined by contraction of hexa- and nitrogen – referring to the presence of six nitrogen atoms in the molecule

hexose
$C_6H_{12}O_6$, derived from hexa- and -ose

Heyns catalyst
Pt/C, named for the German chemist Kurt Fritz Heyns (1908-2005)

HFS
an abbreviation for hyperfine structure

HGH (somatotropin)
an abbreviation for human growth hormone

hibane (beyerane)
$C_{20}H_{34}$, derived from hiba arborvitae (*Thujopsis dolobrata*), from *hiba* (Japanese: hiba arborvitae), and -an(e)

Hilbert-Johnson reaction
named for the US chemists G. E. Hilbert and Treat B. Johnson (1875-1947)

Hildebrand scale (eluotropic series)
named for the US chemist Joel Henry Hildebrand (1881-1983)

Hill system
named for the US chemist Edwin A. Hill (born 1850)

himachalane
$C_{15}H_{28}$, derived from Himalayan deodar (*Cedrus deodara*) and -an(e)

hingganite
$(Yb,Y)_2Be_2Si_2O_8(OH)_2$, named after this mineral's locality Hinggan, China

Hinsberg oxindole and oxiquinoline synthesis
named for the German chemist Oscar Heinrich Daniel Hinsberg (1857-1939)

Hinsberg sulfone synthesis
named for the German chemist Oscar Heinrich Daniel Hinsberg (1857-1939)

Hinsberg test
named for the German chemist Oscar Heinrich Daniel Hinsberg (1857-1939)

Hinsberg thiophene synthesis
named for the German chemist Oscar Heinrich Daniel Hinsberg (1857-1939)

Hinshelwood equation
named for the British chemist Sir Cyril Norman Hinshelwood (1897-1967)

hippadine (pratorine)
$C_{16}H_9NO_3$, derived from the genus name *Hippeastrum* (amaryllis), from *hippeus* (Greek: horseman) and astr(o)-, and -in(e)

hippagine (pancracine)
$C_{16}H_{17}NO_4$, derived from the genus name *Hippeastrum* (amaryllis), from *hippeus* (Greek: horseman), from hipp(o)-, and astr(o)-, and -in(e)

hipp(o)-
derived from *hippos* (Greek: horse)

hippuric acid
$C_9H_9NO_3$, derived from hipp(o)- and urine

hippuricase (histozyme)
derived from hippuric acid and -ase

hirsutic acid
derived from the specific epithet of the species name *Stereum hirsutum* (false turkey tail), from *hirsutus* (Latin: hirsute), ultimately from *horrere* (Latin: to bristle)

hirudin
derived from the genus name *Hirudo* (leeches), from *hirudo* (Latin: leech), and -in(e)

histamine
$C_5H_9N_3$, derived from hist(o)- and amine

histidine
$C_6H_9N_3O_2$, derived from hist(o)- and -idin(e)

hist(o)-
derived from *histos* (Greek: web), from *histanai* (Greek: to cause to stand)

histone
derived from hist(o)- and -one

histopine
$C_9H_{13}N_3O_4$, derived (with contraction) from histidine and opine

histrionicotoxin
$C_{19}H_{25}NO$, derived from the specific epithet

of the species name *Dendrobates histrionicus* (poison arrow frog), ultimately from *histrio* (Latin: actor), of Etruscan origin, and toxin

Hittorf phosphorus
P_n, named for the German chemist and physicist Johann Wilhelm Hittorf (1824-1914)

Hittorf transfer number
named for the German chemist and physicist Johann Wilhelm Hittorf (1824-1914)

HMDI
$C_8H_{12}N_2O_2$, an abbreviation for hexamethylene diisocyanate

HMDS
$C_6H_{19}NSi_2$, an abbreviation for hexamethyldisilazane

HMO
an abbreviation for Hückel molecular orbital

HMPA (HMPT)
$C_6H_{18}N_3OP$, an abbreviation for hexamethylphosphoramide

HMPT (HMPA)
$C_6H_{18}N_3OP$, an abbreviation for hexamethylphosphoric triamide

HMX (octogen)
$C_4H_8N_8O_8$, a British military code for octogen, coined as an abbreviation for high-molecular weight research department explosive

HNE
$C_9H_{16}O_2$, an abbreviation for 4-hydroxynon-2-enal

Hoch-Campbell aziridine synthesis
named for the 20th century French chemist Joseph Hoch and the 20th century US chemist Kenneth N. Campbell

Hock cleavage
named for the German chemist Heinrich Hock (1887-1971)

HOESY
an abbreviation for heteronuclear Overhauser effect spectroscopy

Hofmann degradation (elimination)
named for the German chemist August Wilhelm von Hofmann (1818-1892)

Hofmann isonitrile synthesis
named for the German chemist August Wilhelm von Hofmann (1818-1892)

Hofmann-Löffler-Freytag reaction
named for the German chemists August Wilhelm von Hofmann (1818-1892), Karl Löffler (1878-1910), and Curt Freytag (born 1884)

Hofmann-Martius rearrangement
named for the German chemists August Wilhelm von Hofmann (1818-1892) and Carl Alexander Martius (1838-1920)

Hofmann reaction
named for the German chemist August Wilhelm von Hofmann (1818-1892)

Hofmann rearrangement
named for the German chemist August Wilhelm von Hofmann (1818-1892)

Hofmann-Sand reactions
named for the German chemists August Wilhelm von Hofmann (1818-1892) and J. W. Sand (1878-1917)

Hofmann's rule
named for the German chemist August Wilhelm von Hofmann (1818-1892)

Hofmeister series
named for the German physiologist Franz Hofmeister (1850-1922)

hollandite
$Ba(Mn^{IV},Mn^{II})_8O_{16}$, named for the director of the Indian Geologic Survey, Sir Thomas Henry Holland (1868-1947)

holmia
Ho_2O_3, derived from holmium and -a

holmium
Ho, named for Stockholm, Sweden, derived from *Holmia* (New Latin: Stockholm, Sweden) and -ium

hol(o)-
derived from *holos* (Greek: complete)

holoenzyme
derived from hol(o)- and enzyme

holomycin
$C_7H_6N_2O_2S_2$, a name coined without any stated reason

holophosphoric acid
H_5PO_5, derived from hol(o)- and phosphoric acid

holoprotein
derived from hol(o)- and protein

holoside
derived from hol(o)- and glycoside

holostane
$C_{30}H_{50}O_2$, coined by contraction of holothurin, sterol, and -an(e)

holothurin
$C_{50}H_{82}O_{26}S$, derived from the genus name *Holothuria* (sea cucumbers), from *holothourion* (Greek: a water polyp), and -in(e)

homarine
$C_7H_7NO_2$, derived from the genus name *Homarus* (lobster), ultimately from *humarr* (Old Norse: lobster), and -in(e)

Homberg's alloy
(Pb,Bi,Sn), named for the German physician and chemist Wilhelm Homberg (1652-1715)

homilite
$Ca_2(Fe,Mg)B_2Si_2O_{10}$, derived from *homilein* (Greek: to consort with), from *homilos* (Greek: crowd, assembly), and -ite – referring to this mineral's close association with meliphanite and erdmannite

hominal
coined by contraction of homovicinal

HOMO
an abbreviation for highest occupied molecular orbital

¹homo-
derived from *homos* (Greek: same)

²homo-
a prefix coined by contraction of homologous

homochelidonine
$C_{21}H_{23}NO_5$, derived from ²homo- and chelidonine

homocyclic
derived from ¹homo- and cyclic

homocysteine
$C_4H_9NO_2S$, derived from ²homo- and cysteine

homocystine
$C_8H_{16}N_2O_4S_2$, derived from ²homo- and cystine

homodetic
derived from [1]homo- and *detos* (Greek: bound, linked)

homoeriodictyol
$C_{16}H_{14}O_6$, derived from [2]homo- and eriodictyol

homogeneous
derived from *homogenes* (Greek: of the same kind)

homogentisic acid (alcapton)
$C_8H_8O_4$, derived from [2]homo- and gentisic acid

homoleptic
derived from [1]homo- and -leptic

homologous
derived from *homologos* (Greek: agreeing, of one mind)

homolysis
derived from [1]homo- and lysis

homopoly acid
derived from [1]homo-, poly-, and acid

homoserine
$C_4H_9NO_3$, derived from [2]homo- and serine

homotopic
derived from [1]homo- and -topic

homotrichione
$C_{19}H_{18}O_8$, derived from [2]homo- and trichione

homovanillic acid
$C_9H_{10}O_4$, derived from [2]homo- and vanillic acid

HON
$C_5H_9NO_4$, an abbreviation for 5-hydroxy-4-oxonorvaline

Hooker oxidation
named for the British-US chemist Samuel Cox Hooker (1864-1935)

hopane
$C_{30}H_{52}$, derived from the genus name *Hopea* (tropical trees), after the Scottish physician and botanist John Hope (1725-1786), and -an(e)

hopantenic acid
$C_{10}H_{19}NO_5$, coined by contraction of homopantothenic acid

hopcalite
(MnO_2,CuO,Co_2O_3,Ag_2O), coined as a trademark by contraction of Johns Hopkins University, University of California, and -ite

hopeite
$Zn_3(PO_4)_2 \cdot 4H_2O$, named for the Scottish chemist Thomas Charles Hope (1766-1844)

horde(n)-
derived from the genus name *Hordeum* (grasses), from *hordeum* (Latin: barley)

hordein
derived from horde(n)- and -in(e)

hordenine
$C_{10}H_{15}NO$, derived from horde(n)- and -in(e)

Horecker pathway
named for the US biochemist Bernard Leonard Horecker (born 1914)

hormone
ultimately derived from *horme* (Greek: impulse) and -one

hornblende
$Ca_2(Fe^{II},Mg)_4(Al,Fe^{III})(Si_7Al)O_{22}(OH,F)_2$, derived from horn and blende

Horner reaction
named for the German Chemist Leopold Horner (1911-2005)

Horner-Wadsworth-Emmons reaction
named for the German Chemist Leopold Horner (1911-2005) and the US chemists William Steele Wadsworth, Jr. (born 1927) and William David Emmons (1924-2001)

Hosomi-Sakurai reaction
named for the Japanese chemists Akira Hosomi and Hideki Sakurai (born 1931)

Houben-Fischer synthesis
named for the German chemists Josef Houben (1875-1940) and W. Fischer

Houben-Hoesch synthesis
named for the German chemists Josef Houben (1875-1940) and Kurt Hoesch (1882-1932)

Houdry cracking process
named for the French-US chemist Eugene Jules Houdry (1892-1962)

H-oxime
named for the German chemist Ilse Hagedorn (born 1921)

HPC
an abbreviation for hydroxypropyl cellulose

5-HPETE
$C_{20}H_{32}O_4$, an abbreviation for 5-hydroperoxyeicosa-8,11,14-tetraenoic acid

HPLC
an abbreviation for high-pressure liquid chromatography or high-performance liquid chromatography

HPMC
an abbreviation for 2-hydroxypropyl methyl cellulose

HPTLC
an abbreviation for high-pressure thin layer chromatography or high-performance thin layer chromatography

HQNO
$C_{16}H_{21}NO_2$, an abbreviation for 2-heptylquinolin-4-ol 1-oxide

HREELS (HRELS)
an abbreviation for high-resolution electron energy loss spectroscopy

HSAB
an abbreviation for hard and soft acids and bases

5-HT
$C_{10}H_{12}N_2O$, an abbreviation for 5-hydroxytryptamine

5-HTP
$C_{11}H_{12}N_2O_3$, an abbreviation for 5-hydroxytryptophane

Huang-Minlon reduction
named for the Chinese chemist Huang Minlon (1898-1979)

hübnerite
$MnWO_4$, named for the 19th century German foundry superintendent Adolf Hübner

Hückel-Möbius concept
named for the German chemist Erich Armand Arthur Joseph Hückel (1896-1980) and the German mathematician August Ferdinand Möbius (1790-1868)

Hückel's rule
named for the German chemist Erich Armand Arthur Joseph Hückel (1896-1980)

Hudson's rules
named for the US chemist Claude Silbert

Hudson (1881-1952)

Hughes-Ingold rule
named for the British chemists Edward David Hughes (1906-1963) and Sir Christopher Kelk Ingold (1893-1970)

Hume-Rothery phase
named for the British metallurgist William Hume-Rothery (1899-1968)

humic acid (allomelanin)
derived from *humus* (Latin: soil)

humin
derived from *humus* (Latin: soil) and -in(e)

humite
$(Mg,Fe)_7(SiO_4)_3(F,OH)_2$, named for the British mineral collector Abraham Hume (1749-1838)

hummerite
$KMgV_5O_{14}{\cdot}8H_2O$, named after this mineral's locality Hummer mine, Montrose County, CO, USA

humul-
derived from the genus name *Humulus* (hop), of Germanic origin, cf. *humli* (Old Norse: hop)

humulane
$C_{15}H_{30}$, derived from humul- and -an(e)

humulene
$C_{15}H_{24}$, derived from humul- and -ene

humulone
$C_{21}H_{30}O_5$, derived from humul- and -one

humus
derived from *humus* (Latin: earth, ground, soil)

Hund's rule
named for the German physicist Friedrich Hermann Hund (1896-1997)

Hünig's base
$C_8H_{19}N$, mamed after the German chemist Siegfried Helmut Hünig (born 1921)

Hunsdiecker reaction (Borodin reaction)
named for the German chemists Heinz Hunsdiecker (1904-1981) and Cläre Hunsdiecker (1903-1995)

huperzine A (selagine)
$C_{15}H_{18}N_2O$, derived from the genus name *Huperzia* (mosses), after the German botanist Johann Peter Huperz (1771-1816), and -in(e)

hyacinth
$ZrSiO_4$, ultimately derived from *hyakinthos* (Greek: precious stone, flowering plant) – referring to this mineral's yellow, orange, or red color

hyalin
derived from hyal(o)- and -in(e)

hyalite
$SiO_2{\cdot}nH_2O$, derived from hyal(o)- and -lite – referring to the fact that this mineral is colorless and sometimes clear as glass

hyal(o)-
derived from *hyalos* (Greek: glass)

hyalobiuronic acid
$C_{12}H_{21}NO_{11}$, derived from hyal(o)-, bi-, and -uronic acid

hyalophane
$(K,Ba)Al_2(SiO_4)_2$, derived from hyal(o)- and -phan(e) – referring to this mineral's glass-like appearance

hyaluronic acid
derived from hyal(o)- and -uronic acid

hyaluronidase ([2]kinetin)
derived from hyaluronic acid, [3]-ide, and -ase

hybrid
derived from *hybrida* (Latin: mongrel), probably of non-Indo-European origin

hydantoin
$C_3H_4N_2O_2$, coined by contraction of hydrogen and allantoin

hydnocarpic acid
$C_{16}H_{28}O_2$, derived from the genus name *Hydnocarpus* (trees), from *hydnon* (Greek: truffle) and carp(o)-

hydracrylic acid
$C_3H_6O_3$, derived from [1]hydr(o)- and acrylic acid – referring to this compound's formation by addition of water to acrylic acid

hydrallostane
$C_{21}H_{32}O_5$, derived from [2]hydr(o)-, allo-, sterol, and -an(e)

hydrargillite (gibbsite)
$Al(OH)_3$, derived from [1]hydr(o)- and argillite

hydrargyr(o)-
derived from hydrargyrum

hydrargyrum
Hg, New Latin name for mercury, derived from *hydrargyrum* (Greek: mercury), from [1]hydr(o)- and argyr(o)-

hydrast-
derived from the genus name *Hydrastis* (goldenseal), from [1]hydr(o)-

hydrastine
$C_{21}H_{21}NO_6$, derived from hydrast- and -in(e)

hydrastinine
$C_{11}H_{13}NO_3$, derived from hydrast- and -inin(e)

hydratase
derived from hydratization and -ase

hydrate
derived from [1]hydr(o)- and -ate

hydrazine (diazane)
N_2H_4, derived from [2]hydr(o), az(a)-, and -in(e)

hydrazo-
−NH−NH−, derived from [2]hydr(o)- and az(o)-

hydrazone
coined by insertion of the letter O into hydrazine to reflect the isomerism between hydrazones and azo compounds

hydride
H^-, derived from hydrogen and [2]-ide

hydrindantin
$C_{18}H_{10}O_6$, coined by contraction of 2,2'-dihydroxy[2,2'-biindan]-1,1',3,3'-(2*H*,2'*H*)-tetrone and -in(e)

[1]hydr(o)-
derived from *hydor* (Greek: water)

[2]hydr(o)-
derived from hydrogen

hydrocarbostyril
C_9H_9NO, derived from [2]hydr(o)- and carbostyril

hydrocinnamic acid
$C_9H_{10}O_2$, derived from [2]hydr(o)- and

cinnamic acid

hydrocortisone
$C_{21}H_{30}O_5$, derived from [2]hydr(o)- and cortisone

hydrocotarnine
$C_{12}H_{15}NO_3$, derived from [2]hydr(o)- and cotarnine

hydrogel
deribed from [1]hydr(o)- and gel

hydrogen
H, derived from [1]hydr(o)- and -gen

hydrogenase
derived from hydrogenation and -ase

hydrogenium
H, New Latin name for hydrogen

hydrol
an obsolete name for secondary alcohol, from [2]hydr(o)- and -ol – referring to the presence of a hydrogen atom on the hydroxyl group-bearing carbon atom

hydrolase
derived from hydrolysis and -ase

hydrolyase
derived from [1]hydr(o)- and lyase

hydrolysis
derived from [1]hydr(o)- and lysis

hydron
H^+, derived from [2]hydr(o)- and [3]-on

hydronium (oxonium, oxidanium)
H_3O^+, derived from hydron and -ium

hydrophilic
derived from [1]hydr(o)- and -philic

hydrophilite
$CaCl_2$, derived from hydrophilic and -ite – referring to the hygroscopic nature of this mineral

hydrophobic
derived from [1]hydr(o)- and -phobic

hydroquinidine
$C_{20}H_{36}N_2O_2$, derived from [2]hydr(o)- and quinidine

hydroquinine
$C_{20}H_{36}N_2O_2$, derived from [2]hydr(o)- and quinine

hydroquinone
$C_6H_6O_2$, derived from [2]hydr(o)- and quinone

hydrotropy
derived from [1]hydr(o)- and -tropy

hydroxamic acid
coined by contraction of hydroxylamine and acid

hydroxide
OH^-, derived (with contraction) from hydrogen and oxide

hydroximic acid
coined by contraction of hydroxyimidic acid

hydroxy-
HO–, derived from hydrogen and oxygen

hydroxyl-
HO–, derived from hydrogen, oxygen, and -yl

hydroxylamine
H_3NO, derived (somewhat irregularly) from hydroxyl- and amine

hydroxylapatite (durapatite)
$Ca_5(PO_4)_3(OH)$, derived from hydroxyl- and

apatite

hydroxylapophyllite
$KCa_4(Si_4O_{10})_2(OH)\cdot 8H_2O$, derived from hydroxyl- and apophyllite

hydroxylupanine
$C_{15}H_{24}N_2O_2$, derived from hydroxy- and lupanine

hydroxylysine
$C_6H_{14}N_2O_3$, derived from hydroxy- and lysine

hydroxyproline
$C_5H_9NO_3$, derived from hydroxy- and proline

hydrozincite
$Zn_5(CO_3)_2(OH)_6$, derived from [1]hydr(o)- and zincite

hygrine
$C_8H_{15}NO$, derived from hygr(o)- and -in(e) – referring to the fact that this alkaloid is a liquid at ambient temperature

hygr(o)-
derived from *hygros* (Greek: wet, moist, humid)

hygromycin (homomycin)
derived from the specific epithet of the bacterial species name *Streptomyces hygroscopicus* (Jensen) Waksman & Henrici, from hygroscopic, and -mycin

hygrophylline
$C_{18}H_{27}NO_6$, derived from the specific epithet of the species name *Senecio hygrophilus* Dyer and Sm. (a shrub), from hygr(o)- and -phile, and -in(e)

hygroscope
derived from hygr(o)- and -scope

hygroscopic
derived from hygroscope – referring to the usefulness of water-absorbing substances in hygroscopes

hyo-
derived from *hys* (Greek: swine)

hy(o)-
derived from *hy* (Greek: upsilon)

hyodeoxycholic acid
$C_{24}H_{40}O_4$, derived from hyo-, de(s)-, oxy-, and cholic acid

hyoscyamine (daturine, duboisine)
$C_{17}H_{23}NO_3$, derived from the genus name *Hyoscyamus* (henbane), ultimately from hyo- and *kyamos* (Greek: bean), and -in(e)

hypaphorine
$C_{14}H_{18}N_2O_2$, derived from the genus name *Hypaphorus* Hassk. (bean plants), and -in(e)

hyper-
derived from *hyper* (Greek: over, above, beyond)

hyperforin
$C_{35}H_{52}O_4$, derived (with contraction) from the species name *Hypericum perforatum* (St. John's wort), from *hyperikon* (Greek: St. John's wort), from hyper- and *ereike* (Greek: heath, heather), and *perforatus* (Latin: perforated), from *forare* (Latin: to drill), and -in(e)

hypergolic
derived from hyper- and -ergolic

hypericin
$C_{30}H_{16}O_8$, derived from the genus name *Hypericum* (herbs, shrubs), from *hyperikon* (Greek: St. John's wort), from hyper- and *ereike* (Greek: heath, heather), and -in(e)

hyperconjugation
derived from hyper- and conjugation

hyperoxide
O_2^-, derived from hyper- and oxide

hypersthene (bronzite)
(Fe,Mg)SiO_3, derived from hyper- and -sthene – referring to the fact that this mineral is harder than hornblende with which it is often confused

hypertensin
$C_{49}H_{70}N_{14}O_{11}$, derived from hypertension and -in(e)

hypertonic
derived from hyper- and -tonic

hypervalent
derived from hyper- and -valent

hypho-
derived from *hyphe* (Greek: web)

hypn(o)-
derived from *hypnos* (Greek: sleep)

hypnone (acetophenone)
C_8H_8O, derived from hypn(o)- and -one – referring to the sleep-inducing effects of this compound

hyp(o)-
derived from *hypo* (Greek: under, below, from below)

hypobromic acid
HBrO, derived from hyp(o)- and bromic acid

hypochloric acid
HClO, derived from hyp(o)- and chloric acid

hypoglycine
$C_7H_{11}NO_2$, derived from hypoglycemic and -in(e)

hypoiodic acid
HIO, derived from hyp(o)- and iodic acid

hypon
Uuo, a name suggested for eka-radon, ununoctium, from hyper- and [1]-on

hyponitrous acid
$H_2N_2O_2$, derived from hyp(o)- and nitrous acid

hypophosphoric acid
$H_4O_6P_2$, derived from hyp(o)- and phosphoric acid

hypophosphorous acid
H_3O_2P, derived from hyp(o)- and phosphorous acid

hypotonic
derived from hyp(o)- and -tonic

hypoxanthine
$C_5H_4N_4O$, derived from hyp(o)- and xanthine – referring to the fact that hypoxanthine is the biosynthetic precursor of xanthine

hyprolose
coined by contraction of 2-hydroxypropyl cellulose

hypromellose
coined by contraction of 2-hydroxypropyl methyl cellulose

hypsochrome
derived from *hypsos* (Greek: height) and -chrome

hypusine
$C_{10}H_{23}N_3O_3$, coined by contraction of

hydroxyputrescine and lysine

hysteresis
derived from *hysteresis* (Greek: shortcoming, deficiency, need), from *hysterein* (Greek: to come late, to lag), from *hysteros* (Greek: later)

I

iatro-
derived from *iatros* (Greek: physician), from *iasthai* (Greek: to heal)

iatrochemistry
derived from iatro- and chemistry

iboga-
derived from the specific epithet of the species name *Tabernanthe iboga* Baill. (iboga), from a native plant name in Central Africa

ibogaine
$C_{20}H_{26}N_2O$, derived from iboga- and -in(e)

ibogamine
$C_{19}H_{24}N_2$, derived from iboga- and amine

ibotenic acid
$C_5H_6N_2O_4$, derived from *ibotengutake* (Japanese: the toadstool *Amanita ibotengutake* sp. nov., earlier misidentified as other *Amanita* species)

[1]IC
an abbreviation for internal conversion

[2]IC
an abbreviation for ion chromatography

-ic
derived from *-icus* (Latin: having the character or the form of)

-icetin(e)
variant of -in(e)

ichthy(o)-
derived from *ichthys* (Greek: fish)

-icin
an arbitrary suffix derived by variation of -mycin

-icin(e)
an arbitrary suffix derived by variation of -in(e)

icos(a)-
derived from *eikosi* (Greek: twenty)

icosahedro-
derived from icosahedron

icosane
$C_{20}H_{42}$, derived from icos(a)- and -an(e)

icosanoid
derived from icosane and -oid

ICP
an abbreviation for inductively coupled plasma

ICR
an abbreviation for ion cyclotron resonance

[1]-ide
derived from *-eides* (Greek: -like, -shaped)

[2]-ide
an arbitrary suffix derived from oxide

[3]-ide
an arbitrary suffix derived from anhydride

-idin(e)
an arbitrary suffix patterned after pyridine

idiochromatic
derived from *idios* (Greek: own, particular, personal) and -chromatic

iditol
$C_6H_{14}O_6$, derived from idose and -itol

ido-
derived from *eidos* (Greek: form, shape)

idocrase (vesuvianite)
$(Ca,Na)_{19}(Al,Mg,Fe)_{13}(SiO_4)_{10}(Si_2O_7)_4(OH,F,O)_{10}$, derived from ido- and -crase – referring to the complexity of this mineral

idose
$C_6H_{12}O_6$, coined by contraction of *idem* (Latin: the same) and gulose – referring to the stereoisomerism of idose and gulose

IDP
$C_{10}H_{14}N_4O_{11}P_2$, an abbreviation for inosine 5'-diphosphate

iduronic acid
$C_6H_{10}O_7$, derived from idose and -uronic acid

[1]IEC
an abbreviation for ion exchange chromatography

[2]IEC
an abbreviation for ion exclusion chromatography

IEE (ESCA)
an abbreviation for induced electron emission

IEF
an abbreviation for isoelectric focusing

I-effect
an abbreviation for inductive effect

IEP (IP)
an abbreviation for isoelectric point

IESS
an abbreviation for inelastic ion scattering spectroscopy

IETS
an abbreviation for inelastic electron tunneling spectroscopy

IFA
an abbreviation for immunofluorescence analysis

IFN
an abbreviation for interferon

Ig
an abbreviation for immunoglobulin

IL
an abbreviation for interleukin

-il(e)
ultimately derived from *-ilis* (Latin suffix denoting function or relationship)

ilimaquinone
$C_{22}H_{30}O_4$, derived from ilima (the shrub *Sida fallax*), from *ilima* (Hawaiian: ilima), and quinone – referring to the golden yellow color, rather than the source, of this compound

illinium
Pm, a name unsuccessfully suggested for promethium, after the state of Illinois, USA

illite
$(K,H_3O)Al_2(Si_3Al)O_{10}(H_2O,OH)_2$, named after this mineral's locality, the state of Illinois, USA

illudin
derived from the specific epithet of the species name *Omphalotus illudens* (the mushroom Jack O'-Lantern), from *illudens*

(Latin: mocking), from *illudere* (Latin: to mock), and -in(e)

illudol
$C_{15}H_{24}O_3$, derived from the specific epithet of the fungal species name *Clitocybe illudens*, from *illudens* (Latin: mocking), from *illudere* (Latin: to mock), and -ol

illudosin
$C_{15}H_{24}O_3$, coined by variation of illudol

ilmenite
$FeTiO_3$, named after this mineral's locality, the Ilmen Mountains, Russia

ILS
an abbreviation for ionization loss spectroscopy

ilvaite
$CaFe^{III}Fe^{II}{}_2Si_2O_7(OH)_2{\cdot}H_2O$, named after this mineral's locality, the island of Elba, Italy (Latin: *Ilva*)

imbibition
derived from [2]in- and *bibere* (Latin: to drink)

imene (nitrene)
RN, coined by contraction of imine and carbene

imidazole
$C_3H_4N_2$, derived from imide, az(a)-, and -ol(e)

imidazoline
$C_3H_6N_2$, derived from imidazole and -in(e)

imidazolidine
$C_3H_8N_2$, derived from imidazole and -idin(e)

imide
derived from imine and [2]-ide

imidogen
NH, derived from imide and -gen

imine
coined by alteration of amine

IMMA
an abbreviation for ion microprobe mass analysis

immunoglobulin
derived from immune and globulin

IMP
$C_{10}H_{13}N_4O_8P$, an abbreviation for inosine 5'-monophosphate

imperatorin
$C_{16}H_{14}O_4$, derived from the genus name *Imperatoria* (masterwort), from *imperator* (Latin: emperor), and -in(e)

imperialine
$C_{27}H_{43}NO_3$, derived from the specific epithet of the species name *Fritillaria imperialis* L. (crown imperial), from *imperialis* (Latin: imperial), from *imperator* (Latin: emperor)

Imperial Smelting process
named for the chemical company Imperial Smelting Processes, Cheltenham, UK where this process was developed

[1]**in- (ig-, il-, im-, ir-)**
derived from *in- (ig-, il-, im-, ir-)* (Latin: non-)

[2]**in- (il-, im-, ir-)**
derived from *in* (Latin: into, on)

INAA
an abbreviation for instrumental neutron activation analysis

INADEQUATE
an acronym for incredible natural abundance double quantum transfer experiment

indacene
$C_{12}H_8$, derived from indene and -acene

indaconitine
$C_{34}H_{47}NO_{10}$, derived from ind(o)-, aconit-, and -in(e)

indane
C_9H_{10}, derived from indole and -an(e)

indanthrene
coined as a trademark by contraction of indigo and -anthrene

indazole
$C_7H_6N_2$, derived from indene, az(a)-, and -ol(e)

indene
C_9H_8, derived from indole and -ene

indican
$C_{14}H_{17}NO_6$, derived from indigo and -an(e)

indigane
InH_3, derived from indium (with reference to this element's name root indigo) and -an(e); thus coined to avoid confusion with indane

indigo
$C_{16}H_{10}N_2O_2$, derived from the genus name *Indigofera* (indigo plant), ultimately from *indikon pharmakon* (Greek: literally Indian dye)

indigotin
$C_{16}H_{10}N_2O_2$, derived from indigo and -in(e)

indium
In, derived from indigo – referring to the two indigo-blue lines in this element's spectrum

INDO
an abbreviation for intermediate neglect of differential overlap

ind(o)-
derived from indigo

indole
C_8H_7N, derived from indigo and -ol(e)

indolizine (pyrrocoline)
C_8H_7N, derived from indole; patterned after quinolizine

indolmycin
$C_{14}H_{15}N_3O_2$, derived from indole and -mycin

INDOR
an abbreviation for internuclear double resonance

indoxyl
C_8H_7NO, coined by contraction of indigo and hydroxyl

indulin
derived from indigo, *-ulus* (Latin diminutive suffix), and -in(e)

-in(e)
ultimately derived from *-inus* (Latin suffix meaning belonging to or being similar to)

inesite
$Ca_2Mn_7Si_{10}O_{28}(OH)_2 \cdot 5H_2O$, derived from in(os)- and -ite – referring to the flesh color and fibrous appearance of this mineral

inflammable air
H_2, an archaic name for hydrogen gas, derived from air

INH
$C_6H_7N_3O$, an abbreviation for isonicotinic hydrazide

inhibin
derived from inhibition and -in(e)

inifer
coined by contraction of initiator and transfer agent

iniferter
coined by contraction of initiator, transfer agent, and terminator

initer
coined by contraction of initiator and terminator

INN
an abbreviation for international nonproprietary (drug) name

inorganic
derived from [1]in- and organic

in(os)-
derived from *is* (Greek: fiber, muscle, sinew, tendon)

inosilicate
derived from in(os)- and silicate – referring to these minerals' chain or band structure

inosine
$C_{10}H_{12}N_4O_5$, derived from in(os)- and -in(e)

inosinic acid (IMP)
$C_{10}H_{13}N_4O_8P$, derived from inosine

inositol
$C_6H_{12}O_6$, derived from in(os)-, -ose, and -itol

[1]INS
an abbreviation for inelastic neutron scattering

[2]INS
an abbreviation for ion neutralization spectroscopy

insecticide
derived from insect and -cide

in situ
(Latin: in place)

in statu nascendi
(Latin: in the state of being generated, literally in the state of being born)

insulin
derived from *insulae Langerhans* (New Latin: islets of Langerhans), from *insula* (Latin: island), and -in(e)

integrin
coined by contraction of integral protein

inter-
derived from *inter* (Latin: among, between)

intercalation
derived from *intercalare* (Latin: to insert), from inter- and *calendae, kalendae* (Latin: the first day of the month, month), literally the insertion of a day (or month) into the calendar – referring to the insertion of flat guest molecules into a helical host structure

interferon
derived from interference and [1]-on

interleukin
derived from inter-, leuc(o)-, and -in(e)

intermedin (melanotropin)
derived (with contraction) from *pars intermedia* (New Latin: literally intermediate part) and -in(e)

intermolecular
derived from inter- and -molecular

interstitial
derived from inter- and *stare* (Latin: to stand)

intra- (intro-)
derived from *intra- (intro-)* (Latin: within, inside)

intramolecular
derived from intra- and -molecular

inulase
derived from inulin and -ase

inulin
derived from the genus name *Inula* (elecampane), from *helenion* (Greek: elecampane), ultimately perhaps from *helene* (Greek: wicker basket), from *eilein* (Greek: to wind, to roll), and -in(e)

invertase (saccharase)
derived from to invert and -ase – referring to the fact that this enzyme catalyzes the hydrolysis of a dextrorotatory sucrose solution to a levorotatory mixture of glucose and fructose

iodane
HI, derived from iodine and -an(e)

iodargyrite
AgI, derived from ^{2}iod(o)-, argyr(o)-, and -ite

iodic acid
HIO_3, derived from iodine

iodine
I, derived from ^{1}iod(o)- and -in(e) – referring to the violet color of I_2 vapor

iodinin
$C_{12}H_8N_2O_4$, derived from the specific epithet of the bacterial species name *Chromobacterium iodinum*, from ^{1}iod(o)-, and -in(e)

^{1}iod(o)-
derived from *ioeides* (Greek: purple, violet colored)

^{2}iod(o)-
derived from iodine

iodoform
CHI_3, derived from ^{2}iod(o)- and -form

iodopsin
derived from ^{1}iod(o)-, ops-, and -in(e)

iodous acid
HIO_2, derived from iodine and -ous

iodum
I, New Latin name for iodine, from *Iod* (German: iodine) and -um

iolite (cordierite, dichroite)
$Mg_2Al_4Si_5O_{18}$, derived from *ion* (Greek: the flower violet) and -lite – referring to this mineral's sometimes violet color

ion
derived from *ion* (Greek: going), from *ienai* (Greek: to go) – referring to the mobility of ions in solution

ionium
^{230}Th, unsuccessfully suggested name for a thorium isotope – referring to this species' ionizing action

ion(o)-
derived from ion

ionogen
derived from ion and -gen

ionomer
coined by contraction of ionic polymer

ionomycin
$C_{40}H_{72}O_7$, derived from ion(o)- and -mycin – referring to this antibiotic's ionophoric properties

ionone
$C_{13}H_{20}O$, derived from *ion* (Greek: the flower violet) and -one – referring to the characteristic violet odor of these compounds

ionophore
derived from ion(o)- and -phore

invert sugar
a name referring to the fact that this solution of hydrolysis products of dextrorotatory sucrose is levorotatory

IP (IEP)
an abbreviation for isoelectric point

IPA (EPA)
$C_{20}H_{30}O_2$, an abbreviation for (*all-Z*)-icosa-5,8,11,14,17-pentaenoic acid

IPC
an abbreviation for ion pair chromatography

IPMA
an abbreviation for ion probe microanalysis

IPP
$C_5H_{12}O_7P_2$, an abbreviation for isopentenyl pyrophosphate

ips-
derived from the genus name *Ips* (beetles), from *ips* (Greek: woodworm)

ipsdienol
$C_{10}H_{16}O$, derived from ips-, di-, -ene, and -ol

ipse
derived from *ipse* (Latin: self)

ipsenol
$C_{10}H_{18}O$, derived from ips-, -ene, and -ol

ipso-
derived from ipse

IR
an abbreviation for infrared

-irane
derived from tri- and -an(e)

-irene
derived from tri- and -ene

iridane
$C_{10}H_{20}$, derived from iridoid and -an(e)

iridescent
derived from [1]irid(o)- and -escent

-iridine
derived from tri- and -idin(e)

iridium
Ir, ultimately derived from [1]irid(o)- and -ium – referring to the characteristic colors of iridium salt solutions

[1]irid(o)-
derived from *iris* (Greek: rainbow)

[2]irid(o)-
derived from iridium

iridocene
$C_{10}H_{10}Ir$, derived from [2]irid(o)- and -ocene

iridoid
derived (with contraction) from the genus name *Iridomyrmex* (ants), from [1]irid(o)- and *myrmex* (Greek: ant), and -oid

iridomyrmecin
$C_{10}H_{16}O_2$, derived from the genus name *Iridomyrmex* (ants), from [1]irid(o)- and *myrmex* (Greek: ant), and -in(e)

iridosmine (iridosmium)
(Ir,Os,Rh,Pt), coined by contraction of iridium and osmium

iridosmium (iridosmine)
(Ir,Os,Rh,Pt), coined by contraction of iridium and osmium

irigenin
$C_{18}H_{16}O_8$, derived from the genus name *Iris* (iris), from *iris* (Greek: rainbow), -gen, and -in(e)

-irine
derived from tri- and -in(e)

IRMA
an abbreviation for immunoradiometric assay

iron
Fe, a prehistoric Germanic word, probably of Venetian or Illyrian origin

irone
$C_{14}H_{22}O$, derived from the genus name *Iris* (iris), from *iris* (Greek: rainbow), and -one

Irving-Williams series
named for British 20th century chemists R. J. P. Williams and H. Irving

isanic acid (bolecic acid, erythrogenic acid)
$C_{18}H_{26}O_2$, derived from isano oil, from *isano* (West African native name of the tree *Ongokea klaineana*)

isatin
$C_8H_5NO_2$, derived from the genus name *Isatis* (woad), from *isatis* (Greek: woad), and -in(e)

isatogen
$C_8H_5NO_2$, derived from isatin and -gen

isatoic acid
$C_8H_7NO_4$, derived from isatin

isatoic anhydride
$C_8H_5NO_3$, derived from isatoic acid and anhydride

isatropic acid (diatropic acid)
$C_{18}H_{16}O_4$, derived from is(o)- and atropic acid

ISC
an abbreviation for intersystem crossing

isethionic acid
$C_2H_6O_4S$, derived from is(o)-, eth(a)-, and thi(o)-

is(o)-
derived from *isos* (Greek: equal)

isoalloxazine
$C_{10}H_6N_4O_2$, derived from is(o)- and alloxazine

isoamyl
$C_5H_{11}-$, derived from is(o)- and amyl

isobutane
C_4H_{10}, derived from is(o)- and butane

isochondrodendrine
$C_{36}H_{38}N_2O_6$, derived from is(o)-, the genus name *Chondrodendron* (South American vines), from chondr(o)- and dendr(o)-, and -in(e)

isochore
derived from is(o)- and *choros* (Greek: empty space, room)

isocorybulbine
$C_{21}H_{25}NO_4$, derived from is(o)- and corybulbine

isocorydine
$C_{20}H_{23}NO_4$, derived from is(o)- and corydine

isocorypalmine
$C_{20}H_{23}NO_4$, derived from is(o)- and corypalmine

isocratic
derived from *isokrates* (Greek: equally powerful) or from is(o)- and -crase

isocyclic
derived from is(o)- and cyclic

isodesmic
derived from is(o)- and desm(o)-

isodesmosine
$C_{24}H_{40}N_5O_8$, derived from is(o)- and [1]desmosine

isoelectronic
derived from is(o)- and electron

isoenzyme
coined by contraction of isodynamic enzyme

isoeugenol
$C_{10}H_{12}O_2$, derived from is(o)- and eugenol

isoflavan
$C_{15}H_{14}O$, derived from is(o)- and flavan

isofulminic acid (carboxime)
CHNO, derived from is(o)- and fulminic acid

isoleptic
derived from is(o)- and -leptic

isoleucine
$C_6H_{13}NO_2$, derived from is(o)- and leucine

isolobal
derived from is(o)- and lobe

isolysergic acid
$C_{16}H_{16}N_2O_2$, derived from is(o)- and lysergic acid

isolysine
$C_6H_{14}N_2O_2$, derived from is(o)- and lysine

isomaltol
$C_6H_6O_3$, derived from is(o)- and maltol

isomer
derived from is(o)- and -mer

isomerase
derived from isomerization and -ase

isomerose
$C_6H_{12}O_6$, derived (with contraction) from isomerization and -ose

isomorphous
derived from is(o)- and *morphe* (Greek: shape)

isooctane
C_8H_{18}, derived from is(o)- and octane

isophorone
$C_9H_{14}O$, derived from is(o)- and phorone

isophthalic acid
$C_8H_6O_4$, derived from is(o)- and phthalic acid

isopilosine
$C_{16}H_{18}N_2O_3$, derived from is(o)- and pilosine

isopimaric acid
$C_{20}H_{30}O_2$, derived from is(o)- and pimaric

acid

isopoly acid
derived from is(o)-, poly-, and acid

isoprene
C_5H_8, a name coined without a stated reason

isoprenoid
derived from isoprene and -oid

isopropenyl
C_3H_5-, derived from is(o)- and propenyl

isopropyl
C_3H_7-, derived from is(o)- and propyl

isopropylidene
$C_3H_6<$, derived from is(o)- and propylidene

isoquassin (picrasmin)
$C_{22}H_{28}O_6$, derived from is(o)- and quassin

isoquercitrin
$C_{21}H_{20}O_{12}$, derived from is(o)- and quercitrin

isoquinoline
C_9H_7N, derived from is(o)- and quinoline

isorubijervine
$C_{27}H_{43}NO_2$, derived from is(o)- and rubijervine

isosafrole
$C_{10}H_{10}O_2$, derived from is(o)- and safrole

isosorbide
$C_6H_{10}O_4$, derived from is(o)-, sorbitol, and [3]-ide

isosteric
derived from is(o)- and stere(o)-

isotachophoresis
derived from is(o)-, tach(o)-, and –phoresis

isotactic
derived from is(o)- and -tactic

isothebaine
$C_{19}H_{21}NO_3$, derived from is(o)- and thebaine

isotonic
derived from is(o)- and -tonic

isotope
derived from is(o)- and -tope

isotretinoin
$C_{20}H_{28}O_2$, derived from is(o)- and tretinoin

isotropic
derived from is(o)- and -tropic

isovaleric acid
$C_5H_{10}O_2$, derived from is(o)- and valeric acid

isovaline
$C_5H_{11}NO_2$, derived from is(o)- and valine

isoxerocomic acid
$C_{18}H_{12}O_8$, derived from is(o)- and xerocomic acid

ISS
an abbreviation for ion scattering spectroscopy

itaconic acid
$C_5H_6O_4$, anagrammatically derived from aconitic acid

-ite
derived from *-itus* (Latin suffix denoting origin or relationship)

-itin(e)
variant of -in(e)

-itol
suffix for sugar alcohols, derived from *-it* (German: -itol), as in *Sorbit* (German: sorbitol), and -ol

ITP
$C_{10}H_{15}N_4O_{14}P_3$, an abbreviation for inosine 5'-triphosphate

IUB
an abbreviation for International Union of Biochemistry

IUCr
an abbreviation for International Union of Crystallography

-ium
derived from *-ium* (Latin suffix for adjectives in the neutral gender)

-iun
an arbitrary suffix for trivial names of natural products with unknown structure

IUPAC
an abbreviation for International Union of Pure and Applied Chemistry

IUPAP
an abbreviation for International Union of Pure and Applied Physics

Ivanov reaction
named for the Bulgarian chemist D. P. Ivanov (1894-1975)

ixodin
derived from the genus name *Ixodes* (ticks), from *ixodes* (Greek: like birdlime), from *ixos* (Greek: birdlime), and -in(e)

-izidin(e)
an arbitrary suffix of obscure etymology

-izin(e)
an arbitrary suffix of obscure etymology

J

Jabłoński diagram
named for the Ukrainian-Polish physicist Aleksander Jabłoński (1898-1980)

Jacobsen rearrangement
named for the German chemist Oscar Georg Friedrich Jacobsen (1840-1889)

Jacobsen's catalyst
$C_{36}H_{52}ClMnN_2O_2$, named for the US chemist Eric Niels Jacobsen (born 1960)

jacobsite
$MnFe_2O_4$, named after this mineral's locality Jakobsberg, Sweden

jade
derived from *piedra de ijada* (Spanish: stone for curing pains in the side)

jadeite
$Na(Al,Fe)Si_2O_6$, derived from jade and -ite

Jaffé reaction
named for the German physician Max Jaffé (1841-1911)

Jaffé-Schlesinger reaction
named for the German physician Max Jaffé (1841-1911) and the Austrian physician W. Schlesinger (1869-1947)

Jahn-Teller effect
named for the British mathematician and chemist Hermann Arthur Jahn (1907-1979) and the Hungarian-US physicist Edward Teller (1908-2003)

jamesonite
$Pb_4FeSb_6S_{14}$, named for the Scottish mineralogist Robert Jameson (1774-1854)

Janovsky reaction
named for the Austrian chemist Jaroslav V. Janovsky (1850-1907)

japaconitine
$C_{34}H_{49}NO_{11}$, coined by contraction of Japanese aconitine – referring to the habitat of the japaconitine-producing *Aconitum* species

japonilure
$C_{14}H_{24}O_2$, derived from the specific epithet of the species name *Popillia japonica* (Japanese beetle), from *japonicus* (New Latin: Japanese), and lure

Japp-Klingemann reaction
named for the Scottish chemist Francis Robert Japp (1848-1925) and the German-British chemist Felix Klingemann (1863-1944)

jarosite
$KFe_3(SO_4)_2(OH)_6$, named after this mineral's locality Barranco Jaroso, Spain

jasmo-
derived from the genus name *Jasminum* (jasmine), ultimately from *yasmin*, *yasman* (Persian: jasmine)

jasmolin
derived from jasmo-, -ol(e), and -in(e)

jasmone
$C_{11}H_{16}O$, derived from jasmo- and -one

jasmonic acid
$C_{12}H_{18}O_3$, derived from jasmone

jasper
SiO_2, derived from *iaspis* (Greek: jasper),

from *jaspeh* (Hebrew: jasper)

jatroph(a)-
derived from the genus name *Jatropha* (tropical herbs, shrubs, trees), from iatro- and *trophe* (Greek: food), from *trephein* (Greek: to nourish)

jatrophatrione
$C_{20}H_{26}O_3$, derived from jatroph(a)-, tri- and -one

jatrophone
$C_{20}H_{24}O_3$, derived from jatroph(a)-, and -one

jerv(a)-
derived from *Jervina* (New Latin name for *Veratrum*), ultimately from *ervum* (Latin: vetch, chick-pea)

jerva acid (jervaic acid, jervasic acid, chelidonic acid)
$C_7H_4O_6$, derived from jerv(a)-

jervine
$C_{27}H_{39}NO_3$, derived from jerv(a)- and -in(e)

jesaconitine
$C_{35}H_{49}NO_{12}$, derived from *Jeso* (German: Ezo, the pre-1869 name of the island of Hokkaido, Japan), aconit-, and -in(e) – referring to the habitat of the jesaconitine-producing *Aconitum* species

jet (gagate)
C_n, derived from gagate

joaquinite
$NaBa_2FeTi_2Ce_2(SiO_3)_8O_2(OH){\cdot}H_2O$, named after this mineral's locality Joaquin Ridge, CA, USA

johachidolite
$CaAlB_3O_7$, named after this mineral's type locality Johachido, North Korea

johannsenite
$CaMnSi_2O_6$, named for the US petrologist Albert Johannsen (1871-1962)

joliotium
Db, unsuccessfully suggested name for dubnium, in honor of the French physicist Jean Frédéric Joliot-Curie (1900-1958)

Jones oxidation
named for the Welsh chemist Sir Ewart Ray Herbert Jones (1911-2002)

Jones reagent
$CrO_3{\cdot}H_2SO_4$, named for the Welsh chemist Sir Ewart Ray Herbert Jones (1911-2002)

jordanite
$Pb_{14}(As,Sb)_6S_{23}$, named for the German scientist H. Jordan (1808-1887)

Jorissen's reagent
named for the Belgian chemist A. J. J. Jorissen (1853-1920)

Joule-Thomson effect
named for the British physicists James Prescott Joule (1818-1889) and William Thomson (from 1892 William 1st Baron Kelvin of Largs) (1824-1907)

Jourdan-Ullmann-Goldberg synthesis
named for the 19th century German chemist Friedrich Jourdan and the Swiss chemists Fritz Ullmann (1875-1939) and Irma Goldberg

JSTX
an abbreviation for joro spider toxin

juglone
$C_{10}H_6O_3$, derived from the genus name *Juglandacea* (walnut tree), from *juglans* (Latin: walnut), from the Roman god Jupiter and *glans* (Latin: acorn), and -one

Julia synthesis
named for the French chemist Marc Julia (born 1922)

juniper-
from juniper, derived from *iuniperus* (Latin: juniper)

justicidin
derived from the genus name *Justicia* (water herbs, shrubs), after the Scottish botanist and horticulturist James Justice (1698-1763), and -idin(e)

juvenoid
derived (with contraction) from juvenile hormone and -oid

juxta-
derived from *iuxta* (Latin: next to), ultimately from *iungere* (Latin: to join)

K

κ (kappa)
an affix in systematic inorganic nomenclature, probably derived from coordination (site)

kadsuphilactone
derived (with contraction) from the species name *Kadsura philippinensis* Elmer (kadsura, a Taiwanese shrub), and lactone

kaempferol
$C_{15}H_{10}O_6$, derived from the name of the German physician Engelbert Kämpfer (1651-1716) and -ol

kaersutite
$NaCa_2(Mg_4Ti)Si_6Al_2O_{23}(OH)_2$, named after this mineral's locality Kaersut, Greenland

Kahane's reagent
named for the French biochemist Ernest Kahane (1903-1996)

kainic acid (digenic acid)
$C_{10}H_{15}NO_4$, derived from *kainin-so* (Japanese: the red alga *Digenia simplex*)

kainite
$MgSO_4{\cdot}KCl{\cdot}3H_2O$, derived from *kainos* (Greek: new) and -ite – referring to this mineral's recent origin

kairomone
derived from *kairos* (Greek: right time, right measure, right place) and hormone – referring to the fact that these hormones are beneficial to species other than the producing one

kali
New Latin word for alkali

kalignost (kalibor)
$C_{24}H_{20}BNa$, derived (with contraction) from kalium and diagnostic

kalium
K, New Latin name for potassium, derived from kali and -ium

kalkitoxin
$C_{21}H_{38}N_2OS$, derived from the habitat Playa Kalki, Curaçao of the source organism, the blue-green alga *Lyngbya majuscula,* and toxin

Kalle's acid
$C_{10}H_9NO_6S_2$, named for the German chemist Wilhelm Ferdinand Kalle (1870-1954)

kallidin
$C_{56}H_{85}N_{17}O_{12}$, coined by contraction of kallikrein, peptide, and -in(e

kallikrein
derived from *kallikreas* (Greek: sweetbread, pancreas), from *kallos* (Greek: beauty), from *kalos* (Greek: beautiful), and *kreas* (Greek: flesh), and -in(e)

kalsilite
$KAlSiO_4$, coined by contraction of kalium, aluminum, silicon, and -ite

kamacite
α-(Fe,Ni), derived from *kamax* (Greek: vine pole, shaft) and -ite – referring to this mineral's crystal shape

kanamycin
derived from the specific epithet of the bacterial species name *Streptomyces*

kanamyceticus, from *kana* (Japanese: golden) and *mykes* (Greek: fungus), and -in(e)

kaolin (bolus alba)
derived from a corrupted form of the name of the Chinese locality *Kao-Ling* (Chinese: high ridge), and -in(e)

kaolinite
$Al_2Si_2O_5(OH)_4$, derived from kaolin and -ite

Kaptein-Closs rules
named for the Dutch chemist Robert Kaptein (born 1941) and the German-US chemist Gerhard Ludwig Closs (1928-1992)

karanjin
$C_{18}H_{12}O_4$, derived from the native name *karanja* (Indian beech, *Pongamia glabra*) and -in(e)

Karl Fischer reagent
$I_2 \cdot SO_2 \cdot H_2O$, named for the German chemist Joseph Karl Anton Fischer (1901-1958)

Karplus equation
named for the Austrian-US physicist Martin Karplus (born 1930)

Karstedt catalyst
$C_{16}H_{36}O_2PtSi_4$, named for the 20th century US chemist Bruce D. Karstedt

Kasha's rule
named for the US chemist Michael Kasha (born 1920)

Kasha-Vavilov rule
named for the US chemist Michael Kasha (born 1920) and the Russian chemist Sergei Ivanovich Vavilov (1891-1951)

kasolite
$Pb(UO_2)SiO_4 \cdot 6H_2O$, named after this mineral's locality Kasolo, Zaire

kassinin
$C_{59}H_{95}N_{15}O_{18}S$, derived from the genus name *Kassina* (frogs), and -in(e)

katasulf process
derived from *Katalyse* (German: catalysis) and sulf(o)-

katharometer
derived from *katharos* (Greek: clean, pure) and -meter – referring to the use of this apparatus for the analysis of gas mixtures

katophorite
$Na(Ca,Na)Fe^{II}_4(Al,Fe^{III})Si_7AlO_{22}(OH)_2$, derived from *katapherein* (Greek: to carry down), and -ite – referring to this mineral's volcanic origin

kaurane
$C_{20}H_{34}$, derived from kauri pine (*Agathis australis*), from *kawri* (Maori: kauri pine), and -an(e)

kavatin (kavahin, methysticin)
$C_{15}H_{14}O_5$, derived from kava (*Piper methysticum*), from *kava* (Tongan & Marquesan: kava, literally bitter), and -in(e)

kawain (kavain, gonosan)
$C_{14}H_{14}O_3$, derived from kava (*Piper methysticum* Forst.), from *kava* (Tongan & Marquesan: kava, literally bitter), and -in(e)

keatite
SiO_2, named for the 20th century US chemist Paul Powell Keat

Kedde reaction
named for the 20th century Dutch chemist D. L. Kedde

Keggin anion
$EM_{12}O_{40}^{n-12}$ (n = valency of E), named for the 20th century British chemist J. F. Keggin

kekulene (superbenzene)
$C_{48}H_{24}$, named in honor of the German chemist Friedrich August Kekulé von Stradonitz (1829-1896)

Kekulé structures
named for the German chemist Friedrich August Kekulé von Stradonitz (1829-1896)

Kellogg process
named for the chemical company M. W. Kellogg Co., New York, NY, USA

Kelvin temperature scale
named for the British physicist William Thomson, 1st Baron Kelvin of Largs (1824-1907)

Kemler number
named for the French engineer H. Kemler (born 1910)

Kemp reaction
named for the US chemist Daniel Schaeffer Kemp (born 1936)

Kemp's triacid
$C_{12}H_{18}O_6$, named for the US chemist Daniel Schaeffer Kemp (born 1936)

Kendall-Mattox reaction
named for the US chemists Edward Calvin Kendall (1886-1972) and Vernon R. Mattox

keratin
derived from *keras* (Greek: horn) and -in(e)

kermesic acid
$C_{16}H_{10}O_8$, derived from kermes, ultimately from *qirmiz* (Arabic: vivid red)

kermesite
SbS_2O, ultimately derived from *qirmiz* (Arabic: vivid red) and -ite – referring to this mineral's red color

kernite
$Na_2B_4O_7 \cdot 4H_2O$, named after this mineral's locality Baron, Kern County, CA, USA

kerogen
derived from *keros* (Greek: wax) and -gen

kerosene (kerosine)
derived from *keros* (Greek: wax) and -ene

keryl
$C_{10}H_{21}-$, $C_{11}H_{23}-$, $C_{12}H_{25}-$, $C_{13}H_{27}-$, $C_{14}H_{29}-$, $C_{15}H_{31}-$, $C_{16}H_{33}-$, derived from kerosene and -yl

kestose
$C_{18}H_{32}O_{16}$, derived from the location of the Tate and Lyle Research Laboratories, Ravensbourne, Keston, Kent, UK where this trisaccharide was discovered, and -ose

ketal
derived (with contraction) from ketone and alcohol; patterned after acetal

ketazine
derived from ket(o)- and azine

ketene
C_3H_2O, derived from ket(o)- and -ene

ketide (polyketide)
coined by contraction of polyketide

ketimine
$R_2C=NR$, derived from ket(o)- and imine

ketipic acid
$C_6H_6O_6$, coined by contraction of 2,3-diketoadipic acid

ketitol
derived (with contraction) from ketose and -itol

ket(o)-
derived from ketone

ketol
derived from ket(o)- and -ol

ketone
$R_2C{=}O$, derived from acetone

ketose
derived from ket(o)- and -ose

ketoxime
$R_2C{=}NOR$, derived from ket(o)- and oxime

ketyl
derived from ket(o)- and -yl

kevlar
polyamides, a word invented as a trademark

Kharasch cyclization
named for the Ukrainian-US chemist Morris Selig Kharasch (1895-1957)

khellin
$C_{14}H_{12}O_5$, derived from khellah, from *akhillah* (Arabic: toothpick plant, *Ammi visnaga* Lam.) and -in(e)

kieselguhr
derived from *Kiesel* (German: flint) and *Gu(h)r* (German: fermentation), from *gären* (German: to ferment)

kieserite
$MgSO_4 \cdot H_2O$, named for the German physician and president of the Jena Academy Dietrich Georg Kieser (1779-1862)

Kiliani's reagent
$H_2CrO_4/H_2SO_4/H_2O$, named for the German chemist Heinrich Kiliani (1855-1945)

Kiliani-Fischer synthesis
named for the German chemists Heinrich Kiliani (1855-1945) and Emil Hermann Fischer (1852-1919)

kimberlite
a rock species, named after its locality Kimberley, South Africa

kinase
derived from kinetic and -ase

kin(e)-
derived from *kinein* (Greek: to move, to stimulate, to set in motion)

kinesin
derived from kin(e)- and -in(e)

kinetic
derived from *kinetikos* (Greek: for putting in motion), from kin(e)-

[1]**kinetin**
$C_{10}H_9N_5O$, derived from kinetic and -in(e)

[2]**kinetin (hyaluronidase)**
derived from kinetic and -in(e)

kinin
derived from kin(e)- and -in(e)

kinoite
$Ca_2Cu_2Si_3O_{10} \cdot 2H_2O$, named for the Jesuit explorer Eusebio Francisco Kino (1645-1711)

Kipp apparatus
named for the Dutch pharmacist Petrus Jacobus Kipp (1808-1864)

Kirchhoff's laws
named for the German physicist Gustav Robert Kirchhoff (1824-1887)

Kirkendall effect
named for the US metallurgist Ernest O. Kirkendall (1914-2005)

Kirkwood-Onsager equation
named for the US chemist John Gamble Kirkwood (1907-1959) and the Norwegian-US chemist Lars Onsager (1903-1976)

kirromycin
$C_{43}H_{60}N_2O_{12}$, derived from *kirrhos* (Greek: orange colored) and -mycin – referring to this antibiotic's yellow color

kistrin
derived (with contraction) from the genus name *Agkistrodon* (viper), (irregularly) from *ankistron* (Greek: fishhook), -odon, rhod(o)-, *stoma* (Greek: mouth), and -in(e)

kitol
$C_{40}H_{60}O_2$, derived from *ketos* (Greek: sea monster, whale) and -ol

Kjeldahl determination
named for the Danish chemist Johan Gustav Christoffer Thorsager Kjeldahl (1849-1900)

klado-
derived from *klados* (Greek: branch, slip, sprout)

klaprothium
U, a name unsuccessfully suggested for uranium, in honor of the German chemist Martin Heinrich Klaproth (1743-1817)

Klein's solution
named for the German mineralogist J. F. C. Klein (1842-1907)

knallgas
derived from *Knallgas* (German: knallgas), from *knallen* (German: to detonate) and *Gas* (German: gas)

Knoevenagel condensation
named for the German chemist Emil Knoevenagel (1865-1921)

Knoevenagel diazotization method
named for the German chemist Emil Knoevenagel (1865-1921)

Knoop's rule
named for the German biochemist Franz Knoop (1875-1946)

Knoop synthesis
named for the German biochemist Franz Knoop (1875-1946)

Knorr pyrrole synthesis
named for the German chemist Ludwig Knorr (1859-1921)

Kober reaction
named for the Dutch chemist S. Kober (1903-1945)

Koch-Haaf synthesis
named for the German chemists Herbert Koch (1904-1967) and Wolfgang Haaf (born 1928)

Kochi reaction
named for the US chemist Jay Kazuo Kochi (born 1927)

Koch's acid
$C_{10}H_9NO_9S_3$, named for the US chemist Julius Arnold Koch (1864-1932)

Koch synthesis
named for the German chemist Herbert Koch (1904-1967)

Koelsch radical
$C_{33}H_{21}$, named for the US chemist Charles Frederick Koelsch (1907-1999)

Koenigs-Knorr synthesis
named for the German chemists Wilhelm Koenigs (1851-1906) and Eduard Knorr

Kofler bench
named for the Austrian pharmacognosist Ludwig Kofler (1891-1951)

Kohler reaction
named for the US chemist Elmer Peter Kohler (1865-1938)

Kohler's ketone
$C_{24}H_{24}O$, named for the US chemist Elmer Peter Kohler (1865-1938)

Kohlrausch's law
named for the German physicist Friedrich Wilhelm Georg Kohlrausch (1840-1887)

kojic acid
$C_6H_6O_4$, derived from *koji* (Japanese: a rice enzyme preparation)

Kolbe electrolytic synthesis
named for the German chemist Adolf Wilhelm Hermann Kolbe (1818-1884)

Kölbel-Engelhardt process
named for the German chemists Herbert Kölbel (1908-1995) and Friedrich Engelhardt (1913-1994)

Kolbe-Schmitt reaction
named for the German chemists Adolf Wilhelm Hermann Kolbe (1818-1884) and Rudolf Schmitt (1830-1898)

Kolbe synthesis
named for the German chemist Adolf Wilhelm Hermann Kolbe (1818-1884)

Kondakov reaction
named for the Russian chemist L. L. Kondakov (1957-1931)

Kondakov's rule
named for the Russian chemist L. L. Kondakov (1957-1931)

König's salt
$C_{19}H_{21}ClN$, named for the German chemist Walter Georg König (1878-1964)

Koopmans' theorem
named for the Dutch-US physicist and economist Tjalling Charles Koopmans (1910-1985)

Kopp-Neumann rule
named for the German chemists Hermann Franz Moritz Kopp (1817-1892) and Franz Ernst Neumann (1798-1895)

Kopp's rule
named for the German chemist Hermann Franz Moritz Kopp (1817-1892)

kops-
derived from the genus name *Kopsia* (vinca), after the Dutch botanist Jan Kops (1765-1849)

kopsan
$C_{20}H_{24}N_2$, derived from kops- and -an(e)

kopsine
$C_{22}H_{24}N_2O_4$,derived from kops- and -in(e)

Körner-Contardi reaction
named for the Italian chemists G. Körner and Angelo Contardi (1877-1951)

kornerupine
$Mg_4Al_6(Si,Al,B)_5O_{21}(OH)$, named for the Danish geologist Andreas Nikolaus Kornerup (1857-1881)

Koser's reagent
$C_{13}H_{13}IO_4S$, named for the US chemist Gerald Franklin Koser (born 1942)

kosin
derived from koso (the Ethiopian tree *Brayera anthelmintica*), from *kosso* (Galla: koso), and -in(e)

Kostanecki acylation
named for the German chemist Stanislaus von Kostanecki (1860-1910)

Krafft degradation
named for the German chemist W. L. Friedrich E. Krafft (1852-1923)

Krafft point
named for the German chemist W. L. Friedrich E. Krafft (1852-1923)

Kramer effect
named for the German physicist Johannes Kramer (1905-1975)

Krapcho decarbalkoxylation
named for the US chemist Andrew Paul Krapcho (born 1932)

krauskopfite
$BaSi_2O_5 \cdot 3H_2O$, named for the US geochemist Konrad Bates Krauskopf (1910-2003)

Krebs cycle
named for the German-British biochemist Sir Hans Adolf Krebs (1900-1981)

Krebs-Kornberg cycle
named for the German-British biochemist Sir Hans Adolf Krebs (1900-1981) and the US chemist Arthur Kornberg (born 1918)

Kreis reaction
named for the Swiss chemist H. Kreis (1861-1931)

Kremer-Stein mechanism
named for the 20th century Israeli chemists Mordechai L. Kremer and Gabriel Stein

krennerite
$(Au,Ag)Te_2$, named for the Hungarian mineralogist Joseph S. Krenner (1839-1920)

Kröger-Vink notation
named for the Dutch chemists Ferdinand A. Kröger (born 1915) and Hendrik J. Vink (born 1915)

Kröhnke oxidation
named for the German chemist Fritz Kröhnke (1903-1981)

Kröhnke synthesis
named for the German chemist Fritz Kröhnke (1903-1981)

Kroll process
named for the Luxembourgian metallurgist W. J. Kroll (1889-1973)

krypton
Kr, derived from *kryptos* (Greek: hidden) and [1]-on

Kucherov reaction
named for the Russian-German chemist M. G. Kucherov (1850-1911)

Kuhn-Winterstein reaction
named for the German chemist Richard Kuhn (1900-1967) and the Swiss chemist Alfred Winterstein

Kulinkovich reaction
named for the Belorussian chemist Oleg Grigoryevich Kulinkovich (born 1948)

Kumada cross-coupling reaction
named for the 20th century Japanese chemist Makoto Kumada

kurchatovium
Rf, unsuccessfully suggest name for rutherfordium, in honor of the Russian physicist Igor Vasil'evich Kurchatov (1903-

1960)

Kurrol's salt
$(NaPO_3)_n$, $(KPO_3)_n$, named for the Estonian chemist Julius Kurrol (born 1865)

kutnahorite
$CaMn(CO_3)_2$, named after this mineral's locality Policany, Kutná Hora, Czech Republic

Kutscher-Steudel extraction
named for the German chemists F. Kutscher (1866-1942) and H. F. P. Steudel (1871-1967)

kyanite (cyanite)
$Al_2(O)SiO_4$, derived from cyan(o)- and -ite – referring to this mineral's sometimes blue color

kyanmethin
$C_6H_9N_3$, derived from cyan(o)-, meth(a)-, and -in(e)

kynurenic acid
$C_{10}H_9NO_3$, derived from *kyon*, *kynos* (Greek: dog), ur(o)-, and -in(e)

kynurenine
$C_{10}H_{12}N_2O_3$, derived from kynurenic acid and -in(e)

L

λ (lambda)
an affix in systematic inorganic nomenclature, probably derived from (number of) ligand(s)

L-
derived from *laevus* (Latin: left)

labdane
$C_{20}H_{38}$, derived from the specific epithet of the species name *Cistus labdanum* (rockrose), ultimately from *ledon* (Greek: rockrose), of Semitic origin, and an(e)

laboratory
derived from *laboratorium* (Latin: working place), from *laborare* (Latin: to work), from *labor* (Latin: work), ultimately from *labi* (Latin: to slip, to slide)

labradorite
$(Ca,Na)(Si,Al)_4O_8$, named after this mineral's locality, the Labrador Peninsula, Canada

lacc(a)-
derived from *lacca* (New Latin: lac), ultimately from *laksa* (Sanskrit: lac)

laccaic acid
derived from the specific epithet of the species name *Laccifer lacca* Kerr (a lacquer-producing insect), from lacc(a)-, and -fer

laccase
derived from lacc(a)- and -ase

laccol
$C_{23}H_{36}O_2$, derived from lacc(a)- and -ol

lacmoid
derived from *lacmus* (New Latin: litmus) and -oid

lacmus
New Latin name for litmus, derived from *lakmoes* (Dutch: lacmus), from *leecmos* (Medieval Dutch: lacmus), from *leken* (Medieval Dutch: to drip) and *mos* (Medieval Dutch: green vegetables, mushy foods)

lactam derived from lactic acid and amide

lactamase
derived from lactam and -ase

lactase
derived from lactose and -ase

lactic acid
$C_3H_6O_3$, derived from *lac* (Latin: milk)

lactide
derived from lact(o)- and [3]-ide

lactitol
$C_{12}H_{24}O_{11}$, derived from lactose and -itol

[1]lact(o)-
derived from *lac* (Latin: milk)

[2]lact(o)-
derived from lactic acid

[3]lact(o)-
derived from lactose

lactobacillic acid
$C_{19}H_{36}O_2$, derived from the bacterial genus name *Lactobacillus*, from [1]lact(o)- and the bacterial genus name *Bacillus*, from *bacillus*, diminutive of *baculus* (Latin: staff,

stick)

lactobionic acid
$C_{12}H_{22}O_{12}$, derived from lactobiose

lactobiose (lactose)
$C_{12}H_{22}O_{11}$, derived from ^{3}lact(o)-, bi-, and -ose

lactol
derived from ^{2}lact(o)- and -ol

lactone
derived from ^{2}lact(o)- and -one

lactose (lactobiose)
$C_{12}H_{22}O_{11}$, derived from ^{1}lact(o)- and -ose

lactucin
$C_{15}H_{16}O_{5}$, derived from the genus name *Lactuca* (lettuce), from ^{1}lact(o)- and -in(e)

lactulose
$C_{12}H_{22}O_{11}$, derived from ^{3}lact(o)- and -ulose

Ladenburg benzene (prismane)
C_6H_6, named for the German chemist Albert Ladenburg (1842-1911)

Ladenburg rearrangement
named for the German chemist Albert Ladenburg (1842-1911)

laev(o)- (lev(o)-)
a prefix denoting levorotatory substances, derived from *laevus* (Latin: left)

laevulose
$C_6H_{12}O_6$, an obsolete name for fructose, derived from laev(o)- and -ulose

LAFS
an abbreviation for laser atomic fluorescence spectrometry

laidlomycin
$C_{37}H_{62}O_{12}$, derived from the specific epithet of the bacterial species *Acholeplasma laidlawii*, against which this antibiotic shows activity, and -mycin

Lalancette's reagents
named for the French chemist J.-M. Lalancette (born 1934)

LAL test
an abbreviation for limulus amebocyte lysate test

LAMES
an abbreviation for laser micro emission spectroscopy

laminaran (laminarin)
derived from the genus name *Laminaria* (seaweeds), from *lamina* (Latin: thin plate, scale), and -an

Laming mass
named for the 19th century British chemist R. Laming

laminin
derived from basal lamina and -in(e)

LAMMA (LAMMS, LIMA)
an abbreviation for laser microprobe mass analysis

LAMMS (LAMMA, LIMA)
an abbreviation for laser microprobe mass spectrometry

lampro-
derived from *lampros* (Greek: bright)

lamprophyllite
$Sr_2Na_3Ti_3(Si_2O_7)_2(OH)_4$, derived from lampro-, phyll(o)-, and -ite – referring to the lustrous cleavage of this mineral

LAMS
an abbreviation for laser ablation mass spectrometry

lana philosophica
ZnO, an alchemistic name for threads of zinc oxide formed above melted zinc (New Latin: mysterious wool)

lanatoside
derived from the specific epithet of the species name *Digitalis lanata* Ehrh. (Grecian foxglove), from *lanatus* (Latin: woolly), from *lana* (Latin: wool), -ose, and -ide

Landolt reaction
named for the Swiss chemist Hans Heinrich Landolt (1831-1910)

Landolt vessel
named for the Swiss chemist Hans Heinrich Landolt (1831-1910)

langbeinite
$K_2Mg_2(SO_4)_3$, named for the German 19th century mining official Adalbert Langbein

Langmuir-Blodgett film (Langmuir-Blodgett-Kuhn film)
named for the US chemists Irving Langmuir (1881-1957) and Katharine Burr Blodgett (1898-1979)

Langmuir's equation
named for the US chemist Irving Langmuir (1881-1957)

lankamycin
derived (with contraction) from Sri Lanka, the habitat of the producing bacterium *Streptomyces violaceoniger*, and -mycin

lankavose (chalcose)
$C_7H_{14}O_4$, derived from lankamycin and –ose

lan(o)-
derived from *lana* (Latin: wool)

lanolin
derived from lan(o)-, *oleum* (Latin: oil) and -in(e)

lanostane
$C_{30}H_{54}$, derived from lan(o)-, sterol, and -an(e)

lanosterol
$C_{30}H_{50}O$, derived from lan(o)- and sterol

lanthana
La_2O_3, derived from *lanthanein* (Greek: to escape notice) and -a

lanthanide (lanthanoide)
derived from lanthanum and -ide

lanthanoid (lanthanide)
derived from lanthanum and -oid

lanthanum
La, derived from lanthana and -um – also referring to this element's difficult isolation due to a lack of specific reactions

lanthionine
$C_6H_{12}N_2O_4S$, derived from lan(o)-, thi(o)-, and -in(e)

lantibiotic
coined by contraction of the term lanthionine-containing peptide antibiotic

lapachol
$C_{15}H_{14}O_3$, derived from lapacho (the timber tree genera *Tabebuia* and *Tecoma*), from *lapacho* (Spanish: lapacho), and -ol

lapis (infernalis)
$AgNO_3$, from *lapis infernalis* (New Latin: hellish stone)

lapis lazuli (lazurite)
$Na_3Ca(Si_3Al_3)O_{12}S$, derived from *lapis* (Latin: stone) and, ultimately, Lajward, a place in Turkestan

Laporte's rule
named for the German-US physicist Otto Laporte (1902-1971)

lappaconitine
$C_{32}H_{44}N_2O_8$, derived from Lapp, aconit-, and -in(e)

Lapworth reaction
named for the Scottish chemist Arthur Lapworth (1872-1941)

lariat ether (crown ether)
derived from *la reata* (Spanish: the lasso)

larsenite
$PbZnSiO_4$, named for the US petrologist Esper Signius Larsen, Jr. (1878-1961)

larvicide
derived from larva and -cide

LAS
an abbreviation for linear alkylbenzene-sulfonate

lasalocid
$C_{34}H_{54}O_8$, derived (with contraction) from the specific epithet of the bacterial species name *Streptomyces lasaliensis* and bactericide

laser
an acronym for light amplification by stimulated emission of radiation

laserpitin
$C_{25}H_{38}O_7$, derived from the genus name *Laserpitium* (sermountain), from *lasserpicium* (Latin: name of a now extinct plant, *Ferula historica*), and -in(e)

lasiocarpine
$C_{21}H_{33}NO_7$, derived from the specific epithet of the species name *Heliotropium lasiocarpum* Fish. et C. Mey (an oriental weed), from *lasios* (Greek: shaggy), carp(o)-, and -in(e)

LASMA
an abbreviation for laser mass analyzer

Lassaigne test
named for the French chemist Jean Louis Lassaigne (1800-1859)

laterite
a rock species, derived from *later* (Latin: brick, tile) and -ite

latex
derived from *latex* (Latin: liquid, fluid), from *latax* (Greek: last remnant of a cup of wine)

lathyrogen
derived from the genus name *Lathyrus* (plants), from *lathyros* (Greek: chickling), and -gen

latrotoxin
derived from the genus name *Latrodectus* (venomous spiders), from *latro* (Latin: mercenary soldier, brigand) and *-dektes* (Greek: a biter), from *daknein* (Greek: to bite), and toxin

latrunculin
derived from the genus name *Latrunculia* (marine sponges), from *latrunculus*, diminutive of *latro* (Latin: mercenary soldier, brigand), and -in(e)

laudanidine
$C_{20}H_{25}NO_4$, derived from laudanum and-idin(e)

laudanine
$C_{20}H_{25}NO_4$, derived from laudanum and -in(e)

laudanosine
$C_{21}H_{27}NO_4$, derived from laudanine, -ose, and -in(e)

laudanum
coined by the Swiss-German physician and alchemist Philippus Aureolus Paracelsus (Theophrastus Bombast von Hohenheim) (1493-1541) for an alcoholic solution of opium, ultimately probably from *ladanon* (Greek: a gum resin), of Semitic origin

laughing gas
N_2O, named after its perceived physiological action

laumontite
$Ca(Al_2Si_4)O_{12}\cdot 4H_2O$, named for the French scientist François Pierre Nicolas Gillet de Laumont (1747-1834)

laureline
$C_{19}H_{19}NO_3$, derived from the genus name *Laurelia* (laurel), from *laurus* (Latin: laurel), and -in(e)

Laurell buffer
named for the Swedish biochemist Carl-Bertil Laurell (1919-2001)

Laurent's acid
$C_{10}H_9NO_3S$, named for the French chemist Augustin Laurent (1807-1853)

laurepukine
$C_{18}H_{17}NO_4$, coined by contraction of the species name *Laurelia novae-zelandiae* A. Cunn. (pukatea), pukatea, from *pukatea* (Maori: the forest tree pukatea), and -in(e)

lauric acid
$C_{12}H_{24}O_2$, derived from the genus name *Laurus* (laurel), from *laurus* (Latin: laurel)

laurite
RuS_2, named for Laura R. Joy, the wife of the 19th century US chemist Charles A. Joy

laurocapram
$C_{18}H_{35}NO$, derived (with contraction) from lauryl and caprolactam

laurotetanine (litsoeine)
$C_{19}H_{21}NO_4$, derived from the plant family name *Lauraceae* (shrubs, trees), from *laurus* (Latin: laurel) tetanus, from *tetanos* (Greek: rigid, stretched), from *teinein* (Greek: to stretch), and -in(e) – referring to this alkaloid's tetanizing action

lauroyl
$C_{12}H_{23}O-$, derived from lauric acid and -oyl

lauryl
$C_{12}H_{25}-$, derived from lauric acid and -yl

lauta mass
$Fe(OH)_3$, named after an aluminum plant in the town of Lauta, Germany

lautarite
$Ca(IO_3)_2$, named after this mineral's type locality Oficina Lautaro, Chile

Lauth's violet
$C_{12}H_{10}ClN_3S$, named for the French chemist Charles Lauth (1836-1913)

Lauth's white
$C_{18}H_{23}N_3S$, named for the French chemist Charles Lauth (1836-1913)

låvenite (lavenite)
$Na_2MnZr(Si_2O_7)(O,F)_2$, named after this mineral's locality Låven, Norway

Laves phase
named for the German crystallographer and

mineralogist Fritz H. Laves (1906-1978)

Lavoisier's law
named for the French chemist Antoine Lavoisier (1743-1794)

Lawesson's reagent
$C_{14}H_{14}O_2P_2S_4$, named for the Swedish-Danish chemist Sven-Olov Lawesson (1926-1985)

lawrencium
Lr, named for the US physicist Ernest Orlando Lawrence (1901-1958)

lawsone
$C_{10}H_6O_3$, derived from the genus name *Lawsonia* (tropical shrubs), after the British amateur botanist Isaac Lawson (deceased 1747), and -one

lawsonite
$CaAl_2Si_2O_7(OH)_2{\cdot}H_2O$, named for the Scottish-US geologist Andrew Cowper Lawson (1861-1952)

Lazier catalyst
$Cu_2Cr_2O_5$, named for the US chemist Wilbur Arthur Lazier (born 1900)

lazulite (blue spar, klaprothite)
$(Mg,Fe)Al_2(PO_4)_2(OH)_2$, derived from *azul* (Arabic: sky) and -lite – referring to this mineral's blue color

lazurite (lapis lazuli)
$Na_3Ca(Si_3Al_3)O_{12}S$, derived from *lazur* (Medieval Latin: lapis lazuli), ultimately from Lajward, a place in Turkestan, and -ite – referring to this mineral's blue color

LC
an abbreviation for liquid chromatography

LCAO
an abbreviation for linear combination of atomic orbitals

LDH
an abbreviation for lactate dehydrogenase

LDL
an abbreviation for low-density lipoprotein

LD process
an abbreviation for Linz-Donawitz process; named after the location of the steel mill VOEST (Vereinigte Österreichische Eisen- und Stahlwerke) Alpine AG, Austria

LDPE
an abbreviation for low-density polyethylene

lead
Pb, an indigenous English word

leadhillite
$Pb_4(SO_4)(CO_3)_2(OH)_2$, named after this mineral's locality Leadhills, Lanarkshire, Scotland, UK

LEAFS
an abbreviation for laser excited atomic fluorescence spectroscopy

leaianafulvene
$C_{15}H_{18}O_4$, derived from the specific epithet of the species name *Mycena leaiana* (the mushroom orange mycena), after the 19th century US amateur botanist T. G. Lea, and fulvene

Lebedev process
named for the Russian chemist Sergei Vasilyevich Lebedev (1874-1934)

Leblanc soda process
named for the French physician and chemist Nicolas Leblanc (1742-1806)

LEC
an abbreviation for liquid exclusion chromatography

Le Chatelier's principle (Le Chatelier-Braun priciple)
named for the French chemist Henri Le Chatelier (1850-1936)

lechatelierite
SiO_2, named for the French chemist Henri Le Chatelier (1850-1936)

lecithin
derived from *lekithos* (Greek: egg yolk) and -in(e)

Leclanché element
named for the French chemist Georges Leclanché (1839-1882)

lectin
derived from *lectus* (Latin: selected), from *legere* (Latin: to gather, to select) and -in(e)

ledeburite
(Fe,Fe_3C), named for the German metallurgist Karl Heinrich Adolf Ledebur (1837-1906)

Lederer-Manasse reaction
named for the German chemists L. Lederer and O. Manasse (born 1861)

ledol (ledum camphor)
$C_{15}H_{26}O$, derived from the genus name *Ledum* (evergreen shrubs), from *ledon* (Greek: rockrose), and -ol

LEED
an abbreviation for low-energy electron diffraction

Legal test
named for the German physician E. Legal (1859-1922)

leghemoglobin (legoglobin)
coined by contraction of leguminous, from *leguminosus* (Latin: leguminous), from *legumen* (Latin: legumes), from *legere* (Latin: to gather, to select), and hemoglobin

Lehmstedt-Tanasescu reaction
named for the 20th century German chemist Kurt Lehmstedt and the 20th century Romanian chemist Ioan Tanasescu

LEI
an abbreviation for laser enhanced ionization

leio- (lio-)
derived from *leios* (Greek; smooth)

LEIS
an abbreviation for low-energy ion scattering

Lemieux-Johnson reagent
$NaIO_4/OsO_4$, named for the Canadian chemist Raymond Urgel Lemieux (1920-2000) and the US chemist William Summer Johnson (1914-1995)

Lemieux-von Rudloff reagent
$NaIO_4/KMnO_4$, named for the Canadian chemist Raymond Urgel Lemieux (1920-2000) and the German-Canadian chemist Ernst Max von Rudloff (born 1923)

Lenard phosphor
named for the German physicist Philipp E. A. von Lenard (1862-1947)

Lennard-Jones potential
named for the British chemist Sir John Edward Lennard-Jones (1894-1954)

lenthionine
$C_2H_4S_5$, derived (with contraction) from the genus name *Lentinus* (mushrooms), from *lentus* (Latin: tough), thi(o)-, and -in(e)

lentinan
$(C_5H_{10}O_5)_n$, derived from the genus name *Lentinus* (mushrooms), from *lentus* (Latin: tough), and -an

lentinellic acid
$C_{18}H_{20}O_5$, derived from the genus name *Lentinellus* (mushrooms), diminutive of the genus name *Lentinus* (mushrooms), from *lentus* (Latin: tough)

lentinic acid
$C_{12}H_{22}N_2O_{10}S_4$, derived from the genus name *Lentinus* (mushrooms), from *lentus* (Latin: tough)

leonurine
$C_{14}H_{21}N_3O_5$, derived from the genus name *Leonurus* (herbs), from *leon* (Greek: lion) and *oura* (Greek: tail), and -in(e)

Lepehne tests
named for the German physician Georg Lepehne (1887-1967)

[1]lepidine (cincholepidine)
$C_{10}H_9N$, derived from lepid(o)- and -in(e)

[2]lepidine
$C_{21}H_{20}N_4O_2$, derived from the genus name *Lepidium* (pepperwort), ultimately from lepid(o)- and -in(e)

lepid(o)-
derived from *lepis* (Greek: flake, scale)

lepidocrocite
FeO(OH), derived from lepid(o)-, *kroke* (Greek: thread), and -ite – referring to this mineral's often scaly appearance

lepidolite
$K(Li,Al)_3(Si,Al)_4O_{10}(F,OH)_2$, derived from lepid(o)- and -lite – referring to this mineral's often scaly appearance

lepidomelane (biotite)
$K(Mg,Fe)_3(Si_3Al)O_{10}(OH,F)_2$, derived from lepid(o)- and -melane – referring to this mineral's often scaly appearance and black color

-leptic
derived from *lepsis* (Greek: a seizing), from *lambanein* (Greek: to seize, to take)

leptin
derived from lept(o)- and -in(e)

lept(o)-
derived from *leptos* (Greek: peeled, thin), from *lepein* (Greek: to peel)

leptomycin
derived from lept(o)- and -mycin

Letts nitrile synthesis
named for the British chemist Edmund Albert Letts (1851-1918)

Leuchs anhydride
named for the German chemist Hermann Leuchs (1879-1945)

leucine
$C_6H_{13}NO_2$, derived from leuc(o)- and -in(e)

leucite
$K(AlSi_2)O_6$, derived from leuc(o)- and -ite – referring to this mineral's often white color

Leuckart reaction
named for the German chemist Carl Louis Rudolf Alexander Leuckart (1854-1889)

Leuckart thiophenol reaction
named for the German chemist Carl Louis Rudolf Alexander Leuckart (1854-1889)

Leuckart-Wallach reaction
named for the German chemists Carl Louis Rudolf Alexander Leuckart (1854-1889)

and Otto Wallach (1847-1931)

leuc(o)- (leuk(o)-)
derived from *leukos* (Greek: white)

leucocyanidin
$C_{15}H_{14}O_7$, derived from leuc(o)- and cyanidin

leucodrin
$C_{15}H_{16}O_8$, derived (with contraction) from the genus name *Leucodendron* (trees), from leuc(o)- and dendr(o)-, and -in(e)

leucoglycodrin
$C_{21}H_{26}O_{13}$, derived from leucodrin and glyc(o)-

leucomycin
derived from leuc(o)- and -mycin

leucophanite
$NaCaBeSi_2O_6F$, derived from leuc(o)-, -phan(e), and -ite – referring to this mineral's white color

leucopterin
$C_6H_5N_5O_3$, derived from leuc(o)- and pterin

leuk(o)- (leuc(o)-)
derived from *leukos* (Greek: white)

leukotriene (LT)
derived from leukocyte, tri-, and -ene

leupeptin
derived from leucine, peptide, and -in(e)

lev(o)- (laev(o)-)
a prefix denoting levorotatory substances, derived from *laevus* (Latin: left)

levodopa
$C_9H_{11}NO_4$, derived from lev(o)- and dopa

levopimaric acid
$C_{20}H_{30}O_2$, derived from lev(o)- and pimaric acid

levulin
derived from levulose and -in(e) – referring to the formation of levulose upon hydrolysis of this polysaccharide

levulinic acid
$C_5H_8O_3$, derived from levulin

levulose
$C_6H_{12}O_6$, an obsolete name for D-fructose, derived from lev(o)- and -ulose

Lewis acid and base
named for the US chemist Gilbert Newton Lewis (1875-1946)

lewisite
$C_2H_2AsCl_3$, named for the US chemist Winford Lee Lewis (1878-1943)

LF
an abbreviation for ligand field

LFER
an abbreviation for linear free-energy relation

LFSE
an abbreviation for ligand field stabilization energy

LH
an abbreviation for luteinizing hormone

LH-RH
an abbreviation for luteinizing hormone-releasing hormone

liberin
derived from *liberare* (Latin: to liberate) and -in(e)

libethenite (chinoite)
$Cu_2PO_4(OH)$, named after this mineral's locality Livethen (Lubietova), Czech Republic

lichenin
$(C_6H_{10}O_5)_n$, derived from lichen and -in(e)

Lichtenberg alloy
(Bi,Sn,Pb), named for the German physicist Georg Christoph Lichtenberg (1742-1799)

licochalcone
coined by contraction of licorice (*Glycyrrhiza*) and chalcone

Liebau classification of silicates
named for the 20th century German crystallographer Friedrich Liebau

Lieben iodoform reaction
named for the Austrian chemist Adolf von Lieben (1836-1914)

Liebermann-Burchard reaction
named for the German chemists C. T. Liebermann (1842-1914) and H. R. F. Burchard (1865-1900)

Liebeskind-Srogl coupling
named for the US chemists Lanny Steven Liebeskind (born 1950) and J. Srogl

Liebig condenser
named for Justus Freiherr von Liebig (1803-1873)

Liege nomenclature
named for the 1930 IUPAC congress in Liège (Luik), Belgium

Liesegang rings
named for the German chemist Raphael Eduard Liesegang (1869-1947)

LIF
an abbreviation for laser-induced fluorescence

Lifschitz salt
$Ni^{II}en_2X_2$, named for the 20th century Dutch chemist I. Lifschitz

ligand
derived from *ligandus* (Latin: to be bound), from *ligare* (Latin: to bind, to tie)

ligase
derived from *ligare* (Latin: to tie, to bind) and -ase

lignan
derived from lign(o)- and -an(e)

lignin
derived from lign(o)- and -in(e)

lignite
derived from lign(o)- and -ite

lign(o)-
derived from *lignum* (Latin: wood)

lignoceric acid
$C_{24}H_{48}O_2$, derived from lign(o)- and ^{2}cer(o)-

ligroin
derived from *ligyros* (Greek: light, loud, sharp) and -in(e) – referring to the fact that this hydrocarbon mixture is colorless

lillianite
$Pb_3Bi_2S_6$, named for the Lillian Mining Company, CO, USA

LIMA (LAMMA, LAMMS)
an abbreviation for laser ionization mass analysis

lime
CaO, ultimately derived from *layate*

(Sanskrit: he clings, he sticks)

limestone
$CaCO_3$, a rock species, derived from *limus* (Latin: mud), from *limne* (Greek: marsh, pool)

limettin
$C_{11}H_{10}O_4$, derived from the specific epithet of the species name *Citrus limetta* Auth. (bitter orange), from *limette* (French: lime fruit), and -in(e)

limonene
$C_{10}H_{16}$, derived from *limonum* (Latin: lemon) and -ene

limonin
$C_{26}H_{30}O_8$, derived from *limonum* (Latin: lemon) and -in(e)

limonite (minette)
a rock species, derived from *leimon* (Greek: meadow), from *limne* (Greek: marsh, pool), and -ite

limulus test
derived from the species name *Limulus polyphemus* (horseshoe crab), from *limulus*, diminutive of *limus* (Latin: slanting)

linalool
$C_{10}H_{18}O$, derived from linaloe oil, ultimately from *lignum aloes* (Medieval Latin: wood of the aloe), and –ol

linalyl
$C_{10}H_{17}-$, derived from linalool and -yl

linamarin
$C_{10}H_{17}NO_6$, derived from the genus name *Linum* (flax), from *linum* (Latin: flax), *amarus* (Latin: bitter), and -in(e)

linarin
$C_{28}H_{32}O_{14}$, derived from the genus name *Linaria* (herbs, undershrubs), from *linum* (Latin: flax), and -in(e)

linarite
$CuPbSO_4(OH)_2$, named after this mineral's locality Linares, Spain

linatine
$C_{10}H_{17}N_3O_5$, derived from the genus name *Linum* (flax), from *linum* (Latin: flax), and -in(e)

lincomycin
$C_{18}H_{34}N_2O_6S$, derived from the bacterial species name *Streptomyces lincolnensis*, and -mycin

lindane (BCH, HCH)
$C_6H_6Cl_6$, named for the Belgian chemist Teunis van der Linden (1884-1965)

Linde process
named for the engineering company Linde AG, Wiesbaden, Germany, founded by the German engineer Carl von Linde (1842-1934)

Lindlar's catalyst
$Pd{\cdot}CaCO_3{\cdot}PbO$, named for the Swiss chemist Herbert Lindlar (born 1909)

Lindqvist anion
$Mo_7O_{24}^{6-}$, named for the Swedish chemist Ingvar Lindqvist (born 1921)

lineatin
$C_{10}H_{16}O_2$, derived from the specific epithet of the species name *Trypodendron lineatum* (Olivier) (ambrosia beetle), from *lineatus* (Latin: lined), from *linea* (Latin: line), and -in(e)

Lineweaver-Burk plot
named for the US biochemists Hans Lineweaver (born 1907) and Dean Burk (1904-1988)

linnaeite
$Co^{II}(Co^{III})_2S_4$, named for the Swedish botanist Carl von Linné (latinized Linnaeus) (1707-1778)

linoleic acid
$C_{18}H_{32}O_2$, derived from the genus name *Linum* (flax), from *linum* (Latin: flax), and *oleum* (Latin: oil)

linolenic acid
$C_{18}H_{30}O_2$, derived from the genus name *Linum* (flax), from *linum* (Latin: flax), -ene, and *oleum* (Latin: oil)

Linz-Donawitz process
named after the location of the steel mill VOEST (Vereinigte Österreichische Eisen- und Stahlwerke) Alpine AG, Austria

lio- (leio-)
derived from *leios* (Greek: smooth)

liothyronine
$C_{15}H_{12}I_3NO_4$, derived from L-, ^{2}iod(o)-, and thyronine

lipase
derived from lipid and -ase

lipid
derived from lip(o)-

lip(o)-
derived from *lipos* (Greek: bacon, fat)

lipofuscin
derived from lip(o)-, fusc(o)-, and -in(e)

α-lipoic acid (thioctic acid)
$C_8H_{14}O_2S_2$, derived from lip(o)-

lipophilic
derived from lip(o)- and -philic

lipophobic
derived from lip(o)- and -phobic

lipopolysaccharide
derived from lip(o)- and polysaccharide

lipoprotein
derived from lipo- and protein

lipotropin (LPH)
derived from lip(o)- and -tropin

Lipowitz alloy
(Bi,Pb,Sn,Cd), named for the 19th century chemist A. Lipowitz

lipoxin
coined by contraction of lipoxygenase interaction product and -in(e)

lipoxygenase
coined by contraction of lipid oxygenase

liptinite (exinite)
derived from *leipein* (Greek: to leave, to remain) and -ite

liquid
derived from *liquidus* (Latin: liquid), from *liquere* (Latin: to be liquid)

liquidus curve
derived from *liquidus* (Latin: liquid), from *liquere* (Latin: to be liquid)

-lite
ultimately derived from *lithos* (Greek: stone)

-lith
derived from *lithos* (Greek: stone)

litharge
PbO, derived from *lithargyros* (Greek: literally stone silver), from *lithos* (Greek: stone) and *argyros* (Greek: silver) – the

Greek name was originally used by Dioscorides for a material obtained in the process of separating lead from silver by pyrometallurgy

lithergolic
derived from -lith and -ergolic

lithiophilite (lithiophylite)
$LiMnPO_4$, derived from lithium, phil(o)-, and -ite

lithium
Li, derived from *lithos* (Greek: stone) and -ium – referring to the fact that this element was first found in minerals while its cognates sodium and potassium were first found in plants

lith(o)-
derived from l*ithos* (Greek: stone)

lithocholic acid
$C_{24}H_{40}O_3$, derived from lith(o)- and cholic acid

lithopone
($BaSO_4$,ZnS), derived from lith(o)- and *ponos* (Greek: work, artifact)

lithos magnetis
Fe_3O_4, an archaic name for magnetite, derived from [1]Magnesian stone

litmus
ultimately derived from *litmosi* (Old Norse: herbs used in dyeing), from *litr* (Old Norse: color) and *mosi* (Old Norse: moss)

litsoeine (laurotetanine)
$C_{19}H_{21}NO_4$, derived from the genus name *Litsea* (litsea), from *lei tsai* (Chinese: cherry, small plum), and -in(e)

livetin
anagrammatically derived from vitellin

[1]**livingstonite**
$HgSb_4S_7$, named for the Scottish missionary and explorer David Livingstone (1813-1873)

[2]**livingstonite**
$HgSb_4S_8$, named for the Scottish missionary and explorer David Livingstone (1813-1873)

lizardite
$Mg_3Si_2O_5(OH)_4$, derived from lizard and -ite – referring to this mineral's often green or green-blue color

LLC
an abbreviation for liquid-liquid chromatography

LLDPE
an abbreviation for linear low-density polyethylene

LLPC
an abbreviation for liquid-liquid partition chromatography

LNG
an abbreviation for liquefied natural gas

lobeline
$C_{22}H_{27}NO_2$, derived from the genus name *Lobelia* (lobelia), after the Flemish botanist Matthias de Lobel (1538-1616), and -in(e)

Lobry de Bruyn-van Ekenstein transformation
named for the Dutch chemists Cornelius Adrian Lobry de Bruyn (1857-1904) and W. A. van Ekenstein

lochnericine
$C_{21}H_{24}N_2O_3$, derived from the genus name *Lochnera* (periwinkle), after the German botanist and physician Michael Friedrich Lochner von Hummelstein (1662-1720),

and -icin(e)

locustol
$C_9H_{12}O_2$, derived from locust and -ol

loganin
$C_{17}H_{26}O_{10}$, derived from the family name *Loganiaceae* (herbs, shrubs, trees), after the Irish-US botanist James Logan (1674-1751), and -in(e)

-logy
derived from *logos* (Greek: word, reason, speech, account)

Lohmann transphosphorylation reaction
named for the German biochemist Karl Heinrich Adolf Lohmann (1898-1978)

LOI
an abbreviation for limiting oxygen index

loline (festucine)
$C_8H_{14}N_2O$, derived from the genus name *Lolium* (lolium), from *lolium* (Latin: darnel), and -in(e)

löllingite
$FeAs_2$, named after this mineral's locality Lölling, Austria

London force
named for the German-US physicist Fritz Wolfgang London (1900-1954)

longifolene
$C_{15}H_{24}$, derived from the specific epithet of the species name *Pinus longifolia* Roxb. (chir pine), from *longus* (Latin: long) and *folium* (Latin: leaf), and -ene

Longwell-Maniece method
named for the 20th century British chemists J. Longwell and W. D. Maniece

lonsdaleite
C_n, named for the British crystallographer Dame Kathleen Lonsdale (1903-1971)

looplure
$C_{14}H_{26}O_2$, derived from looper moth (*Trichoplusia ni* (Hübner)) and lure

loparite
$(Ce,Na,Ca)_2(Ti,Nb)_2O_6$, derived from *lopar'* (Russian: Lapp) and -ite

loph(o)-
derived from *lophos* (Greek: crest)

lophotoxin
$C_{22}H_{24}O_8$, derived from the genus name *Lophogorgia* (gorgonians), from loph(o)-, and toxin

lorandite
$TlAsS_2$, named for the Hungarian physicist Eötvös Lorand (1848-1919)

lorenzenite (ramsayite)
$Na_2Ti_2O_3(Si_2O_6)$, named for the Danish mineralogist Johannes Theodor Lorenzen (1855-1884)

Loschmidt's number (Avogadro's number)
named for the Austrian high school chemistry teacher Josef Loschmidt (1821-1895)

Lossen rearrangement
named for the German chemist Wilhelm Lossen (1838-1906)

lost (mustard gas)
$C_4H_8Cl_2S$, derived from the German code name *Lost*, coined by using the first two letters each of the names of the German chemists W. Lommel (1878-1962) and W. Steinkopf (1879-1949)

love (MDA)
$C_{10}H_{13}NO_2$, the nickname of this hallucinogenic drug refers to its psychopharmacological effects

Lowry concept
named for the British chemist Thomas Martin Lowry (1874-1936)

Lowry method
named for the US biochemist Oliver H. Lowry (1910-1996)

lox
an abbreviation for liquid oxygen

LPES
an abbreviation for laser photodetachment electron spectroscopy

LPG
an abbreviation for liqufied petroleum gas

LSC
an abbreviation for liquid-solid chromatography

LSD (lysergide)
$C_{20}H_{25}N_3O$, an abbreviation for *Lysergsäure*-N,N-*diethylamid* (German: lysergic acid *N,N*-diethylamide)

LT
an abbreviation for leukotriene

lubricant
derived from *lubricare* (Latin: to make smooth), from *lubricus* (Latin: slippery)

Lucas test
named for the US chemist Howard Johnson Lucas (1885-1963)

lucensomycin
$C_{36}H_{53}NO_{13}$, derived from the specific epithet of the bacterial species name *Streptomyces lucensis*, from *lucensis* (New Latin: of Lucca, Italy), and -mycin

luciferase
derived from luciferin and -ase

luciferin
derived from *lucifer* (Latin: light bearing), from *lux* (Latin: light) and *ferre* (Latin: to bear, to carry), and -in(e)

Luff solution
named for the 20th century German chemist G. Luff

luliberin (lutropin-releasing factor)
coined by contraction of lutropin and liberin

lumichrome
$C_{12}H_{10}N_4O_2$, derived from lumi(n)- and -chrome

lumiflavin
$C_{13}H_{12}N_4O_2$, derived from lumi(n)- and flavin

lumi(n)-
derived from luminescence

luminescence
ultimately derived from *lumen* (Latin: light) and -escence

luminol
$C_8H_7N_3O_2$, derived from lumi(n)- and -ol

lumisterol
$C_{28}H_{44}O$, derived from lumi(n)- and sterol

LUMO
an abbreviation for lowest unoccupied molecular orbital

lunacridine
$C_{17}H_{23}NO_4$, derived from the genus name *Lunasia* (tropical trees and shrubs), from

lunas (Tagalog: lunasia), and acridine

luna cornea
AgCl, alchemistic name for horn silver, from *luna* (Latin: moon), the alchemistic name for silver, and *corneus* (Latin: of horn), from *cornu* (Latin: horn)

lunacrine
$C_{16}H_{19}NO_3$, derived from the genus name *Lunasia* (tropical trees and shrubs), from *lunas* (Tagalog: lunasia), and -acrine

lunar caustic
$AgNO_3$, alchemistic name for silver nitrate, *luna* (Latin: moon) being the alchemistic name for silver

lunarine
$C_{25}H_{31}N_3O_4$, derived from the genus name *Lunaria* (henbane), from *lunarius* (Latin: of the moon), from *luna* (Latin: moon), and -in(e)

lunasine
$C_{17}H_{22}NO_3^+$, derived from the genus name *Lunasia* (plants), from *lunas* (Tagalog: lunasia), and -in(e)

Lunge's reagent
named for the German chemist G. Lunge (1839-1923)

lunularic acid
$C_{15}H_{14}O_4$, derived from the genus name *Lunularia* (liverwort), from *lunula* (Latin: crescent-shaped ornament worn by a woman), diminutive of *luna* (Latin: moon)

lupane
$C_{30}H_{52}$, derived from lup(e)- and -an(e)

lupanine
$C_{15}H_{25}N_2O$, derived from lupane and -in(e)

lup(e)-
derived from the genus name *Lupinus* (lupine), from *lupinum* (Latin: lupine, literally of the wolf), from *lupus* (Latin: wolf)

lupeol
$C_{30}H_{50}O$, derived from lup(e)- and -ol

lupinine
$C_{10}H_{19}NO$, derived from lup(e)- and -inin(e)

lupi spuma (spuma lupi)
$(Fe,Mn)WO_4$, alchemistic name for wolframite, derived from *lupi spuma* (New Latin: wolf's foam)

lupone
$C_{31}H_{46}O_4$, derived (with contraction) from the specific epithet of the species name *Humulus lupulus* (hop plant), from *lupulus*, diminutive of *lupus* (Latin: hop plant), and -one

lupulone
$C_{26}H_{38}O_4$, derived from the specific epithet of the species name *Humulus lupulus* (hop plant), from *lupulus*, diminutive of *lupus* (Latin: hop plant), and -one

lutein (xanthophyll)
$C_{40}H_{56}O_2$, derived from lute(o)- and -in(e)

lute(o)-
derived from *luteus* (Latin: yellow), from *lutum* (Latin: dyer's rocket)

luteolin
$C_{15}H_{10}O_6$, derived from the specific epithet of the species name *Reseda luteola* (dyer's rocket), from *luteolus*, diminutive of *luteus* (Latin: yellow), from *lutum* (Latin: dyer's rocket), -ol, and -in(e)

luteomycin
$C_{22}H_{27}NO_9$, derived from lute(o)- and

-mycin – referring to this antibiotic's orange-yellow color

lutetia
Lu_2O_3, derived from lutetium and -a

lutetium
Lu, named for Paris, France, derived from *Lutetia* (New Latin: Paris, France), ultimately from *Lutetia Parisiorum* (Latin: Paris, Gaul)

lutidine
C_7H_9N, anagrammatically derived from toluidine

lutropin (luteinizing hormone)
derived from corpus luteum, from lute(o)-, and -tropin

lyase
derived from lysis and -ase

lyc(o)-
derived from *lykos* (Greek: wolf)

lycoctonine
$C_{25}H_{41}NO_7$, derived from the specific epithet of the species name *Aconitum lycoctonum* L. (wolfsbane), from *lykoktonos* (Greek: wolf-slaying), from lyc(o)- and *kteinein* (Greek: to kill), and -in(e)

lycodine
derived (with contraction) from lycopod- and -in(e)

lyco(p)-
derived from the genus name *Lycopersicon* (tomato plants), from *lykopersion* (Greek: an Egyptian plant), from lyc(o)-

lycopene
$C_{40}H_{56}$, from lyco(p)- and -ene

lycophyll
$C_{40}H_{56}O_2$, derived from lyco(p)- and -phyll

lycopod-
derived from the genus name *Lycopodium* (lycopodium), from lyc(o)- and -podium

lycopodane
$C_{15}H_{25}N$, derived from lycopod- and -an(e)

lycopodine
$C_{16}H_{25}O$, derived from lycopod- and -in(e)

lycor-
derived from the genus name *Lycoris* (surprise lilies), from *Lycoris* (Latin: a woman's name)

lycoramine
$C_{17}H_{23}NO_3$, derived from lycor- and amine

lycorenan
$C_{15}H_{17}NO$, derived from lycor-, -ene, and -an(e)

lycorine
$C_{16}H_{17}NO_4$, derived from lycor- and -in(e)

lycoxanthine
$C_{40}H_{56}O$, derived from lyco(p)-, xanth(o)-, and -in(e)

lyddite (picric acid)
$C_6H_3N_3O_7$, named after the town of Lydd, Kent, UK where this compound was manufactured in WWI

lymphokine
derived from lymph and -kine

lymphotoxin
derived from lymph and -toxin

Lynen cycle
named for the German biochemist Feodor Felix Konrad Lynen (1911-1979)

lyo-
derived from lysis

lyophilic
derived from lyo- and -philic

lyophobic
derived from lyo- and -phobic

lypressin
$C_{46}H_{65}N_{13}O_{12}S_2$, coined by contraction of lysine and vasopressin

lysergamide
$C_{16}H_{17}N_3O$, coined by contraction of lysergic acid amide

lysergic acid
$C_{16}H_{16}N_2O_2$, derived from lys(o)- and ergot

lysergide (LSD)
$C_{20}H_{25}N_3O$, coined by contraction of lysergic acid *N,N*-diethylamide

lysidine
$C_4H_8N_2$, derived from lys(o)- and amidine – referring to the relatively high water solubility of this compound's urate

lysine
$C_6H_{14}N_2O_2$, derived from lys(o)- and -in(e)

lysis
derived from *lysis* (Greek: loosening, dissolution), from *lyein* (Greek: to loosen, to dissolve)

lys(o)-
derived from lysis

lysopine
$C_9H_{18}N_2O_4$, derived (with contraction) from lysine and opine

lysostaphin
derived from the specific epithet of the bacterial species name *Staphylococcus staphylolyticus*, from staphyl(o)- and -lytic, and -in(e)

lysozyme (muramidase, muraminidase)
derived (with contraction) from lysis and enzyme

lythran
$C_{24}H_{27}NO$, derived from lythr(o)- and -an(e)

lythranidine
$C_{26}H_{35}NO_4$, derived from lythr(o)-, -an(e), and -idin(e)

lythr(o)-
derived from the genus name *Lythrum* (herbs, subshrubs), from *lythron* (Greek: gore)

lyxitol
$C_5H_{12}O_5$, derived from lyxose and -itol

lyxoflavin
$C_{17}H_{20}N_4O_6$, derived from lyxose and flavin

lyxose
$C_5H_{10}O_5$, anagrammatically derived from xylose

M

μ (mu)
an affix denoting bridging complex ligands in systematic inorganic nomenclature, suggested and adopted without any stated reason, but probably the similarity between the letter m and a Roman bridge was deemed too obvious to deserve explicit mention

maclurin
$C_{13}H_{10}O_6$, derived from the genus name *Maclura* (fustic), after the Scottish-US geologist William Maclure (1763-1840), and -in(e)

macro-
derived from *makros* (Greek: long, large)

macrolide
coined by contraction of macrocyclic and -olide

Maddrell's salt
$(NaPO_3)_n$, named for the 19th century British chemist R. Maddrell

Madelung constant
named for the German physicist Erwin Madelung (1881-1972)

Madelung synthesis
named for the 20th century German chemist W. Madelung

maduramicin
$C_{47}H_{83}NO_{17}$, derived from the bacterial genus name *Actinomadura*, derived from actin(o)- and the city of Madura, India, and -micin

maduramycin
$C_{28}H_{22}O_{10}$, derived from the bacterial genus name *Actinomadura*, derived from actin(o)- and the city of Madura, India, and -mycin

mafite
group name of rocks, coined by contraction of magnesium, ferrum, and -ite

magadiite
$NaSi_7O_{13}(OH)_3{\cdot}4H_2O$, named after this mineral's locality Lake Magadi, Kenya

magainin
derived from *maghen* (Hebrew: shield), and -in(e) – referring to these polypeptides antimicrobial properties

magenta (fuchsin)
$C_{20}H_{20}ClN_3$, named for the battle of Magenta, Italy in 1859 – referring to the time of the discovery of this dye shortly after

maghemite
$Fe_{2.67}O_4$, coined by contraction of magnetite and hematite

magic acid
(FSO_3H,SbF_5) – referring to this mixture's extreme acidic strength

magic methyl
CH_4FO_3S – referring to this ester's extreme alkylating power

magic numbers
2, 8, 20, 28, 50, 82, 126 – referring to the exceptional stability of atomic nuclei where the number of protons and/or neutrons corresponds to a magic number or a sum of magic numbers

magisterium bismuti (bismutum subnitricum)
$BiO(NO_3)\cdot H_2O$, an archaic name for bismuth subnitrate, derived from *magisterium* (New Latin: precipitate) and *bismutum* (New Latin: bismuth)

magma
derived from *magma* (Greek: kneaded mass, ointment), from *massein* (Greek: to knead)

magmatite
derived from magma and -ite

magnalium
(Al,Mg), coined by contraction of magnesium and aluminum

Magnéli phase
named for the Swedish crystallographer Arne Magnéli (1914-1996)

[1]magnes
Fe_3O_4, a corruption of [1]Magnesian stone

[2]magnes
MnO_2, a corruption of [2]Magnesian stone

[1]magnesia (magnesia usta)
MgO, derived from magnesia alba

[2]magnesia
Fe_3O_4, derived from [1]Magnesian stone

[3]magnesia
MnO_2, derived from [3]Magnesian stone

magnesia alba
$MgCO_3$, derived from [3]Magnesian stone and *albus* (Latin: white)

magnesia nigra
MnO_2, derived from [2]Magnesian stone and *niger* (Latin: black)

[1]Magnesian stone
Fe_3O_4, an archaic name for magnetite, ultimately derived from *Magnetis lithos* (Greek: Magnesian stone) – Magnesian refers either to the ancient Greek province Thracian Magnesia, the ancient town of Magnesia ad Maeandrum, Asia Minor, the ancient town of Magnesia ad Sipylum, Asia Minor, or to the ancient shepherd Magnes who according to Nicander (as cited by Pliny the Elder) discovered this magnetic mineral on Mount Ida, Crete, Greece

[2]Magnesian stone
MnO_2, an archaic name for manganese dioxide, coined by confusion of [2]Magnesian stone with [1]Magnesian stone because of their similar appearance

[3]Magnesian stone
$MgCO_3$, ultimately derived from *Magnetis lithos* (Greek: Magnesian stone) – Magnesian refers either to the ancient Greek province Thracian Magnesia, the ancient town of Magnesia ad Maeandrum, Asia Minor, the ancient town of Magnesia ad Sipylum, Asia Minor

magnesia usta (magnesia)
MgO, derived from [3]magnesia and *ustus* (Latin: burnt), from *urere* (Latin: to burn)

magnesiochromite
$MgCr_2O_4$, derived from magnesium and chromite

magnesioferrite
$MgFe_2O_4$, derived from magnesium and ferrite

magnesite (bitter spar, giobertite)
$MgCO_3$, derived from magnesium and -ite

magnesium
Mg, derived from [1]magnesia and -ium

magnesocene
$C_{10}H_{10}Mg$, derived from magnesium and -ocene

magneson
$C_{12}H_9N_3O_4$, derived from magnesium and [1]on – referring to this compound's usefulness for the detection of magnesium (and molybdenum)

magnetite
Fe_3O_4, derived from [1]Magnesian stone and -ite

magno-
derived from the genus name *Magnolia* (magnolia), after the French botanist Pierre Magnol (1638-1715)

magnoflorine
$C_{20}H_{24}NO_4^+$, derived (with contraction) from the species name *Magnolia grandiflora* L. (southern magnolia), from *grandis* (Latin: large) and *florere* (Latin: to flower), and -in(e)

magnolol
$C_{18}H_{18}O_2$, derived from magno- and -ol

Magnus's green salt
$Cl_4H_{12}N_4Pt_2$, named for the German chemist Heinrich Gustav Magnus (1802-1870)

Maillard browning reaction
named for the French chemist Louis Camille Maillard (1878-1936)

makatite
$Na_2Si_4O_8(OH)_2{\cdot}4H_2O$, derived from *emakat* (Masai: soda) and -ite – referring to this mineral's high sodium content

malachite
$Cu_2(CO_3)_2(OH)_2$, derived from *malache* (Greek: mallow) and -ite – referring to this mineral's green color

malachite green
$C_{23}H_{25}ClN_2$, derived from malachite and green – referring to this dye's color

Malaprade reaction
named for the French chemist L. A. J. Malaprade (1903-1982)

maleamic acid
$C_4H_5NO_3$, coined by contraction of maleic acid monoamide

maleanilic acid
$C_{10}H_9NO_3$, coined by contraction of maleic acid monoanilide

maleic acid
$C_4H_4O_4$, coined by modification of malic acid – referring to a preparation of maleic acid by distillation of malic acid

maleuric acid
$C_5H_6N_2O_4$, coined by contraction of maleic acid monoureide

malic acid
$C_4H_6O_5$, derived from *malum* (Latin: apple)

mallardite
$MnSO_4{\cdot}7H_2O$, named for the French crystallographer Ernest Mallard (1833-1894)

malonic acid
$C_3H_4O_4$, coined by modification of malic acid – referring to a preparation of malonic acid by oxidation of malic acid

malt
ultimately derived from *mel* (Indo-European: to grind)

maltase
derived from maltose and -ase

maltol
$C_6H_6O_3$, derived from malt and -ol

maltose
$C_{12}H_{22}O_{11}$, derived from malt and -ose

malv-
derived from the genus name *Malva* (mallow), from *malva* (Latin: mallow)

malvidin
$C_{17}H_{15}O_7^+$, derived from malv- and -idin(e)

malvin
$C_{19}H_{35}O_{17}^+$, derived from malv- and -in(e)

mammotropin (prolactin)
derived from *mamma* (Latin: female breast) and -tropin

mancude
an abbreviation for maximum number of noncumulative double bonds

mandelic acid
$C_8H_8O_3$, derived from *Mandelsäure* (German: mandelic acid), from *Mandel* (German: almond), ultimately from *amygdale* (Greek: almond)

manganate
derived from manganese and -ate

manganaxinite
$Ca_2MnAl_2(BO_3)Si_4O_{12}(OH)$, derived from manganese and axinite

manganes
MnO_2, an obsolete name for manganese dioxide coined by variation of magnes under the influence of *manganizein* (Greek: to purify) – referring to manganese dioxide's ability to decolorize iron-containing green glass

manganese
Mn, derived from manganesium

manganesium
Mn, an obsolete name for manganese, derived from manganes – discarded in order to avoid confusion with magnesium

manganite
MnO(OH), derived from manganese and -ite

manganium (manganum)
Mn, New Latin name for manganese, coined by contraction of manganesium

mangan(o)-
derived from manganese

manganocene
$C_{10}H_{10}Mn$, derived from mangan(o)- and -ocene

manganocolumbite
$(Mn,Fe)(Nb,Ta)_2O_6$, derived from mangan(o)- and columbite

manganosite
MnO, derived from mangan(o)-, oxide, and -ite

manganotantalite
$(Mn,Fe)(Ta,Nb)_2O_6$, derived from mangan(o)- and tantalite

manganum (manganium)
Mn, New Latin name for manganese, coined by contraction of manganesium

mangostin
$C_{24}H_{26}O_6$, derived from the specific epithet of the species name *Garcinia mangostana* L. (mangosteen tree), from *mangustan* (Malay: mangosteen), and -in(e)

mannan
derived from mannose and –an

Mannich base
named for the German chemist Carl Ulrich Franz Mannich (1877-1947)

Mannich reaction
named for the German chemist Carl Ulrich Franz Mannich (1877-1947)

mannitol
$C_6H_{14}O_6$, derived from mannose and -itol

mannose
$C_6H_{12}O_6$, derived from manna, from *man* (Hebrew: gift), and -ose

manometer
derived from *manos* (Greek: thin) and meter

MAO
an abbreviation for monoamine oxidase

marasmin
derived from *marasmos* (Greek: a wilting), from *marainein* (Greek: to waste away), and -in(e) – referring to these compound's wilting effect on plants

marble
$CaCO_3$, a rock species, ultimately derived from *marmaros* (Greek: white glistening stone)

marcasite (hydropyrite)
FeS_2, derived from *marcasita* (Latin: pyrite), from *maqqashita* (Aramaic: pyrite), ultimately derived from Karkhashi, a region perhaps located in northeast Persia

Marcus equation
named for the Canadian-US chemist Rudolph Arthur Marcus (born 1923)

[1]margaric acid
$C_{17}H_{34}O_2$, derived from margarine

[2]margaric acid
$(C_{16}H_{32}O_2,C_{18}H_{36}O_2)$, a mixture of palmitic and stearic acids erroneously believed to be a single acid, derived from margarine

margarin
$C_{20}H_{40}O_4$, derived from [1]margaric acid and -in(e)

margarine
derived from *margaron* (Greek: pearl) and -in(e)

margarite
$CaAl_2(Al_2Si_2)O_{10}(OH)_2$, derived from *margaron* (Greek: pearl) and -ite – referring to this mineral's pearly luster

margarone
derived from [2]margaric acid and -one – referring to the formation of this mixture of ketones by condensation of [2]margaric acid

margatoxin
$C_{178}H_{286}N_{52}O_{50}S_7$, derived from the specific epithet of the species name *Centruroides margaritatus* (a scorpion), ultimately from *margaron* (Greek: pearl), and toxin

marialite (wernerite)
$(Na,Ca)_4(Si,Al)_{12}O_{24}(Cl,CO_3,SO_4)$, named by its discoverer in honor of his wife Maria Rosa von Rath (1830-1888)

Mark-Houwink relationship
named for the Austrian-US chemist Herman Francis Mark (1895-1992) and the Dutch chemist R. Houwink (1897-1988)

Markovnikov's rule
named for the Russian chemist Vladimir Vasilevich Markovnikov (1838-1904)

Markush formula
named for the Hungarian-US chemist Eugene Armand Markush (1888-1968)

marl
a rock species, ultimately derived from *margila*, Medieval Latin diminutive of *marga* (Latin: clay, earth), of Gaulish origin

Marquis reaction
named for the French chemist R. Marquis (1872-1950)

marrubiin
$C_{20}H_{28}O_4$, derived from the genus name *Marrubium* (horehound), from *marrubium* (Latin: horehound), and -in(e)

Marschalk reaction
named for the Swiss chemist Charles Henri Marschalk (1885-1968)

Marshall's acid
$H_2S_2O_8$, named for the 19th century Scottish chemist Hugh Marshall

Marsh test
named for the British chemist James Marsh (1794-1846)

martensite
(Fe,Fe_3C), named for the German metallurgist Adolf Martens (1850-1914)

Martinet dioxindole synthesis
named for the 20th century French chemist J. Martinet

Martius yellow
$C_{10}H_5N_2NaO_5$, named for the German chemist Carl Alexander von Martius (1838-1920)

MAS
an abbreviation for magic angle spinning

maser
an abbreviation for microwave amplification by stimulated emission of radiation

massicotite (massicot)
PbO, derived from *marzacotta* (Italian: potter's glaze), ultimately from *sabb qubti* (Arabic: Egyptian alum, i.e. iron and/or aluminum sulfate), and -ite

masurium
Tc, a name unsuccessfully suggested for technetium, after the Masuria region, East Prussia, Germany, now Poland

matri(car)-
derived from the genus name *Matricaria* (feverfew), from *matricaria* (Latin: feverfew), from *matrix* (Latin: womb, uterus)

matricarin
$C_{17}H_{20}O_5$, derived from matri(car)- and -in(e)

matridine
$C_{15}H_{26}N_2$, derived from matri(car)- and -idin(e)

matrine
$C_{15}H_{24}N_2O$, derived from matri(car)- and -in(e)

Mattauch's rules
named for the Austrian-German physicist Josef Mattauch (1895-1976)

MBAS
an abbreviation for methylene blue active substance

MBK
$C_6H_{12}O$, an abbreviation for methyl butyl ketone

MC
an abbreviation for methyl cellulose

McFadyen-Stevens reaction
named for the British chemists John S.

McFadyen and Thomas Stevens Stevens (1900-2000)

McLafferty rearrangement
named for the US chemist Fred Warren McLafferty (born 1923)

McMurry's reagent
$TiCl_3 \cdot LiAlH_4$, named for the US chemist John Edward McMurry (born 1942)

MCR
an abbreviation for multi-component reaction

McReynolds constants
named for the US chemist W. O. McReynolds (deceased 1976)

MCT
an abbreviation for medium chain triglycerides

MDA (love)
$C_{10}H_{13}NO_2$, an abbreviation for 3,4-methylenedioxyamphetamine

MDE (MDEA, Eve)
$C_{12}H_{17}NO_2$, an abbreviation for 3,4-methylenedioxyethamphetamine

MDI
$C_{15}H_{10}N_2O_2$, an abbreviation for the unsystematic name methylene diphenyl isocyanate

MDMA (Adam, E, ecstasy, XTC)
$C_{11}H_{15}NO_2$, an abbreviation for 3,4-methylenedioxymethamphetamine

mecon-
derived from *mekon* (Greek: poppy)

meconic acid
$C_7H_4O_7$, derived from mecon-

meconin
$C_{10}H_{10}O_4$, derived from mecon- and -in(e)

mecrylate
$C_5H_5NO_2$, coined by contraction of methyl 2-cyanoacrylate

medic(ag)-
derived from the genus name *Medicago* (medic), from *medica* (Latin: medic), from the ancient country of Media, and *plantago* (Latin: plantain)

medicagol
$C_{16}H_8O_6$, derived from medic(ag)- and -ol

medicarpin
$C_{16}H_{14}O_4$, derived (with contraction) from medic(ag)- and pterocarpin

medullipin
coined by contraction of medullary lipid and -in(e)

meerschaum (sepiolite)
$Mg_4(OH)_2(Si_2O_5)_3 \cdot 6H_2O$, derived from *Meerschaum* (German: meerschaum, literally sea froth) – referring to the fact that this mineral floats in water

Meerwein arylation
named for the German chemist Hans Leberecht Meerwein (1879-1965)

Meerwein-Ponndorf-Verley reduction
named for the German chemists Hans Leberecht Meerwein (1879-1965) and Wolfgang Ponndorf (1894-1948), and the French chemist A. Verley

Meerwein rearrangement
named for the German chemist Hans Leberecht Meerwein (1879-1965)

Meerwein salt
$C_3H_9BF_4O$, named for the German chemist

Hans Leberecht Meerwein (1879-1965)

Meerwein's ester
$C_{17}H_{20}O_{18}$, named for the German chemist Hans Leberecht Meerwein (1879-1965)

M-effect (R-effect)
an abbreviation for mesomeric effect

mega-
derived from *megas* (Greek: great, large, mighty)

megacin
derived from the specific epithet of the bacterial species name *Bacillus megaterium*, from mega-, *therion* (Greek: beast), and -icin(e)

megaphone
$C_{22}H_{30}O_6$, derived from the specific epithet of the species name *Aniba megaphylla* (a tropical tree), from mega- and phyll(o)-, and -one

meio-
derived from *meion* (Greek: less)

meionite (wernerite)
$(Ca,Na)_4(Si,Al)_{12}O_{24}(CO_3,SO_4,Cl)$, derived from meio- and -ite – referring to this mineral's less acute pyramidal shape compared to vesuvianite

Meisenheimer complex
named for the German chemist Jakob Meisenheimer (1876-1934)

Meisenheimer rearrangement
named for the German chemist Jakob Meisenheimer (1876-1934)

meitnerium
Mt, named for the Austrian-Swedish physicist Lise Meitner (1878-1968)

MEK
C_4H_8O, an abbreviation for methyl ethyl ketone

Meker burner
named for the 20th century US chemist George Meker

melam
$C_6H_9N_{11}$, an arbitrary name

melamine (cyanuramide)
$C_3H_6N_6$, derived from melam and amine

-melane
derived from *melas* (Greek: black) and -an(e)

melanin
derived from mela(no)- and -in(e)

mela(no)-
derived from *melas* (Greek: black)

melanoidin
derived from melanin, -oid, and -in(e)

melanoliberin (melanotropin-releasing factor, melanotropin releasing hormone)
derived (with contraction) from melanotropin and liberin

melanophlogite
$SiO_2 \cdot n(C,H,O,S)$, derived from melan(o)-, phlog(o)-, and -ite – referring to the fact that this mineral's organic inclusions burn up upon heating and turn it black

melanostatin (melanotropin release inhibiting factor, melanotropin release inhibiting hormone)
derived (with contraction) from melanotropin and -statin

melanotropin (melanocyte stimulating hormone)
derived from mela(no)- and -tropin

melanterite
$FeSO_4 \cdot 7H_2O$, derived from *melanteria* (Greek: copperas, pigment used to blacken shoes), from mela(no)- and *terein* (Greek: to watch, to preserve, to keep), and -ite

melatonin
$C_{13}H_{16}N_2O_2$, derived from mela(no)- and -tonin

Meldola blue
$C_{18}H_{15}Cl_3N_2OZn$, named for the British chemist Raphael Meldola (1849-1915)

Meldrum's acid
$C_6H_8O_4$, named for the Scottish chemist Andrew Norman Meldrum (1876-1934)

melem
$C_6H_6N_{10}$, coined by variation of melam

melezitose
$C_{18}H_{32}O_{16}$, derived from *mélèze* (French: larch) and -ose; patterned after melitose

melibiose
$C_{12}H_{22}H_{11}$, derived from mel(l)i-, bi-, and -ose

melilite
$(Ca,Na)_2(Al,Mg)(Si,Al)_2O_7$, derived from mel(l)i- and -lite – referring to this mineral's honey color

meliphanite
$Na(Na,Ca)BeSi_2O_6F$, derived from mel(l)i-, -phan(e), and -ite – referring to this mineral's honey color

melissic acid
$C_{30}H_{60}O_2$, derived from *melissa* (Greek: bee)

melissyl alcohol
$C_{30}H_{62}O$, derived from melissic acid

melitose (melitriose, raffinose)
$C_{18}H_{32}O_{16}$, derived from mel(l)i- and -ose

melitoxin (dicoumarol, dicumarol)
$C_{19}H_{12}O_6$, derived from the genus name *Melilotus* (sweetclover), from *melilotos* (Latin: melilot), from *meli* (Greek: honey) and *lotos* (Greek: clover), and toxin

melitriose (melitose, raffinose)
$C_{18}H_{32}O_{16}$, derived from mel(l)i-, tri-, and -ose

melittin
$C_{131}H_{229}N_{39}O_{31}$, derived from the specific epithet of the species name *Apis mellifera* (honey bee), from mel(l)i-, and -in(e)

mel(l)i-
derived from *mel* (Latin: honey)

mellite
$Al_2C_6(COO)_6 \cdot 16H_2O$, derived from mel(l)i- and -lite – referring to this mineral's honey color

mellitic acid
$C_{12}H_6O_{12}$, derived from mellite, a salt of mellitic acid

melon
$C_6H_3N_9$, coined by variation of melam

mellose
coined by contraction of methyl cellulose

membrane
derived from *membrana* (Latin: fine skin, literally that which covers the members of the body)

menadiol
$C_{11}H_{10}O_2$, coined by contraction of 2-

methylnaphthalene-1,4-diol

menadione
$C_{11}H_8O_2$, coined by contraction of 2-methylnaphthalene-1,4-dione or 2-methyl-1,4-naphthoquinone

menaquinone
coined by contraction of (3-substituted) 2-methyl-1,4-naphthoquinone

mendelevium
Md, named for the Russian chemist Dmitry Ivanovich Mendeleyev (1834-1907)

Menshutkin reaction
named for the Russian chemist Nikolai Aleksandrovich Menshutkin (1842-1907)

menthane
$C_{10}H_{20}$, derived from menth(o)- and -an(e)

menthene
$C_{10}H_{18}$, derived from menth(o)- and -ene

menthiafolin
$C_{26}H_{36}O_{12}$, derived (with contraction) from the species name *Menyanthes trifoliata* (buckbean), from *Menyanthes* (bog plants), from *menyein* (Greek: to disclose) and anth(o)- – referring to this plant's successively opening raceme – and fol-, and -in(e)

menth(o)-
derived from *mentha* (Latin: mint plant)

menthol
$C_{10}H_{20}O$, derived from menth(o)- and -ol

menthone
$C_{10}H_{18}O$, derived from menth(o)- and -one

menthyl
$C_{10}H_{19}-$, derived from menth(o)- and -yl

mepartricin
coined by contraction of partricin methyl ester

mephit-
derived from *mephitis* (Latin: noxious exhalation from the earth), from Oscan

mephitic acid
CO_2, an archaic name for carbon dioxide, derived from mephit-

mephitic air
N_2, CO_2, an archaic name, derived from mephit- and air

mer-
derived from *meridionalis* (Latin: meridional), from *meridies* (Latin: noon, south), from *medius* (Latin: middle) and *dies* (Latin: day)

-mer
derived from *meris*, *meros* (Greek: part, share)

mercaptal
coined by contraction of mercaptan and acetal

mercaptan
coined by contraction of the term *corpus mercurium captans* (New Latin: a substance striving to seize mercury)

mercapto- (sulfanyl)
$HS-$, derived from mercaptan

mercaptol (thioacetal)
coined by contraction of mercaptan and -ol

mercapturic acid (*N*-acetyl-L-cysteine)
$C_5H_9NO_3S$, derived from mercaptan, urine, and acid

mercerization
named for the British chemist John Mercer (1791-1866)

mercuric
Hg^{2+}, derived from mercury and -ic

mercurium
Hg, New Latin name for mercury, after the Roman god Mercurius whose name is possibly related to *merx* (Latin: merchandise)

mercur(o)-
derived from mercurium

mercurocene
$C_{10}H_{10}Hg$, derived from mercur(o)- and -ocene

mercurous
Hg_2^{2+}, derived from mercury and -ous

mercury
Hg, named for the Roman god Mercurius whose name is possibly related to *merx* (Latin: merchandise)

meri- (mero-)
derived from *meris*, *meros* (Greek: part, share)

meriquinone
derived from meri- and quinone – referring to the only partially quinoid character of these compounds

mero- (meri-)
derived from *meris*, *meros* (Greek: part, share)

merocyanin
derived from mero- and cyanin

merotropy (tautomerism)
derived from mero- and –tropy

merrifield solid-phase peptide synthesis
named for the US chemist Robert Bruce Merrifield (1921-2006)

MES
$C_6H_{13}NO_4S$, an abbreviation for morpholine-4-(ethanesulfonic acid)

mesaconic acid
$C_5H_6O_4$, probably coined by contraction of mesocitraconic acid

mescaline
$C_{11}H_{17}NO_3$, derived from mescal (*Lophophora williamsii*), ultimately from *metl* (Nahuatl: maguey) and *ixcalli* (Nahuatl: stew, decoction), and -in(e)

mesembrine
$C_{17}H_{23}NO_3$, derived from the genus name *Mesembryanthemum* (herbs), from *mesembria* (Greek: midday, noon), from mes(o)- and *hemera* (Greek: day), and anth(o)-, and -in(e)

mesit-
derived from *mesites* (Greek: mediator)

[1]mesityl
C_3H_5, derived from mesit- and -yl – referring to the obsolete notion of acetone as mesityl hydroxide with a boiling point between those of diethyl ether and ethanol

[2]mesityl
$C_9H_{11}-$, derived from mesitylene and -yl

mesitylene
C_9H_{12}, derived from [1]mesityl and -ene

mesityl oxide
$C_6H_{10}O$, derived from [1]mesityl and oxide

mesna
$C_2H_5NaO_3S_2$, an abbreviation for *2-Mercaptoethansulfonsäurenatriumsalz*

(German: 2-mercaptoethanesulfonic acid sodium salt)

mes(o)-
derived from *mesos* (Greek: middle)

mesogen
derived from mes(o)- and -gen

mesoionic
derived from mes(o)- and ionic

mesolite (cotton stone)
$Na_2Ca_2(Al_6Si_9)O_{30}\cdot 8H_2O$, derived from mes(o)- and -lite – referring to this mineral's equal relationship with natrolite and scolecite

mesomerism
derived from mes(o)- and -mer

mesomorphism
derived from mes(o)- and *morphe* (Greek: form, shape)

mesoperiodic acid
$H_6I_2O_{10}$, derived from mes(o)- and periodic acid – referring to this acid's intermediate position, with regard to hydration, between orthoperiodic acid (H_5IO_6) and metaperiodic acid (HIO_4)

mesoperrhenic acid
$H_6Re_2O_{10}$, derived from mes(o)- and perrhenic acid – referring to this acid's intermediate position, with regard to hydration, between orthoperrhenic acid (H_5ReO_6) and metaperrhenic acid ($HReO_4$)

mesophase
derived from mes(o)- and phase

mesothorium I
^{228}Ra, an obsolete name for a radium isotope, derived from mes(o)- and thorium

mesothorium II
^{228}Ac, an obsolete name for an actinium isotope, derived from mes(o)- and thorium

mesoxalic acid
$C_3H_2O_5$, derived from mes(o)- and oxalic acid

mesyl
CH_3SO_2-, coined by contraction of methanesulfonyl

met(a)-
derived from *meta* (Greek: between, with, after)

metabolism
derived from *metabole* (Greek: change), from *metaballein* (Greek: to change), from met(a)- and *ballein* (Greek: to throw)

metaboric acid
$(HBO_2)_n$, derived from met(a)- and boric acid

metacinnabarite
HgS, derived from met(a)- and cinnabarite

metal
derived from *metallon* (Greek: mine, metal), from *metallan* (Greek: to search after, to inquire about)

metaldehyde
$(C_2H_4O)_n$, derived (with contraction) from met(a)- and acetaldehyde

metalepsis (substitution)
derived from *metalepsis* (Greek: exchange, transposition), from met(a)- and -leptic

metalliferous
derived from metal and *ferre* (Latin: to carry)

metallocene
$C_{10}H_{10}M^{II}$, derived from metal and -ocene; patterned after ferrocene

[1]**metalloid (half-metal)**
derived from metal and -oid

[2]**metalloid (nonmetal)**
derived (improperly) from metal and -oid

metallurgy
derived from metal and -urgy

metamerism
derived from met(a)- and -mer

metametal
derived from met(a)- and metal

metanilic acid
$C_6H_7NO_3S$, coined by contraction of *m*-anilinesulfonic acid

metaperiodic acid (periodic acid)
HIO_4, derived from met(a)- and periodic acid

metaperrhenic acid
$HReO_4$, derived from met(a)- and perrhenic acid

metaphanine
$C_{19}H_{23}NO_5$, coined by contraction of metastephanine, from met(a)- and stephanine

metaphosphoric acid
$(HPO_3)_n$, derived from met(a)- and phosphoric acid

metasilicic acid
$(H_2SiO_3)_n$, derived from met(a)- and silicic acid

metastable
derived from met(a)- and stable

metastable radio-technetium
^{99m}Tc, derived from radio- and technetium

metathesis
derived from *metathesis* (Greek: change of position), from met(a)- and *tithenai* (Greek: to place, to set)

metatungstate
$H_2W_{12}O_{40}{}^{6-}$, derived from met(a)- and tungstate

metavanadate
$(VO_3^-)_n$, derived from met(a)- and vanadate

meteloidine
$C_{13}H_{21}NO_4$, derived from the specific epithet of the species name *Datura meteloides* DC. (Jimson weed, sacred datura), from *meteloides* (New Latin: resembling *Datura metel*, garden thorn apple), from metel nut, from *jouz mathal* (Arabic: metel nut), and -idin(e)

meteorite
derived from meteor, from *meteoron* (Greek: astronomical phenomenon), from met(a)- and, ultimately, *airein* (Greek: to lift, to raise), and -ite

-meter
derived from *metron* (Greek: measure)

meth(a)-
derived from methane

methacrylic acid
$C_4H_6O_2$, coined by contraction of α-methylacrylic acid

methacrylonitrile
C_4H_5N, derived from methacrylic acid and nitrile

methadone
$C_{21}H_{27}NO$, coined by contraction of 6-

(dimethylamino)-4,4-diphenylheptan-3-one

methamphetamine
$C_{10}H_{15}N$, coined by contraction of *N*-methylamphetamine

methane
CH_4, derived from meth(a)-, ultimately from methylene, and -an(e)

methanol
CH_4O, derived from methane and -ol

methemoglobin
coined by contraction of metahemoglobin

methenamine (urotropin)
$C_6H_{12}N_4$, derived from methene and amine

methene
CH_2, derived from meth(a)- and -ene

methionine
$C_5H_{11}NO_2S$, derived from meth(a)-, thi(o)-, and -in(e)

methyl
derived from methylene and -yl

methyl blue (Helvetia blue)
$C_{37}H_{27}N_3Na_2O_9S_3$, derived from methyl and blue – probably referring to this dye's triphenylmethane core

methylene
derived from 'methylene bihydrate', the first structure-based name of methanol, and 'methylene hydrate', the analogous name for dimethyl ether, derived from *methy* (Greek: wine) and *hyle* (Greek: wood)

methylene blue (Swiss blue)
$C_{16}H_{18}ClN_3S$, derived from methylene and blue – referring to the formal insertion of four methylene groups into the four N–H bonds of the parent diamine

methylene green
$C_{16}H_{17}ClN_4O_2S$, derived from methylene and green – referring to the formal insertion of four methylene groups into the four N–H bonds of the parent diamine

methylene red
$C_{17}H_{19}ClN_2S$, derived from methylene and red – referring to the formal insertion of four methylene groups into the four N–H bonds of the parent diamine

methylenomycin
derived from methylene and -mycin – referring to the presence of an exomethylene group in these antibiotics

methyl green (ethyl green)
$C_{27}H_{35}BrClN_3$, derived from methyl and green – probably referring to this dye's triphenylmethane core

methyl orange (helianthine)
$C_{14}H_{14}N_3NaO_3S$, derived from methyl and orange – referring to this dye's orange color and to the presence of methyl groups

methyl red
$C_{15}H_{15}N_3O_2$, derived from methyl and red – referring to this dye's red color and to the presence of methyl groups

methyl violet
derived from methyl and violet – referring to this dye's violet color and to the presence of methyl groups

methyl yellow (butter yellow)
$C_{14}H_{15}N_3$, derived from methyl and yellow – referring to this dye's yellow color and to the presence of methyl groups

methysticin (kavahin, kavatin)
$C_{15}H_{14}O_5$, derived from the specific epithet of the species name *Piper methysticum* Forst. (kava), from *methystos* (Greek:

drunk), from *methyskein* (Greek: to make drunk), from *methyein* (Greek: to be drunk), from *methy* (Greek: wine), and -in(e)

-metry
derived from -meter

mevaldic acid
$C_6H_{10}O_4$, derived from mevalonic acid and aldehyde

mevalonic acid
$C_6H_{12}O_4$, coined by contraction of dihydroxymethylvalerolactone and acid

mevinoid
derived (with contraction) from mevalonic acid and -oid

mevastatin
$C_{23}H_{34}O_5$, derived from mevalonic acid and -statin – referring to this compound's inhibition of the biosynthesis of mevalonic acid and, thus, lipid biosynthesis

mexicain
derived from the specific epithet of the species name *Pileus mexicanus* (wild papaya), from Mexico, and -in(e)

meyerhofferite
$Ca_2B_6O_6(OH)_{10}\cdot 2H_2O$, named for the German chemist Wilhelm Meyerhoffer (1864-1906) who synthesized this mineral

Meyer reaction
named for the 19th century German chemist G. Meyer

Meyer-Ronge catalyst
supported copper, named for the 20th century German chemists F. R. Meyer and Grethe Ronge

Meyer-Schuster rearrangement
named for the German chemists Karl Heinrich Meyer (1883-1952) and Kurt Eduard Schuster (1892-1990)

Meyer synthesis
named for the German chemist Victor Meyer (1848-1897)

Meyers aldehyde synthesis
named for the US chemist Albert Irving Meyers (born 1933)

MFO
an abbreviation for mixed function oxidase

miargyrite
$AgSbS_2$, derived from *meion* (Greek: less), argyr(o)-, and -ite – referring to this mineral's relatively low silver content

MIBK
$C_6H_{12}O$, an abbreviation for methyl isobutyl ketone

mica
derived from *mica* (Latin: crumb), also influenced by unrelated *micare* (Latin: to glisten)

micaceous
derived from mica

Michael addition
named for the US chemist Arthur Michael (1853-1942)

Michaelis-Arbuzov reaction (Arbuzov-Michaelis reaction)
named for the German chemist August Karl Arnold Michaelis (1847-1916) and the Russian chemist Aleksandr Erminingeldovich Arbuzov (1877-1968)

Michaelis-Menten equation
named for the German-US biochemist Leonor Michaelis (1875-1949) and the Canadian biochemist Maud Leonora

Menten (1879-1960)

Michler's base
$C_{17}H_{22}N_2$, named for the German chemist Wilhelm Traugott Michler (1846-1889)

Michler's ketone
$C_{17}H_{20}N_2O$, named for the German chemist Wilhelm Traugott Michler (1846-1889)

-micin
coined by variation of -mycin

micranthine
$C_{34}H_{32}N_2O_5$, derived from the specific epithet of the species name *Daphnandra micrantha* (Til.) Benth. (socket sassafras), from micr(o)- and anth(o)-, and -in(e)

micr(o)-
derived from *mikros* (Greek: small)

microcline
$KAlSi_3O_8$, derived from micr(o)- and -cline – referring to the triclinic nature of this mineral's crystals

micrococcin
derived from the bacterial genus name *Micrococcus*, from micr(o)- and, ultimately, *kokkos* (Greek: grain, kernel, kermes berry), and -in(e)

microcosmic salt
NH_4NaHPO_4, an archaic name for ammonium sodium hydrogenphosphate – referring to this salt's origin in man

microlite
$(Ca,Na)_2Ta_2O_6(O,OH,F)$, derived from micro- and -lite – referring to the small crystals of this mineral

micron
derived from micro-

micronomycin
$C_{20}H_{41}N_5O_7$, derived (with contraction) from the bacterial genus name *Micromonospora*, from micr(o)-, mon(o)-, and spore, and -mycin

Midland's reagent
$C_{18}H_{31}B$, named for the US chemist Michael Mark Midland (born 1946)

Miescher degradation
named for the Swiss chemist Karl Miescher (1892-1974)

Mignonac reaction (Moureu-Mignonac reaction)
named for the 20th century French chemist Georges Mignonac

milarite
$(K,Na)Ca_2(Be,Al)_3Si_{12}O_{30}\cdot H_2O$, named after this mineral's locality Val Milar, Switzerland

Milas hydroxylation of olefins
named for the US chemist Nicholas A. Milas (1897-1971)

milbemycin
derived from *Milbe* (German: mite) and -mycin – referring to this antibiotic's miticidal activity

mildiomycin
$C_{19}H_{30}N_8O_9$, derived from mildew and -mycin – referring to this antibiotic's anti-mildew activity

milky quartz
SiO_2, named after its appearance

millerite
NiS, named for the British mineralogist William Hallowes Miller (1801-1880)

Millon base
HHg_2NO, named for the French physician and chemist N. A. Eugène Millon (1812-1867)

Millon reaction
named for the French physician and chemist N. A. Eugène Millon (1812-1867)

Milori blue (Berlin blue, Paris blue, Prussian blue)
$C_6Fe_2KN_6$, named for the 19th century French dye manufacturer A. Milori

Milori green (chrome green)
named for the 19th century French dye manufacturer A. Milori

mimetite
$Pb_5(AsO_4)_3Cl$, derived from *mimetes* (Greek: imitator), from *mimesthai* (Greek: to imitate), and -ite – referring to this mineral's similarity to pyromorphite

mimosine
$C_8H_{10}N_2O_4$, derived from the genus name *Mimosa* (mimosa), from *mimos* (Greek: mime, imitator) – referring to the notion that these plants imitate animal life by their visible reactions to touch – and -in(e)

MINDO
an abbreviation for modified intermediate neglect of differential overlap

mineral
derived from *mineralis* (Medieval Latin: of the ore, of the mine), from *minera* (Medieval Latin: ore, mine)

minette (limonite)
a rock species, derived from *minette*, diminutive of *mine* (French: mine)

minium
Pb_3O_4, derived from *minium* (Latin: cinnabar), probably of Iberian origin

minnesotaite
$(Fe,Mg)_3Si_4O_{10}(OH)_2$, named after this mineral's locality, the state of Minnesota, USA

mirabilite (Glauber's salt)
$Na_2SO_4 \cdot 10H_2O$, derived from *sal mirabile* (Latin: wonderful salt) and -ite – referring to this mineral's perceived anticonstipatory effect

miraculin
derived from miracle fruit (*Synsepalum dulcificum* (Schum. Daniell.) and -in(e)

mirbane oil (oil of Myrbane, nitrobenzene)
$C_6H_5NO_2$, probably a trademark derived from *mirobolan*, *mirobalan* (Middle French: aromatic nuts used in perfumery), from *myron* (Greek: unguent, perfume) and *balanos* (Greek: acorn)

mirex
$C_{10}Cl_{12}$, originally coined as a trademark, probably by contraction of pismire, from *pissemyre* (Danish: pismire), from *pisse* (Danish: to pee) and *myre* (Danish: ant), and exterminator – referring to this compound's insecticidal properties

misch metal
derived from *Mischmetall* (German: mixed metal), from *mischen* (German: to mix)

Mislow-Evans rearrangement
named for the US chemists Kurt Mislow (born 1923) and David A. Evans (born 1941)

miticide
derived from mite and -cide

Mitis green (Schweinfurt green)
named for the Austrian industrialist Ignaz Edler von Mitis (1771-1842)

mitomycin
$C_{15}H_{18}N_4O_5$, derived from mitosis, from *mitos* (Greek: thread), and -mycin – referring to this antibiotic's antimitotic activity

mitragynine
$C_{23}H_{30}N_2O_4$, derived from the species name *Mitragyna speciosa* (kratom), from *mitra* (Greek: cap) and *gyne* (Greek: woman), and *speciosus* (Latin: good-looking, showy), and -in(e)

Mitscherlich's rule
named for the German chemist Eilhard Mitscherlich (1794-1863)

Mitscherlich's test
named for the German chemist Eilhard Mitscherlich (1794-1863)

Mitsunobu reaction
named for the Japanese chemist Oyo Mitsunobu (born 1934)

Mitsunobu reagent
named for the Japanese chemist Oyo Mitsunobu (born 1934)

mizoribine (bredinin)
$C_9H_{13}N_3O_6$, derived (with contraction) from the name of the discoverer, the 20th century Japanese chemist Kimio Mizuno, ribose, and -in(e)

MMA
$C_5H_8O_2$, an abbreviation for methyl methacrylate

MMT
$C_9H_7MnO_3$, an abbreviation for (methylcyclopentadienyl)manganese tricarbonyl

MNDO
an abbreviation for modified neglect of diatomic overlap

MO
an abbreviation for molecular orbital

Möbius aromaticity
named for the German mathematician August Ferdinand Möbius (1790-1868)

Möbius compound
named for the German mathematician August Ferdinand Möbius (1790-1868)

Möbius electrolysis
named for the German chemist Hans-Heinrich Möbius (born 1929)

mocimycin (delvomycin, kirromycin)
$C_{43}H_{60}N_2O_{12}$, derived (with contraction) from the specific epithet of the bacterial species name *Streptomyces ramosissimus*, from *ramosissimus*, superlative of *ramosus* (Latin: branched), from *ramus* (Latin: branch), and -mycin

Mohr's salt
$(NH_4)_2Fe(SO_4)_2 \cdot 6H_2O$, named for the German chemist Carl Friedrich Mohr (1806-1879)

Mohs hardness
named for the German chemist and mineralogist Friedrich Mohs (1773-1839)

molal
derived from mole

molar
derived from mole

molasses
ultimately derived from *mellaceus* (Late

Latin: resembling honey), from mel(l)i-

mole
coined by contraction of gram-molecule

molecule
derived from *molecula* (New Latin: molecule), diminutive of *moles* (Latin: mass)

Molisch reaction
named for the German botanist Hans Molisch (1856-1937)

molluscicide
derived from mollusc and –cide

moloxide
derived from molecule and oxide – referring to such compounds' formation by 1:1 reaction of a substrate with oxygen

molozonide
derived from molecule and ozonide – referring to such compounds' formation by 1:1 reaction of a substrate with ozone

molybdenite
MoS_2, derived from *molybdos* (Greek: lead, graphite, molybdenite) and -ite

molybdenum
Mo, derived from molybdenite and -um

molybdocene
$C_{10}H_{10}Mo$, derived from molybdenum and -ocene

molybdomenite
$PbSeO_3$, derived from *molybdos* (Greek: lead, graphite, molybdenite), *mene* (Greek: moon) – referring to selenium's naming after the Greek moon goddess Selene – and -ite – referring to this mineral's lead and selenium content

monacetin
$C_5H_{10}O_4$, coined by contraction of *Monoacetylglycerin* (German: monoacetylglycerol)

monactin
$C_{41}H_{66}O_{12}$, coined by contraction of monohomononactin, from mon(o)-, homo-, and nonactin

-monas
derived from *monas* (Late Latin: unit), from mon(o)-

monazite
(Ce,La,Th,Y)PO_4, derived from *monazein* (Greek: to live alone), from mon(o)- and -ite – referring to the isolated occurrence and rarity of this mineral

Mond metal
(Ni,Cu,Mn), named for the German-British inventor and industrialist Ludwig Mond (1839-1909)

Mond nickel
Ni, named for the German-British inventor and industrialist Ludwig Mond (1839-1909)

Mond process
named for the German-British inventor and industrialist Ludwig Mond (1839-1909)

monel
(Ni,Cu), coined as a trademark, after the US industrialist and philanthropist Ambrose Monell (1873-1921)

monellin
named for the Monell Chemical Senses Center, Philadelphia, PA, ISA where this natural polypeptide was isolated and characterized, ultimately after the US industrialist and philanthropist Ambrose Monell (1873-1921)

monensin
$C_{36}H_{62}O_{11}$, derived (with contraction) from the specific epithet of the bacterial species name *Streptomyces cinnamonensis*, from cinnamon, and -in(e)

monergolic
derived from mon(o)- and -ergolic

mon(o)-
derived from *monos* (Greek: alone)

monobactam
coined by contraction of monocyclic β-lactam

monobutyrin
$C_7H_{14}O_4$, derived from *Monobutyrylglycerin* (German: monobutyrylglycerol)

monoclinic
derived from mon(o)- and *klinein* (Greek: to bend, to incline)

monocrotaline
$C_{16}H_{23}NO_6$, derived from mon(o)-, the genus name *Crotalaria* (showy rattlebox), from crota(lo)-, and -in(e)

monokine
derived from monozyte and -kine

monomer
derived from mon(o)- and -mer

monooxygenase
derived from mon(o)- and oxygenase – referring to the fact that these enzymes only transfer one oxygen atom of an oxygen molecule to their substrates

monopalmitin
$C_{19}H_{38}O_4$, coined by contraction of *Monopalmitoylglycerin* (German: monopalmitoylglycerol)

monorden (radicicol)
$C_{18}H_{17}ClO_6$, derived (with contraction) from the fungal species name *Monosporium bonorden*

monosaccharide
derived from mon(o)- and saccharide

monostearin
$C_{21}H_{42}O_4$, coined by contraction of *Monostearoylglycerin* (German. Monostearoylglycerol)

monoterpene
$(C_5H_8)_2$, derived from mono- and terpene

Monsanto process
named for the chemical company Monsanto Company, St. Louis, MO, USA

montanic acid
$C_{28}H_{56}O_2$, derived from montan wax

montan wax
derived from *Montanwachs* (German: mountain wax), ultimately from *mons* (Latin: mountain)

Mont-Cenis process
named for the Mont-Cenis coal mine, Herne, Germany – referring to the first customer to acquire an ammonia plant designed for this process

montebrasite
$LiAlPO_4(OH)$, named after this mineral's locality Montebras, France

monticellite
$CaMgSiO_4$, named for the Italian mineralogist Teodoro Monticelli (1759-1846)

montmorillonite
$(Al,Mg)_2Si_4O_{10}(OH)_2{\cdot}nH_2O$, named after this mineral's locality Montmorillon,

Département Vienne, France

Moore cyclization
named for the US chemist Harold Wesley Moore (born 1936)

Moore's ketene
C_7H_9NO, named for the US chemist Harold Wesley Moore (born 1936)

Moore-Stein analysis
named for the US biochemists Stanford Moore (1913-1982) and William Howard Stein (1911-1980)

MOPS
$C_7H_{15}NO_4S$, an abbreviation for morpholine-4-(propanesulfonic acid)

mordenite (clinoptilolite, ptilolite)
$(K,Ca,Na)_6(A_{19}Si_{39})O_{96}\cdot 29H_2O$, named after this mineral's locality Morden, King's County, Nova Scotia, Canada

morganite (rose beryl)
$Be_3Al_2Si_6O_{18}$, named for the US industrialist John Pierpont Morgan (1837-1913)

Morgan-Walls reaction (Pictet-Hubert reaction)
named for the 20th century British chemists Gilbert T. Morgan and Leslie Percy Walls

morin
$C_{15}H_{10}O_7$, derived from the genus name *Morus* (mulberry trees), from *morus* (Latin: mulberry tree), and -in(e) – referring to this compound's occurrence in the fustic tree (*Chlorophora tinctoria*) which once was erroneously believed to belong to the genus *Morus*

morindin
$C_{27}H_{30}O_{14}$, derived from the genus name *Morinda* (tropical trees, shrubs), from *morus inda* (Latin: Indian mulberry tree), from *morus* (Latin: mulberry tree) and *indus* (Latin: Indian), and -in(e)

morphinan
$C_{16}H_{21}N$, derived from morphine and -an(e)

morphine (morphium)
$C_{17}H_{19}NO_3$, named for Morpheus, the Greek god of dreams, from *morphe* (Greek: form, shape) – referring to this compound's analgesic and sedative properties

morph(o)-
derived from *morphe* (Greek: form, shape)

morpholine
C_4H_9NO, coined by contraction of morphine and quinoline – referring to an erroneous belief that morphine contained a morpholine ring as part of its structure

[1]mortar
(material), derived from *mortarium* (Latin: [2]mortar)

[2]mortar
(tool), derived from *mortarium* (Latin: [2]mortar)

mosandrite
$Na(Na,Ca)_2(Ca,Ce,Y)_4(Ti,Nb,Zr)(Si_2O_7)_2(O,F)_2F_3$, named for the Swedish chemist and mineralogist Carl Gustav Mosander (1797-1858)

Moseley's law
named for the British physicist Henry George Jeffreys Moseley (1887-1915)

Mosher's acid (Mosher's reagent)
$C_{10}H_9F_3O_3$, named for the US chemist Harry Stone Mosher (1915-2001)

moss agate
SiO_2, named after the mosslike or dendritic markings of this mineral

Mössbauer effect
named for the German physicist Rudolf Ludwig Mössbauer (born 1929)

Mössbauer spectroscopy
named for the German physicist Rudolf Ludwig Mössbauer (born 1929)

motilin
derived from motility and -in(e)

mottramite
$PbCu(VO_4)(OH)$, named after its locality Mottram, St. Andrews, Cheshire, England, UK

Moureu-Mignonac reaction (Mignonac reaction)
named for the 20th century French chemists Charles Moureu and Georges Mignonac

MPIC
an abbreviation for mobile phase ion chromatography

MPS
an abbreviation for mucopolysaccharide polysulfate

MPTP
$C_{12}H_{15}N$, an abbreviation for 1-methyl-4-phenyl-1,2,3,6-tetrahydropyridine

MS
an abbreviation for mass spectroscopy

MSH
an abbreviation for melanocyte-stimulating hormone

MTB (MTBE)
$C_5H_{12}O$, an abbreviation for methyl *tert*-butyl ether

MTG process
an abbreviation for methanol to gasoline

mucic acid (galactaric acid)
$C_6H_{10}O_8$, derived from muc(o)-

mucin
derived from muc(o)- and -in(e)

mucinase (hyaluronidase)
derived from mucin and -ase

muc(o)-
derived from *mucus* (Latin: nasal mucus)

mucochloric acid
$C_4H_2Cl_2O_3$, derived (with contraction) from pyromucic acid and ^{2}chlor(o)-

mucochloric anhydride
$C_8H_2Cl_4O_5$, derived from mucochloric acid and anhydride

muconic acid
$C_6H_6O_4$, coined by variation of mucic acid; patterned after itaconic acid

mucopolysaccharide
derived from muc(o)- and polysaccharide

mucoprotein
derived from muc(o)- and protein

Muetterties' rule
named for the US chemist Earl Leonard Muetterties (1927-1984)

Mukaiyama aldol reaction
named for the Japanese chemist Teruaki Mukaiyama (born 1927)

Mukaiyama-Michael reaction
named for the Japanese chemist Teruaki Mukaiyama (born 1927) and the Michael reaction

Mukaiyama's reagent
C_6H_7ClIN, named for the Japanese chemist Teruaki Mukaiyama (born 1927)

mullite
$Al_4Si_2O_{10}$, named after this mineral's locality, the island of Mull, Scotland, UK

multi-
derived from *multus* (Latin: much, many)

münchnone
named for the city of *München* (Munich, Germany), where these compounds were prepared for the first time; patterned after sydnone

muonium
Mu, derived from muon, from the Greek letter μ (mu), and -ium

Murahashi reaction
named for the Japanese chemist Shin-Ichi Murahashi (born 1937)

muramic acid
$C_9H_{17}NO_7$, derived (with contraction) from *murus* (Latin: wall) and glucosamine – referring to this compound's occurrence in bacterial cell walls

muramidase (muraminidase, lysozyme)
derived from muramic acid, amide, and -ase

murein
derived from *murus* (Latin: wall) and -ein(e)

murex-
derived from *murex* (Latin: purple shell), from *myax* (Greek: sea-mussel)

murexide
$C_8H_8N_6O_6$, derived from murex- and ¹-ide – referring to the purple color of aqueous solutions of this compound

murexine
$C_{11}H_{18}N_3O_2^+$, derived from the genus name *Murex* (molluscs), from murex-, and -in(e)

muriate
Cl^-, an obsolete name for chloride, derived from muriatic acid

muriatic acid
HCl, an obsolete name for aqueous hydrochloric acid, derived from *muriaticus* (Latin: pickled in brine), from *muria* (Latin: brine)

murmanite
$Na_3(Ti,Nb)_4O_4(Si_2O_7)_2{\cdot}4H_2O$, most likely named after this mineral's locality Murmansk, Russia

muscalure
$C_{23}H_{46}$, derived from the genus name *Musca* (fly), from *musca* (Latin: fly), and lure

musc(ar)-
derived from the specific epithet of the species name *Amanita muscaria* (fly agaric), from *muscarius* (Latin: of a fly), from *musca* (Latin: fly)

muscarine
$C_9H_{20}NO_2$, derived from musc(ar)- and -in(e)

muscazone
$C_5H_6N_2O_4$, derived from musc(ar)-, az(a)-, and -one

muscimol (agarin)
$C_4H_6N_2O_2$, derived from musc(ar)-, imine, and -ol

muscone
$C_{16}H_{30}O$, derived from *muscus* (Latin: musk) and -one

muscovite (Muscovy glass)
$KAl_2(Si,Al)_4O_{10}(OH,F)_2$, named after this mineral's locality, the province of Muscovy, Russia, from *Muscova* (New Latin: Moscow, Russia), and –ite

mustard
(as in aniline mustard), derived from mustard gas

mustard gas (lost, yperite)
$C_4H_8Cl_2S$, named for the characteristic mustard odor of this compound which, however, at ambient temperature is a heavy oil, not a gas

mutarotase (aldose 1-epimerase)
derived from mutarotation and -ase

mutarotation
derived from *mutare* (Latin: to change, to exchange) and rotation

mutase
derived (with contraction) from dismutation and -ase

mutein
derived from mutation and protein

mycaminose
$C_8H_{17}NO_4$, derived from myc(o)-, amine, and -ose

Muthmann's liquid
$C_2H_2Br_4$, named for the German chemist Friedrich Wilhelm Muthmann (1861-1913)

mycarose
$C_7H_{14}O_4$, probably derived from a contracted anagram of carbomycin and -ose

mycelianamide
$C_{22}H_{28}N_2O_5$, derived from mycelium, from myc(o)- and *helos* (Greek: nail, wart, callus), and amide

-myces
derived from *mykes* (Greek: fungus)

mycetin
derived from myc(o)- and -etin(e)

-mycin
derived from the bacterial order name *Actinomyces*, from myc(o)-, and -in(e)

myc(o)-
derived from *mykes* (Greek: fungus)

mycobacidin (acidomycin, cinnamonin)
$C_9H_{15}NO_3S$, derived (with contraction) from myc(o)-, bacillus, and -idin(e)

mycobacillin
$C_{65}H_{85}N_{13}O_{30}$, derived from myc(o)-, bacillus, and -in(e)

mycobactin
derived from the bacterial genus name *Mycobacterium*, from myc(o)- and bacterium, and -in(e)

mycolic acid
derived from the bacterial genus name *Mycobacterium*, from myc(o)- and bacterium, and -ol

mycomycin
$C_{13}H_{10}O_2$, derived from myc(o)- and -mycin

mycophenolic acid
$C_{17}H_{20}O_6$, derived from myc(o)- and phenol

mycosamine
$C_6H_{13}NO_4$, derived from myc(o)-, -ose, and amine

mycosterol
derived from myc(o)- and sterol

mycotoxin
derived from myc(o)- and toxin

myelin
derived from myel(o)- and -in(e)

myel(o)-
derived from *myelos* (Greek: marrow),

ultimately from *mys* (Greek: muscle)

Myers cyclization
named for the 20th century US chemist Andrew G. Myers

my(o)-
derived from *mys* (Greek: muscle)

myoglobin
derived from my(o)- and -globin

myosin
derived from my(o)- and -in(e)

myrcene
$C_{10}H_{16}$, derived from the genus name *Myrcia* (tropical trees, shrubs), ultimately from *myrtos* (Greek: myrtle), probably of Semitic origin, and -ene

myricetin
$C_{15}H_{10}O_8$, derived from the genus name *Myrica* (aromatic shrubs), from *myrike* (Greek: tamarisk), probably of Semitic origin, and -etin(e)

myricyl alchol
$C_{30}H_{62}O$, derived from the genus name *Myrica* (myrtle), from *myrike* (Greek: tamarisk), probably of Semitic origin, and -yl

myristic-
derived from the genus name *Myristica* (nutmeg tree), ultimately from *myron* (Greek: sweet oil, unguent, perfume)

myristic acid
$C_{14}H_{28}O_2$, derived from myristic-

myristicin
$C_{11}H_{12}O_3$, derived from myristic- and -in(e)

myristoyl
$C_{14}H_{27}O-$, derived from myristic acid and -oyl

myristyl
$C_{14}H_{29}-$, derived from myristic acid and -yl

myronic acid
$C_{10}H_{17}NO_9S_2$, derived from *myron* (Greek: sweet oil, unguent, perfume)

myrosin (myrosinase)
derived from *myron* (Greek: sweet oil, unguent, perfume), -ose, and -in(e)

myrosinase (myrosin)
derived from myrosin and -ase

myxin
$C_{13}H_{10}N_2O_4$, derived from the bacterial genus name *Myxobacter*, from myx(o)- and -bacter, and -in(e)

myx(o)-
derived from *myxa* (Greek: nasal slime)

myxothiazol
$C_{25}H_{33}N_3O_3S_2$, derived from the bacterial genus name *Myxococcus*, from myx(o)-, and thiazol

N

NAA
an abbreviation for neutron activation analysis

nacrite
$Al_2Si_2O_5(OH)_4$, derived from *nacre* (French: mother-of-pearl), ultimately from *naqqarah* (Arabic: bowl), and -ite

NAD (NAD$^+$, DPN)
$C_{21}H_{27}N_7O_{14}P_2$, an abbreviation for nicotinamide-adenine dinucleoside

NADH (NADH$_2$)
$C_{21}H_{29}N_7O_{14}P_2$, an abbreviation for 1,4-dihydronicotinamide-adenine dinucleoside

NADP (NADP$^+$, TPN)
$C_{21}H_{28}N_7O_{17}P_3$, an abbreviation for nicotinamide-adenine dinucleotide phosphate

NADPH (NADPH$_2$, TPN)
$C_{21}H_{30}N_7O_{17}P_3$, an abbreviation for 1,4-dihydronicotinamide-adenine dinucleotide phosphate

Nagata hydrocyanation
named for the 20th century Japanese chemist Wataru Nagata

nagyagite
$Sb_{1.5}AuPb_5(S_{6.7}Te_{2.55})$, named after this mineral's locality Nagyag mine, Hungary (now Sacaramb, Romania)

naled
$C_4H_7Br_2Cl_2O_4P$, originally a trademark, of undisclosed origin

nalidixic acid
$C_{12}H_{12}N_2O_3$, originally a trademark, probably derived from naphthyridine and carboxylic acid

Nametkin rearrangement
named for the Russian chemist Sergei Semyonovich Nametkin (1876-1950)

NANA
$C_{11}H_{19}NO_9$, an abbreviation for *N*-acetylneuraminic acid

nandinine
$C_{19}H_{19}NO_4$, derived from the genus name *Nandina* (nandina), from *nandin* (Japanese: nandina), and -in(e)

nano-
derived from *nanos* (Greek: dwarf)

napadisilate
$C_{10}H_6O_6S_2^{2-}$, coined by contraction of naphthalene-1,5-disulfonate

napalm
coined by contraction of naphthenic acid and palmitic acid

napelline
$C_{22}H_{33}NO_3$, derived from the specific epithet of the species name *Aconitum napellus* L. (monkshood), from *napellus*, diminutive of *napus* (Latin: turnip), and -in(e)

naphtha
derived from *naphtha* (Greek: naphtha), ultimately from *neft* (Persian: naphtha)

naphthacene (tetracene)
$C_{18}H_{12}$, derived from naphth(o)- and -acene

naphthalene
$C_{10}H_8$, derived from naphtha and -ene

naphthene
derived from naphtha and -ene – referring to these hydrocarbons' molecular formula C_nH_{2n} and, thus, analogy with alkenes

naphthenic acid
derived from naphthene

naphth(o)-
derived from naphthalene

naphthoic acid
$C_{11}H_8O_2$, derived from naphth(o)-; patterned after benzoic acid

naphthyl
$C_{10}H_7-$, derived from naphth(o)- and -yl

naphthyridine
$C_8H_6N_2$, coined by contraction of naphth(o)- and pyridine

naphthol AS
$C_{17}H_{13}NO_2$, coined as a trademark, from *Naphtol AS* (German: naphthol AS), from *Naphtol* (obsolete German: naphthol) and *Amid einer Säure* (German: amide of an acid)

Naples yellow
basic lead antimonate(V), named for the city of Naples, Italy

napsyl
$C_{10}H_7O_2S-$, coined by contraction of naphthalenesulfonyl

narbomycin
$C_{28}H_{47}NO_7$, derived from the specific epithet of the bacterial species name *Streptomyces narbonensis*, from *narbonensis* (New Latin: Provençal), from *Provincia narbonensis* (Latin: Provence) – referring to the habitat of this species – and -mycin

narceine
$C_{23}H_{27}NO_8$, coined by variation of narcotine

narcissistic reaction
derived from the Greek mythical figure Narkissos (Latin: Narcissus), a beautiful youth who according to ancient myth fell in love with his own image – referring to the fact that the product of this reaction is the mirror image of the starting material

narcotine (noscapine)
$C_{22}H_{23}NO_7$, derived from *narkotikos* (Greek: benumbing, narcotic), from *narkoun* (Greek: to benumb), from *narke* (Greek: numbness), and -in(e)

narcotoline
$C_{21}H_{21}NO_7$, coined by variation of narcotine

naringenin
$C_{15}H_{12}O_5$, derived from naringin, -gen, and -in(e)

naringin
$C_{27}H_{32}O_{14}$, ultimately derived from *naranga* (Sanskrit: orange tree, *Citrus*) and -in(e)

narsarsukite
$Na_2(Ti,Fe,Zr)Si_4(O,F)_{11}$, named after this mineral's locality Narssarssuk, Greenland

nascent
derived from *nascens* (Latin: being born, being generated), from *nasci* (Latin: to be born, to be generated)

natamycin (pimaricin)
$C_{33}H_{47}NO_{13}$, derived from the specific epithet of the bacterial species name *Streptomyces natalensis*, from *natalensis*

(New Latin: of the province of Natal, South Africa), and –mycin

natrium
Na, New Latin name for sodium, derived (with contraction) from natron and -ium

natrocalcite (gaylussite)
$Na_2Ca(CO_3)_2{\cdot}5H_2O$, derived from natrium, calcium, and -ite

natrochalcite
$NaCu_2(OH)(SO_4)_2{\cdot}H_2O$, derived from natrium, chalc(o)-, and -ite

natrolite
$Na_2Al_2Si_3O_{10}{\cdot}2H_2O$, derived from natrium and -lite

[1]**natron**
$NaHCO_3$, ultimately derived from *ntry* (Old Egyptian: niter)

[2]**natron**
$Na_2CO_3{\cdot}10H_2O$, ultimately derived from *ntry* (Old Egyptian: niter)

naujakasite
$Na_6(Fe,Mn)Al_4Si_8O_{26}$, named after this mineral's locality Naujakasik, Greenland

Nazarov cyclization
$C_7H_{10}O_3$, named for the Russian chemist Ivan Nikolaevich Nazarov (1906-1957)

Nazarov's reagent
$C_7H_{10}O_3$, named for the Russian chemist Ivan Nikolaevich Nazarov (1906-1957)

NBS
$C_4H_4BrNO_2$, an abbreviation for *N*-bromosuccinimide

NC (CN)
an abbreviation for nitrocellulose

NDDO
an abbreviation for neglect of diatomic differential overlap

neamine
$C_{12}H_{26}N_4O_6$, derived (with contraction) from neomycin and amine

Neber rearrangement
named for the German chemist Peter W. Neber (1883-1960)

nebramycin (tenebrimycin, tenemycin)
derived from the specific epithet of the bacterial species name *Streptomyces tenebrarius*, from *tenebrae* (Latin: darkness), ultimately from *temos* (Latin: darkness), and -mycin

nebularine
$C_{10}H_{12}N_4O_4$, derived from the specific epithet of the species name *Clitocybe nebularis* (Batsch.) Quel. (a mushroom), from *nebularis* (Latin: foggy), from *nebula* (Latin: mist, cloud), and -in(e)

nebulin
derived from *nebula* (Latin: mist, cloud), and -in(e) – referring to this protein's histological localization in the N_2 line of the sarcomere, a nebulous striation within the I band

nebulium
name of a hypothetical element inferred from certain lines in the spectra of nebulae later related to oxygen; derived from *nebula* (Latin: mist, cloud) and -ium

necic acid
derived from necine alkaloids

necine (amabiline)
$C_{15}H_{25}NO_4$, derived from necine alkaloids and -in(e)

necine alkaloids
coined by contraction of senecio alkaloids, derived from the genus name *Senecio* (herbs, shrubs, trees), from *senecio* (Latin: groundsel), from *senex* (Latin: old man)

Néel temperature
named for the French physicist Louis Eugene Felix Néel (1904-2000)

Nef reaction
named for the Swiss-US chemist John Ulric Nef (1862-1915)

Nef synthesis
named for the Swiss-US chemist John Ulric Nef (1862-1915)

Negishi cross coupling
named for the Japanese-US chemist Ei-ichi Negishi (born 1935)

nekoite
$Ca_3Si_6O_{15}\cdot 7H_2O$, coined as an anagram of okenite – referring to the similarity of nekoite and okenite

nemadectin
$C_{36}H_{52}O_8$, probably coined by contraction of nematocidal and avermectin

nematic
derived from nemat(o)-

nematicide (nematocide)
derived from nematode, from nemat(o)-, and -cide

nemat(o)-
derived from *nema* (Greek: thread)

nematocide (nematicide)
derived from nematode, from nemat(o)-, and -cide

Nencki reaction
named for the German chemist Marcel von Nencki (1847-1901)

Nenitzescu indole synthesis
named for the Romanian chemist Costin D. Nenitzescu (1902-1970)

Nenitzescu reductive acylation
named for the Romanian chemist Costin D. Nenitzescu (1902-1970)

neo-
derived from *neos* (Greek: new)

neocembrene
$C_{20}H_{32}$, derived from neo- and cembrene

neocupferron
$C_{10}H_{11}N_3O_2$, derived from neo- and cupferron

neocuproine
$C_{14}H_{12}N_2$, derived from neo- and cuproine

neodidymium
Nd, now obsolete, derived from neo- and *didymos* (Greek: twin)

neodymia
Nd_2O_3, derived from neodymium and -a

neodymium
Nd, coined by alteration of obsolete neodidymium

neoflavan
$C_{15}H_{14}O$, derived from neo- and flavan

neomycin
derived from neo- and -mycin

neon
Ne, derived from *neon* (Greek: the new)

neopentane
C_5H_{12}, derived from neo- and pentane

neopentyl
$C_5H_{11}-$, derived from neopentane and -yl

neophyl
$C_{10}H_{13}-$, coined by contraction of neopentyl and phenyl – referring to the substitution of a methyl group in neopentyl with a phenyl group

neopine
$C_{18}H_{21}NO_3$, derived (with contraction) from neo-, opium, and -in(e)

neoprene
a trademark coined by contraction of neo- and chloroprene

neopterin
$C_9H_{11}N_5O_4$, derived (with contraction) from neo- and biopterin

neoquassin
$C_{22}H_{30}O_6$, derived from neo- and quassin

neovitamin A
$C_{20}H_{30}O$, derived from neo- and vitamin A

nepetalactone
$C_{10}H_{14}O_2$, derived from the genus name *Nepeta* (mints), from *nepeta* (Latin: catnip), and lactone

nephelauxetic effect
derived from nephel(o)- and *auxesis* (Greek: increase), from *auxanein*, *auxein* (Greek: to increase)

nepheline
(K,Na)$AlSiO_4$, derived from nephel(o)- and -in(e) – referring to the observation that this mineral becomes cloudy when immersed in strong acid

nephel(o)-
derived from *nephele* (Greek: cloud)

nephelometry
derived from nephel(o)- and -metry

nephrite
$Ca_2(Mg,Fe)_5Si_8O_{22}(OH)_2$, derived from *nephron* (Greek: kidney) and -ite – referring to the traditional use of this stone as a charm against diseases of the kidney

neptunite (carlosite)
$KNa_2Li(Fe,Mg,Mn)_2Ti_2Si_8O_{24}$, named for Neptunus, the Roman god of the sea, of Etruscan origin – referring to the fact that this mineral was found associated with aegirine, named for Ægir, the Old Norse god of the sea

neptunium
Np, named for the planet Neptune, ultimately from Neptunus, the Roman god of the sea, of Etruscan origin

neral (citral b)
$C_{10}H_{16}O$, derived from nerol and -al

neriifolin
$C_{30}H_{46}O_8$, derived from the specific epithet of the species name *Thevetia neriifolia* Juss. (yellow oleander), from *neriifolius* (New Latin: oleander-leaved), from *Nerium* (oleander), from *nerion* (Greek: oleander), and *folium* (Latin: leaf), and -in(e)

Nernst equation
named for the German chemist Hermann Walther Nernst (1864-1941)

Nernst rule
named for the German chemist Hermann Walther Nernst (1864-1941)

nerol
$C_{10}H_{18}O$, derived from neroli oil, named for

Anna Maria de la Tremoille, princess of Nerola (1642-1722), and -ol

nerolidol
$C_{15}H_{26}O$, derived from nerol, -ide, and -ol

nervon(e)
$C_{48}H_{91}NO_8$, derived from *nervus* (Latin: nerve, sinew) and -on(e)

nervonic acid
$C_{24}H_{46}O_2$, derived from nervon(e)

Nesmeyanov reaction
named for the Russian chemist Aleksandr Nikolaevich Nesmeyanov (1899-1980)

nesosilicate
derived from *nesos* (Greek: island) and silicate

Nessler's reagent
named for the German chemist Julius Neßler (1827-1905)

netropsin
$C_{18}H_{26}N_{10}O_3$, originally coined as a trademark, derived from the specific epithet of the bacterial species name *Streptomyces netropsis* and -in(e)

Neuberg ester (fructose 6-phosphate)
named for the German biochemist Carl Neuberg (1877-1956)

Neumann-Kopp rule
named for the 19th century German chemists Franz Neumann and Hermann Kopp

neuraminic acid
$C_9H_{17}NO_8$, derived from neur(o)- and amine

neuraminidase
coined by contraction of neuraminic acid and glycosidase

neurine
$C_5H_{13}NO$, derived from neur(o)- and -in(e)

neur(o)-
derived from *neuron* (Greek: nerve, sinew)

neuromedin
coined by contraction of neur(o)-, mediator, and -in(e)

neurophysin
derived from neurohypophysis, from neur(o)- and *hypophysis* (Greek: outgrowth), and -in(e)

neurotensin
derived from neur(o)- and -tensin

neurotoxin
derived from neur(o)- and toxin

neurotrophin
derived from neur(o)- and -trophin

neutral
derived from neutr(o)- – referring to the property of being neither acidic nor basic

neutral red
$C_{15}H_{17}ClN_4$ – referring to this indicator dye's color change at pH 6.8–8.0

neutr(o)-
derived from *neuter* (Latin: neither)

Newman projection
named for the US chemist Melvin Spencer Newman (1908-1993)

newtonium (etherion)
a name suggested for a hypothetical element, identical with [4]ether, in honor of the British physicist Sir Isaac Newton (1642-1727)

Newtonian liquid
named for the British physicist Sir Isaac Newton (1642-1727)

Newton metal
(Bi,Pb,Sn), named for the British physicist Sir Isaac Newton (1642-1727)

NEXAFS
an abbreviation for near-edge extended X-ray absorption fine structure

NGF
an abbreviation for nerve growth factor

niacin (nicotinic acid, vitamin B_5)
$C_6H_6N_2O$, derived (with contraction) from nicotinic acid and -in(e)

niccol(o)- (nickel(o)-)
derived from niccolum

niccolum
Ni, New Latin name for nickel, a latinization of nickel

Nicholas reaction
named for the 20th century US chemist Kenneth M. Nicholas

nichrome
(Ni,Cr), coined by contraction of the names nickel and chromium

nickel
Ni, ultimately derived from *Kupfernickel* (German: nickel), from *Kupfer* (German: copper) and *Nickel*, nickname for *Nikolaus* (German: Nicholas) – the name *Kupfer-nickel*, equivalent to mock copper, refers to medieval miners' disappointment of having mined an apparent copper ore which then failed to yield any copper metal

nickeline (kupfernickel, niccolite)
NiAs, derived from nickel and -in(e)

nickel(o)- (niccol(o)-)
derived from nickel

nickel vitriol
$NiSO_4 \cdot 7H_2O$, derived from nickel and vitriol

nickelocene
$C_{10}H_{10}Cr$, derived from nickel(o)- and -ocene

nickel-skutterudite (chloanthite)
$(Ni,Co)As_{3-x}$, derived from nickel and skutterudite – referring to this mineral's nature as a nickel-rich skutterudite

nicotine
$C_{10}H_{14}N_2$, derived from the genus name *Nicotiana* (tobacco plants), after the French diplomat Jean Nicot (1530-1600), and -in(e)

nicotinamide
$C_6H_6N_2O$, derived from nicotinic acid and amide

nicotinic acid (niacin, vitamin B_5)
$C_6H_5NO_2$, derived from nicotine – referring to this compound's formation upon oxidation of nicotine

nicotinyl
$C_6H_6N–$, derived from nicotinic acid and -yl

***nido*-**
derived from *nidus* (Latin: nest)

nielsbohrium
Bh, name unsuccessfully suggested for bohrium, in honor of the Danish physicist Niels Henrik David Bohr (1885-1962) – the suggested name was intended to avoid confusion between boron and bohrium in derived names such as borate and bohrate

Niementowski quinazoline synthesis
named for the 19th century Polish chemist

Stefan von Niementowski

Niementowski quinoline synthesis
named for the 19th century Polish chemist Stefan von Niementowski

Nierenstein reaction
named for the 20th century British chemist Maximilian Nierenstein

Nieuwland enyne synthesis
named for the Belgian-US chemist Julius Arthur Nieuwland (1878-1936)

nigelline (damascenine)
$C_{10}H_{13}NO$, derived from the genus name *Nigella* (European herbs), from *nigella* (Late Latin: black caraway), from *nigellus* (Latin: blackish, dark), diminutive of *niger* (Latin: black), and -in(e)

Niggli formula
named for the Swiss mineralogist Paul Niggli (1888-1953)

nigr(o)-
derived from *niger* (Latin: black)

nigrosine
derived from nigr(o)-, -ose, and -in(e) – referring to the black color of these dyes

NIH shift
named for the National Institutes of Health, Bethesda, MD, USA where this reaction was discovered

nil-
derived from *nihil*, *nil* (Latin: nothing)

nimbi-
derived from *nimba* (Sanskrit: neem tree, *Azadirachta indica* A. Juss.)

nimbin
$C_{30}H_{36}O_9$, derived from nimbi- and -in(e)

nimbiol
$C_{18}H_{24}O_2$, derived from nimbi- and -ol

ninhydrin
$C_9H_6O_4$, originally coined as a trademark, probably by manipulation of the obsolete name triketohydrindene hydrate

niobia
Nb_2O_5, derived from niobium and -a

niobium
Nb, named for Niobe, Greek mythical figure – Niobe was the daughter of Tantalos, thus this name refers to the close relationship of niobium with tantalum

niobocene
$C_{10}H_{10}Nb$, derived from niobium and -ocene

nipecotic acid
$C_6H_{11}NO_2$, probably coined by contraction of nicotinic acid and piperidine

NIR
an abbreviation for near infrared

nisin
$C_{143}H_{230}N_{42}O_{37}S_7$, derived from a partial abbreviation for (streptococci) group N inhibitory substance and -in(e)

nitarsone
$C_6H_6AsNO_5$, coined by contraction of 4-nitrophenylarsonic acid

niter (nitre)
KNO_3, $NaNO_3$, ultimately derived from *ntry* (Old Egyptian: niter)

nitinol
(Ni,Ti), coined by contraction of Ni, Ti, and Naval Ordnance Laboratory where this alloy was developed

niton
Rn, a name unsuccessfully suggested for radon, derived from *nitere* (Latin: to shine) and [1]-on

nitramide
$H_2N_2O_2$, derived from nitr(o)- and amide

nitramine (tetryl)
$C_7H_5N_5O_8$, coined by contraction of *N*-methyl-*N*,2,4,6-tetranitrobenzenamine

nitranilic acid
$C_6H_2N_2O_8$, derived from nitr(o)- and anil

nitrate
NO_3^-, derived from nitrogen and -ate

nitratine
$NaNO_3$, derived from nitrate and -in(e)

nitrene
HN, RN, derived from nitrogen and -ene

nitric
derived from nitrogen

nitric acid
HNO_3, derived from nitric

nitric oxide
NO, derived from nitric and oxide

nitride
N^{3-}, derived (with contraction) from nitrogen and –ide

nitrification
derived from nitrogen and *facere* (Latin: to make)

nitrile
RCN, derived from nitr(o)- and -il(e)

nitrin
$C_{13}H_{13}N_3$, derived from nitrite and -in(e) – referring to this compound's usefulness as an analytical reagent for nitrite

nitrite
NO_2^-, derived from nitrogen and -ite

nitr(o)-
O_2N-, derived from nitrogen

nitrogen
N, derived from niter and -gen

nitrogenase
derived from nitrogen and -ase

nitrogenium
N, New Latin name for nitrogen, derived from nitrogen and -ium

nitroglycerin
$C_3H_5N_3O_9$, derived from *Nitroglycerin* (German: nitroglycerol)

nitroglycerol
$C_3H_5N_3O_9$, derived from nitro- and glycerol – this compound is not a nitro compound, but rather a nitrate

nitrokalite
KNO_3, derived from nitrate, kalium, and -ite

nitrolic acid
$RC(=NOH)NO_2$, derived from nitr(o)- and -ol

nitron
$C_{20}H_{16}O_4$, derived from nitrate and [1]-on – referring to this compound's usefulness as analytical reagent for nitrate

nitrone
$R_2C=N(O)R$, derived from nitr(o)- and -one

nitronic acid
$R_2C=N(O)OH$, derived from nitrone

nitroprusside
$Fe(CN)_5NO^{2-}$, derived (with contraction) from [1]nitrosyl and [1]prusside

nitroric acid (orthonitric acid)
H_3NO_4, derived from nitr(o)- and -oryl

nitros(o)-
ON−, derived from nitrosus

nitrosolic acid
RC(=NOH)NO, derived from nitros(o)- and -ol

nitrosus
New Latin name for nitrous

[1]nitrosyl
NO^+, derived from nitrosus and -yl

[2]nitrosyl
ON−, derived from nitrosus and -yl

nitrous
derived from nitrogen and -ous

nitrous acid
HNO_2, derived from nitrous

nitrous air
NO, an archaic name for nitric oxide, derived from nitrous and air

nitrous oxide
N_2O, derived from nitrous and oxide

nitroxide
R_2NO, probably derived (with contraction) from nitric oxide

nitryl
NO_2^+, derived from nitr(o)- and -yl

nivalenol
$C_{15}H_{20}O_7$, derived from the specific epithet of the fungal species name *Fusarium nivale*, from *nivalis* (Latin: snow white), from *nix* (Latin: snow), -ene, and -ol

NMDA
$C_5H_9NO_4$, an abbreviation for *N*-methyl-D-aspartic acid

NMN
$C_{11}H_{15}N_2O_8P$, an abbreviation for nicotinamide mononucleotide

[1]NMP
C_5H_9NO, an abbreviation for *N*-methyl-2-pyrrolidone

[2]NMP
an abbreviation for nucleoside 5'-monophosphate

NMR
an abbreviation for nuclear magnetic resonance

nobelium
No, named for the Swedish chemist, industrialist, and philanthropist Alfred Bernhard Nobel (1833-1896)

noble gas
a name referring to the monatomic nature of these elements

noble metal
a name referring to the resistance of these elements to oxidation

nocard-
derived from the bacterial genus name *Nocardia*, after the French veterinarian and biologist Edmond I. E. Nocard (1850-1903)

nocardamine
$C_{27}H_{48}N_6O_9$, derived from nocard- and amine

nocardicin
derived from nocard- and -icin(e)

nociceptin (orphanin)
$C_{79}H_{129}N_{27}O_{22}$, derived from nociceptive, from *nocere* (Latin: to harm) and *receptare* (Latin: to receive), and -in(e)

nodakenin
$C_{20}H_{24}O_9$, derived from *nodake* (Japanese: peucedanum, hogfennel root, *Peucedanum decursivum* (Miq.) Maxim.), and -in(e)

nodulisporic acid
$C_{43}H_{53}NO_6$, derived from the fungal genus name *Nodulisporium*, from *nodulus*, diminutive of *nodus* (Latin: knot), and spore

NOE
an abbreviation for nuclear Overhauser effect

NOEL
an abbreviation for no observed effect level

NOESY
an abbreviation for nuclear Overhauser enhancement spectroscopy

noformicin
$C_8H_{15}N_5O$, derived (with contraction) from the bacterial species name *Nocardia formica*, from nocard- and *formica* (Latin: ant), and -icin(e)

nogalamycin
$C_{39}H_{49}NO_{16}$, derived from the specific epithet of the bacterial species name *Streptomyces nogalater*, from *nogalater* (New Latin: walnut black), from *nogal* (Spanish: walnut tree) and *ater* (Latin: black), and -mycin

nomenclature
derived from *nomenclatura* (Latin: act of calling by name), from *nomenclator* (Latin: a house slave who called the names of visitors), from *nomen* (Latin: name) and *calare* (Latin: to call, to mention)

nomilin
$C_{28}H_{34}O_9$, coined as an anagram of limonin

non-
derived from *non* (Latin: not)

nona-
derived from *nonus* (Latin: the ninth), from *novem* (Latin: nine)

nonaconta-
a New Latin construction derived from nona- and the Greek numeral suffix *-konta* (-conta) – meaning ninety

nonactic acid
$C_{10}H_{18}O_4$, derived from nonactin

nonactin
$C_{40}H_{64}O_{12}$, derived from non-, active, and -in(e) – referring to this compound's lack of antibiotic activity

nonadeca-
a New Latin construction from nona- and *deka* (Greek: ten) – meaning nineteen

nonakis-
a New Latin construction from nona- and the Greek numeral suffix -akis – meaning nine times

nonoxynol
$(C_2H_4O)_nC_{15}H_{24}O$, coined by contraction of nonylphenoxypolyethoxyethanol

nootkatone
$C_{15}H_{22}O$, derived from the specific epithet of the species name *Chamaecyparis nootkatensis* (yellow cedar), from Nootka, an American Indian people

nopaline
$C_{11}H_{20}N_4O_6$, derived from *nopal* (French: Barbary fig, common prickley pear, *Opuntia vulgaris*) and -in(e)

nopinene
$C_{10}H_{16}$, derived from nopinic acid and -ene

nopinic acid
$C_{10}H_{16}O_3$, coined as an anagram of pinonic acid

nor-
derived from normal in the sense of reduced to its core structure by removal of a peripheral carbon atom

noradrenaline (arterenol, norepine phrine)
$C_8H_{11}NO_3$, derived from nor- and adrenaline

norbergite
$Mg_3(SiO_4)(F,OH)_2$, named after this mineral's locality Norberg, Västmanland, Sweden

norbixin
$C_{24}H_{28}O_4$, derived from nor- and bixin

norbornane
C_7H_{12}, derived from nor- and bornane

norbornyl cation
$C_7H_{11}^+$, derived from norbornane and -yl

norcarane
C_7H_{12}, derived from nor- and carane

nordstrandite
$Al(OH)_3$, named for the US chemist Robert A. van Nordstrand (1917-2000) who first synthesized this mineral

norephedrine
$C_9H_{13}NO$, derived from nor- and ephedrine

norepinephrine (arterenol, noradrenaline)
$C_8H_{11}NO_3$, derived from nor- and epinephrine

[1]**Norge saltpeter (Norway saltpeter)**
$Ca(NO_3)_2$, derived from *Norge* (Norwegian: Norway) and saltpetre

[2]**Norge saltpeter (Norway saltpeter)**
NH_4NO_3, derived from *Norge* (Norwegian: Norway) and saltpeter

Norit
C_n, a trademark

[1]**norite**
C_n, derived from Norit

[2]**norite**
a rock species, derived from *Norge* (Norwegian: Norway) and -ite

norleucine
$C_6H_{13}NO_2$, derived from nor- and leucine

Normann process
named for the German chemist and industrialist Wilhelm Normann (1870-1939)

nornicotine
$C_9H_{12}N_2$, derived from nor- and nicotine

norpseudoephedrine
$C_9H_{13}NO$, derived from nor- and pseudoephedrine

Norrish reactions
named for the British chemist Ronald George Wreyford Norrish (1897-1978)

Northern blot (Northern blotting)
jocularly so named with reference to Southern blot

norvaline
$C_5H_{11}NO_2$, derived from nor- and valine

nosean (noselite)
$Na_8Al_6Si_6O_{24}(SO_4)$, named for the German mineralogist Karl Wilhelm Nose (1753-1835)

nosiheptide (multhiomycin)
$C_{51}H_{43}N_{13}O_{12}S_6$, a name coined without any stated reason

notatin (glucose oxidase)
derived from the specific epithet of the microbiological species name *Penicillium notatum*, from *notatus* (Latin: marked with spots or lines), from *nota* (Latin: mark) and -in(e)

nourseothricin
derived from the specific epithet of the bacterial species name *Streptomyces noursei*, *thrix* (Greek: hair), and -in(e)

novi-
derived from *novies* (Latin: nine times)

novo-
derived from *novus* (Latin: new)

novobiocin
$C_{31}H_{36}N_2O_{11}$, coined by contraction of novo-, bio-, and -mycin

novocain(e)
$C_{13}H_{20}N_2O_2$, coined by contraction of novo- and cocaine

Noyori hydrogenation
named for the Japanese chemist Ryoji Noyori (born 1938)

Noyori reagent
$C_{22}H_{18}AlLiO_3$, named for the Japanese chemist Ryoji Noyori (born 1938)

Nozaki-Hiyama coupling reaction (Nozaki-Hiyama-Kishi reaction)
named for the 20th century Japanese chemists Hitosi Nozaki and Timejiro Hiyama

NQR
an abbreviation for nuclear quadrupole resonance

nRNA
an abbreviation for nuclear ribonucleic acid

NTA
$C_6N_9NO_6$, an abbreviation for nitrilotriacetic acid

nuclear Overhauser effect
named for the US physicist Albert W. Overhauser (born 1925)

nuclease
derived (with contraction) from nucleic acid and –ase

nucleic acid
derived from nucle(o)-

nuclein
coined by contraction of nucleoprotein

nucle(o)-
derived from *nucleus* (Latin: kernel, nut)

nucleocidin
$C_{10}H_{13}FN_6O_6S$, derived from nucle(o)-, -cide, and -in(e)

nucleofuge
derived from nucle(o)- and -fuge

nucleophile
derived from nucle(o)- and -phile

nucleosidase
derived from nucleoside and –ase

nucleoside
derived from nucle(o)-, -ose, and [3]-ide

nucleotidase
derived from nucleotide and -ase

nucleotide
coined by variation of nucleoside, probably in analogy with phosphatide

nuclide
derived from nucle(o)- and [1]-ide

nudic acid
derived from the specific epithet of the species name *Tricholoma nudus* (a mushroom), from *nudus* (Latin: nude)

nupharidine
$C_{15}H_{23}NO_2$, derived from the genus name *Nuphar* (water lilies), ultimately from *nilotpala* (Sanskrit: nenuphar), from *nila* (Sanskrit: dark blue) and *utpala* (Sanskrit: nenuphar blossom), and -idin(e)

nybomycin
$C_{16}H_{14}N_2O_4$, derived (with contraction) from New York Botanical Garden and -mycin

Nyholm-Gillespie model (VSEPR model)
named for the Australian chemist Sir Ronald Sydney Nyholm (1917-1971) and the Canadian chemist Sir Ronald James Gillespie (born 1924)

Nyholm-Gillespie rules
named for the Australian chemist Sir Ronald Sydney Nyholm (1917-1971) and the Canadian chemist Sir Ronald James Gillespie (born 1924)

Nylander's test
named for the Swedish chemist C. W. G. Nylander (1835-1907)

nylon (Carothers silk)
polyamides, a word invented as a trademark, coined by reshuffling of 'no run' – referring to the legendary sturdiness of the first nylon stockings produced

nystatin
named for New York State, USA where these antibiotics were developed

O

-oate
coined by euphonic variation of -ate

ob- (o-, oc-, of-, op-)
derived from *ob- (o-, oc-, of-, op-)* (Latin: against, opposed)

Obermayer's test
named for the Austrian physician Friedrich Obermayer (1861-1925)

obsidian
a rock species, derived from *Obsidianus lapis*, a corrupted form of *Obsianus lapis* (Latin: stone of Obsius)

-ocane
derived (with contraction) from octa- and -ane

occlusion
derived from *occludere* (Latin: to shut up, to close up), from *claudere* (Latin: to shut, to close)

-ocene
derived from ferrocene

ochratoxin
derived from the specific epithet of the fungal species name *Aspergillus ochraceus* Wilh., from *ochraceus* (Latin: ocher colored), from *ochra* (Latin: ocher), and toxin

ocimene
$C_{10}H_{16}$, derived from the genus name *Ocimum* (mints), ultimately from *okimon* (Greek: basil), and -ene

-ocine
derived (with contraction) from octa- and -in(e)

oct(a)-
derived from *okto* (Greek: eight)

octacont(a)-
derived from *oktakonta* (Greek: eighty)

octacos(a)-
derived from *oktakaikosi* (Greek: twenty-eight)

octadec(a)-
derived from *oktakaideka* (Greek: eighteen)

octahedro-
derived from octahedron

octakis-
derived from *oktakis* (Greek: eight times)

octane
C_8H_{18}, derived from oct(a)- and -an(e)

octi-
derived from *octies* (Latin: eight times)

octogen (HMX)
$C_4H_8N_8O_8$, derived (with contraction) from oct(a)- and nitrogen – referring to the presence of eight nitrogen atoms in this molecule

octop-
derived from the genus name *Octopus* (octopus), from *oktopous* (Greek: octopod), from oct(a)- and *pous* (Greek: foot)

octopamine
$C_8H_{11}NO_2$, derived from octop- and amine

octopine
$C_9H_{18}N_4O_4$, derived from octop- and -in(e)

octopinic acid
$C_8H_{16}N_2O_4$, derived from octopine

ODMR
an abbreviation for optically detected magnetic resonance

-odon
derived from *odon* (Greek: tooth)

OE
an abbreviation for Overhauser effect

OES
an abbreviation for optical emission spectroscopy

-oid
ultimately derived from *-oeides* (Greek: -like), from *eidos* (Greek: form, shape)

oil
derived from *oleum* (Latin: oil), from *elaion* (Greek: olive oil)

-oin
an arbitrary suffix patterned after [1]benzoin

okadaic acid
$C_{44}H_{68}O_{13}$, derived from the specific epithet of the species name *Halichondria okadai* (a marine black sponge)

okenite
$Ca_{10}Si_{18}O_{46}\cdot 18H_2O$, named for the German naturalist Lorenz Oken (1779-1851)

-ol
an arbitrary suffix derived from alcohol

Olah's reagent
HF/C_5H_5N, named for the Hungarian-US chemist George Andrew Olah (born 1927)

olation
derived from hydroxyl group

-ol(e)
a suffix derived from *oleum* (Latin: oil)

oleanane
$C_{30}H_{52}$, derived from olean(dro)- and -an(e)

oleandomycin
$C_{35}H_{61}NO_{12}$, derived from oleandrose and -mycin

oleandrin
$C_{12}H_{48}O_9$, derived from olean(dro)- and -in(e)

olean(dro)-
derived from oleander (*Nerium*), from *oleander* (Medieval Latin: oleander), probably an alteration of *rhododendron* (Latin: rhododendron), from rhod(o)- and dendr(o)-

oleandrose
$C_7H_{14}O_4$, derived from olean(dro)- and -ose

oleanolic acid
$C_{30}H_{48}O_3$, derived from the genus name *Olea* (olive trees), from *olea* (Latin: olive tree), and -ol

olefiant gas
C_2H_4, an archaic name for ethene, derived from *gaz oléfiant* (French: oil making gas), ultimately from *oleum* (Latin: oil) and *facere* (Latin: to make) – referring to the fact that addition of halogens to ethene leads to heavy liquid products

olefin
derived (with contraction) from olefiant gas and -in(e)

oleic acid
$C_{18}H_{34}O_2$, derived from *oleum* (Latin: oil)

olein (triolein)
$C_{57}H_{104}O_6$, coined by contraction of *Trioleylglycerin* (German: trioleylglycerol)

ole(o)-
derived from *oleum* (Latin: oil)

oleogel
derived from ole(o)- and gel

oleoyl
$C_{18}H_{33}O-$, derived from oleic acid and -oyl

oleum
(H_2SO_4,SO_3), derived from *oleum* (Latin: oil)

oleuropein
$C_{25}H_{32}O_{13}$, derived (with contraction) from the species name *Olea europea* L. (European olive tree), from *olea* (Latin: olive tree) and Europe, and -in(e)

oleyl
$C_{18}H_{35}-$, derived from oleic acid and -yl

olfactory
derived from *olfacere* (Latin: to smell), from *olere* (Latin: to emit a smell) and *facere* (Latin: to make)

-olide
derived from glycolide

oligase
coined (with contraction) from oligo-saccharide and -ase

oligo-
derived from *oligos* (Greek: few, scanty)

oligoclase
(Na,Ca)(Si,Al)$_4$$O_8$, derived from oligo- and -clase – referring to the fact that this mineral has a less perfect cleavage than albite

oligomer
derived from oligo- and -mer

oligomycin
a name coined without any stated reason, derived from oligo- and -mycin

oligopeptide
derived from oligo- and peptide

oligosaccharide
derived from oligo- and saccharide

olivacine
$C_{17}H_{14}N_2$, derived from the specific epithet of the species name *Aspidosperma olivaceum* Müll. Arg. (guatambú), from *olivaceus* (Latin: olive green), from *oliva* (Latin: olive, olive tree), and -in(e)

olivanic acid
derived from the specific epithet of the bacterial species name *Streptomyces olivaceus*, from *olivaceus* (Latin: olive green), from *oliva* (Latin: olive, olive tree)

olivenite
$Cu_2AsO_4(OH)$, named after its olive color

olivil
$C_{20}H_{24}O_7$, derived from the genus name *Olea* (olive trees), from *olea* (Latin: olive tree), and -il(e)

olivine (chrysolite, peridot)
(Fe,Mg)$_2$$SiO_4$, named after its olive color

olivomycin
derived from the specific epithet of the bacterial species name *Streptomyces olivoreticuli*, from *oliva* (Latin: olive, olive tree), and -mycin

[1]**-olone**
derived from -ol and -one; an obsolete non-systematic suffix

[2]-olone
derived from -ol(e) and -one

ommatin
derived from omm(ato)- and -in(e)

omm(ato)-
derived from *omma* (Greek: eye)

ommatochrome (ommochrome)
derived from omm(ato)- and -chrome

ommochrome (ommatochrome)
derived from omm(ato)- and -chrome

OMP
$C_{10}H_{13}N_2O_{11}P$, an abbreviation for orotidine 5'-monophosphate

omphacite
$(Ca,Na)(Mg,Fe,Al)Si_2O_6$, derived from *omphakites* (Greek: green stone), from *omphax* (Greek: unripe grape), and -ite – referring to this mineral's green color

[1]-on
derived from the Greek adjective neuter suffix *-on*

[2]-on
in the names boron and silicon by analogy with carbon

[3]-on
in the names of small particles by analogy with proton and electron

-onane
derived (with contraction) from non(a)- and -an(e)

-one
derived from the Greek feminine patronymic suffix *-one*

-onic acid
derived from gluconic acid

-onine
derived (with contraction) from non(a)- and -in(e)

-onium
derived from ammonium

onyx
SiO_2, derived from *onyx* (Greek: finger nail, toe nail) – referring to this mineral's beneficial properties for skin, hair, and nails perceived in ancient mythology

o(o)-
derived from *oion* (Greek: egg)

oosporein
$C_{14}H_{10}O_8$, derived from the fungal genus name *Oospora*, from o(o)- and spore, and -in(e)

opal (girasol)
$SiO_2{\cdot}nH_2O$, ultimately derived from *upala* (Sanskrit: (precious) stone)

opalescence
derived from opal and -escence

ophi(o)-
derived from *ophis* (Greek: snake)

ophiobolane
$C_{25}H_{46}$, derived from the fungal genus name *Ophiobolus*, from ophi(o)- and *-bolos* (Greek: throwing, casting), from *ballein* (Greek: to throw, to cast), and -an(e)

opianic acid
$C_{10}H_{10}O_5$, derived from opium and -an(e)

opiate (opioid)
derived from opium and -ate

opine
a class name probably coined by contraction of octopine-type compound

opioid (opiate)
derived from opium and -oid

opium
derived from *opion* (Greek: poppy juice), from *opos* (Greek: plant juice)

Oppenauer oxidation
named for the Austrian chemist Rupert Viktor Oppenauer (1910-1969)

-ops
derived from *ops* (Greek: eye)

opsin
derived from rhodopsin

-opsis
derived from *opsis* (Greek: sight)

opsonin (fibronectin)
derived from *opsonium* (Latin: relish), from *opsonein* (Greek: to purchase victuals), and -in(e)

-orane
an arbitrary suffix patterned after phosphorane

orbital
derived from *orbita* (Latin: wheel track), from *orbis* (Latin: circle, ring)

orc(e)-
derived from *orcine* (French: orcin)

orcein
derived from orc(e)- and -in(e)

orcin (orcinol)
$C_7H_8O_2$, derived from orc(e)- and -in(e)

orcinol (orcin)
$C_7H_8O_2$, derived from orc(e)-, -in(e), and -ol

ORD
an abbreviation for optical rotatory dispersion

ore
ultimately derived from *aer* (Old English: brass)

orexin
derived from *orexis* (Greek: appetite), from *oregein* (Greek: to stretch), and -in(e)

[1]organic
(in the sense of carbon-containing) derived from [2]organic

[2]organic
(in the sense of generated by a living organism) derived from *organon* (Greek: bodily organ)

organyl
derived from [1]organic and -yl

oripavine
$C_{18}H_{19}NO_3$, coined by contraction of the species name *Papaver orientale* L. (oriental poppy), from papaver- and *orientalis* (Latin: oriental), and -in(e)

orlean (annatto)
named after a gallicized form of the name of the Spanish discoverer Francisco de Orellana (1511-1546)

orlon
polyacrylates, a word invented as a trade-mark; patterned after nylon

ormocer
coined by contraction of organically modified ceramics

ormos-
derived from the genus name *Ormosia* (shrubs, trees), from *hormos* (Greek: chain, necklace)

ormosanine (piptamine)
$C_{20}H_{35}N_3$, derived from ormos-, -an(e), and -in(e)

ormosinine
$C_{40}H_{66}N_6$, derived from ormos- and -inin(e)

ornipressin
$C_{45}H_{63}N_{13}O_{12}S_2$, coined by contraction of ornithine and vasopressin

ornithine
$C_5H_{12}N_2O_2$, derived from *ornis* (Greek: bird) and -in(e)

ornaline (nopalinic acid)
$C_{10}H_{18}N_2O_6$, derived (with contraction) from ornithine and nopaline

orotic acid
$C_5H_4N_2O_4$, probably derived from *oros* (Greek: whey)

orotidine
$C_{10}H_{12}N_2O_8$, derived from orotic acid and -idin(e)

oroxyline
derived from the genus name *Oroxylum* (midnight horror), from *oros* (Greek: mountain) and xyl(o)-, and -in(e)

orpiment
As_2S_3, derived from *auripigmentum* (Latin: gold paint), from *aurum* (Latin: gold) and *pigmentum* (Latin: coloring matter), from *pingere* (Latin: to paint)

Orsat apparatus
named for the 19th century French engineer L. H. Orsat

orsellinic acid
$C_8H_8O_4$, derived from *orselle* (Old French: archil)

ORTEP
an abbreviation for Oak Ridge thermal ellipsoid plot

orthanilic acid
$C_6H_7NO_3S$, coined by contraction of *o*-anilinesulfonic acid

orth(o)-
derived from *orthos* (Greek: straight, right, true)

orthoboric acid
H_3BO_3, derived from orth(o)- and boric acid

orthocarbonic acid
H_4CO_4, derived from orth(o)- and carbonic acid

orthocarboxylic acid
$RC(OH)_3$, derived from orth(o)- and carboxylic acid

orthochrysotile
$Mg_3Si_2O_5(OH)_4$, derived (with contraction) from orthorhombic and chrysotile

orthoclase (adularia)
$KAlSi_3O_8$, derived from orth(o)- and -clase – referring to this mineral's right angle of good cleavage

orthohelium
He, derived from orth(o)- and helium

orthohydrogen
H_2, derived from orth(o)- and hydrogen

orthonitric acid (nitroric acid)
H_3NO_4, derived from orth(o)- and nitric acid

orthonitrous acid
H_3NO_3, derived from orth(o)- and nitrous acid

orthoperiodic acid
H_5IO_6, derived from orth(o)- and periodic acid

orthoperrhenic acid
H_5ReO_6, derived from orth(o)- and perrhenic acid

orthophosphoric acid
H_3PO_4, derived from orth(o)- and phosphoric acid

orthosilicic acid
H_4SiO_4, derived from orth(o)- and silicic acid

orthosomycin
derived from orthoester and -mycin – referring to a characteristic substructure in these antibiotics

Orton rearrangement
named for the British chemist Kennedy Joseph Previté Orton (1872-1930)

-oryl
(O)X≡, a suffix derived from phosphoryl

oryza-
ultimately derived from *oryza* (Greek: rice) of non-Indo-European origin, akin to *vrihi* (Sanskrit: rice)

oryzacidin
$C_8H_{13}NO_5$, derived from the specific epithet of the fungal species name *Aspergillus oryzae*, from oryza-, -cide, and -in(e)

oryzanol
derived from oryza- and -ol

osajin
$C_{25}H_{24}O_5$, derived (with contraction) from osage orange (*Maclura pomifera* (Raf.) Schneid.) and -in(e)

-osan (-an)
a suffix derived from -ose and -an

osazone
derived (with contraction) from -ose and hydrazone

Osborne fractions
named for the US biochemist T. B. Osborne (1859-1935)

-ose
an arbitrary suffix derived from glucose

osm(e)-
derived from *osme* (Greek: smell), from *ozein* (Greek: to smell)

osmiridium
(Os,Ir), derived from osmium and iridium

osmium
Os, derived from osm(e)- and -ium – referring to the strong characteristic smell of osmium tetroxide which emanates from air oxidized osmium surfaces

^{1}osm(o)-
derived from *osmos* (Greek: a thrusting), from *othein* (Greek: to push)

^{2}osm(o)-
derived from osm(e)-

^{3}osm(o)-
derived from osmium

osmocene
$C_{10}H_{10}Os$, derived from ^{3}osm(o)- and -ocene

osmogen
derived from [2]osm(o)- and -gen

osmolality
derived from osmosis and molality

osmometry
derived from osmosis and -metry

osmophore
derived from [2]osm(o)- and -phore

osmosis
derived from [1]osmo-

osone (ketoaldose)
derived from -ose and -one

OSPE
an abbreviation for octahedral site preference energy

ossein
derived from *os* (Latin: bone) and -in(e)

oste(o)-
derived from *osteon* (Greek: bone)

osteocalcin
derived from oste(o)-, calcium, and -in(e)

osthole
$C_{15}H_{16}O_3$, derived, with contraction, from ostruth- and -ol(e)

ostruth-
derived from the specific epithet of the species name *Peucedanum ostruthium* (L.) Koch (masterwort, hogfennel), from *ostruthius* (New Latin: born under a lucky star, lucky, useful), from *astruc* (Old Occitanic: born under a lucky star, lucky, useful)

ostruthin
$C_{19}H_{22}O_3$, derived from ostruth- and -in(e)

ostruthol
$C_{21}H_{22}O_7$, derived from ostruth- and -ol

Ostwald dilution law
named for the German chemist Friedrich Wilhelm Ostwald (1853-1932)

Ostwald maturation
named for the German chemist C. W. W. Ostwald (1883-1943)

Ostwald rule
named for the German chemist Friedrich Wilhelm Ostwald (1853-1932)

Ostwald-Volmer rule
named for the German chemists Friedrich Wilhelm Ostwald (1853-1932) and Max Volmer (1885-1965)

osumilite
$K(Fe,Mg)_2(Al,Fe)_3(Si,Al)_{12}O_{30}$, named after this mineral's locality, the province of Osumi, Kyushu, Japan

otavite
$CdCO_3$, named after this mineral's locality, the Tsumeb mines, Otavi, Namibia

otobain (otobite)
$C_{20}H_{30}O_4$, derived from the specific epithet of the species name *Dialyanthera otoba* (Humb. & Bonpl.) (cuangare, otoba, saosa), from *otoba* (Spanish: *Dialyanthera otoba*), and -in(e)

ouabagenin (G-strophantidin)
$C_{23}H_{34}O_8$, derived from ouabain, -gen, and -in(e)

ouabain (acocantherin, Gratus strophantin)
$C_{29}H_{44}O_{12}$, derived from the specific epithet of the species name *Acokanthera ouabaio* (Wll. & Hock.) (ouabaio), from *waba yo* (Somali: ouabaio), and -in(e)

Ouchterlony's test
named for the Swedish microbiologist Örjan Ouchterlony (1914-2004)

-ous
ultimately derived from *-osus* (Latin: having, full of)

Outokumpu process
named for the metallurgical company Outokumpu Oy, Espoo, Finland

ovalbumin
derived from ov(o)- and albumin

ovalene
$C_{32}H_{14}$, derived from oval and -ene – referring to the oval shape of this molecule

Overhauser effect
named for the US physicist Albert W. Overhauser (born 1925)

Overman rearrangement
named for the US chemist Larry E. Overman (born 1943)

ovicide
derived from ov(o)- and -cide

ov(o)-
derived from *ovum* (Latin: egg)

ovoglobulin
derived from ov(o)- and globulin

ovotransferrin
derived from ov(o)- and transferrin

ox(a)-
derived from oxygen

oxalacetic acid
$C_4H_4O_5$, derived from oxal(o)- and acetic acid

oxalic acid
$C_2H_2O_4$, derived from the genus name *Oxalis* (herbs), from *oxalis* (Latin: garden sorrel), ultimately from *oxys* (Greek: acidic)

oxal(o)-
C_2HO_3-, derived from oxalic acid

oxalyl
$-C_2O_2-$, derived from oxalic acid

oxamic acid
$C_2H_3NO_3$, derived from oxalic acid and amide

oxamide
$C_2H_4N_2O_2$, derived from oxalic acid and amide

oxane (tetrahydropyran)
$C_5H_{10}O$, derived from ox(a)- and -an(e)

oxenin
$C_{20}H_{30}O_2$, derived from ox(o)-, -ene, and -in(e)

oxidane (water)
H_2O, systematic name for water, derived (irregularly) from oxide, and -an(e) – thus coined to avoid confusion with oxane, $C_5H_{10}O$

oxidase
derived (with contraction) from oxidation and -ase

oxidation
derived from oxide

oxide
derived from *oxyde* (French: oxide), coined by contraction of *oxygène* (French: oxygen) and *acide* (French: acid)

oxidoreductase
derived (with contraction) from oxidation,

reduction, and -ase

oxime
derived (with contraction) from oxygen and imide

ox(o)-
derived from oxygen

oxolane (tetrahydrofuran)
C_4H_8O, derived from ox(a)-, -ol(e), and -an(e)

oxolation
derived from ox(o)-

oxole (furan)
C_4H_4O, derived from ox(a)- and -ol(e)

oxonic acid
$C_4H_3N_3O_4$, derived from ox(o)- and -one

oxonium (oxidanium, hydronium)
H_3O^+, derived from ox(o)- and -onium

oxonol dye (cyanine dye)
derived from ox(o)- and -ol

oxo synthesis (Roelen reaction)
derived from ox(o)- – referring to the formation of oxo compounds from alkenes, carbon monoxide, and hydrogen

oxy-
an obsolete name for hydroxy-, derived from oxygen

oxyacanthan
$C_{32}H_{30}N_2O_2$, derived from oxyacanth(o)- and -an(e)

oxyacanthine
$C_{37}H_{40}N_2O_6$, derived from oxyacanth(o)- and -in(e)

oxyacanth(o)-
derived from the specific epithet of the species name *Crataegus oxyacantha* (hawthorn), from *oxys* (Greek: sharp) and acanth(o)-

oxycodone
$C_{18}H_{21}NO_4$, coined by contraction of dihydrohydroxycodeinone

oxygen
O, derived from *oxys* (Greek: sharp, acidic) and -gen – referring to the acidic nature of many combustion products

oxygenase
derived from oxygen and -ase

oxygenium
O, New Latin name for oxygen, derived from oxygen and -ium

oxymorphone
$C_{17}H_{19}NO_4$, coined by contraction of dihydrohydroxymorphinone

oxymuriatic acid
Cl_2, an archaic name for chlorine, from oxy- and muriatic acid – referring to the formation of chlorine by oxidation of hydrogen chloride

oxytetracycline
$C_{22}H_{24}N_2O_9$, derived from oxy- and tetracycline

oxytocin
$C_{43}H_{66}N_{12}O_{12}S_2$, derived from *oxys* (Greek: sharp), *tokos* (Greek: childbirth), and -in(e)

-oyl
RCO–, an arbitrary suffix derived from -yl; patterned after benzoyl

ozo-
derived from *ozein* (Greek: to smell)

ozocerite (ozokerite)
derived from ozo-, *keros* (Greek: beeswax), and -ite – referring to this mineral's characteristic smell and waxy appearance

ozone
O_3, derived from ozo- and -one – referring to this compound's characteristic unpleasant smell

ozonide
derived from ozone and [2]-ide

P

Paal-Knorr pyrrole synthesis
named for the Austrian chemist Carl Paal (1860-1935) and the German chemist Ludwig Knorr (1859-1921)

pachnolite
$NaCaAlF_6 \cdot H_2O$, derived from *pachne* (Greek: hoarfrost), akin to *pegnynai* (Greek: to fasten together, to coagulate), and -lite – referring to the gel-like appearance of this mineral

pactamycin
$C_{28}H_{38}N_4O_8$, derived from the specific epithet of the bacterial species name *Streptomyces pactum* var *pactum*, from *pactus* (Latin: fastened), from *pangere* (Latin: to fasten), and -mycin

paddlane
derived from paddle and -an(e) – referring to the paddle wheel molecular shape of these hydrocarbons

PAF
an abbreviation for platelet activating factor

PAGE
an abbreviation for polyacrylamide gel electrophoresis

PAH
an abbreviation for polycyclic aromatic hydrocarbon

paktong (packfong)
(Ni,Zn,Cu), derived from *paak t'ung* (Cantonese: paktong), from *paak* (Cantonese: white) and *t'ung* (Cantonese: copper)

palitantin
$C_{14}H_{22}O_4$, derived from the specific epithet of the fungal species name *Penicillium palitans*, from *palitans* (Latin: vagary), from *palitari* (Latin: to wander), and -in(e)

palladin
cell phosphoproteins, named for the Italian architect Andrea Palladio (Andrea di Pietro) (1508-1580), referring to the localization of these proteins to architectural elements of the cell

palladium
Pd, named for the asteroid Pallas, ultimately derived from Pallas Athene, the Greek goddess of wisdom, from *pallas* (Greek: maiden)

palladocene
$C_{10}H_{10}Pd$, derived from palladium and -ocene

palmatine (calystigine)
$C_{21}H_{22}NO_4^+$, derived from the specific epithet of the species name *Jatrorrhiza palmata* (DC.) Miers (calumba), from *palmatus* (Latin: hand-shaped), and -in(e)

palmidrol
$C_{18}H_{37}NO_2$, derived (with contraction) from palmitic acid, amide, hydroxy-, and -ol

palmitic acid
$C_{16}H_{32}O_2$, derived from palmitin

palmitin (tripalmitin)
$C_{51}H_{98}O_6$, derived from *palmite* (French: pith of the palm tree), ultimately from *palma* (Latin: palm tree), and -in(e)

palustric acid
$C_{20}H_{30}O_2$, derived from the specific epithet

of the species name *Pinus palustris* (longleaf pine), from *paluster* (Latin: marshy), from *palus* (Latin: marsh)

palygorskite (attapulgite)
$(Mg,Al)_2Si_4O_{10}(OH)\cdot 4H_2O$, named after this mineral's locality Palygorsk, Russia

palytoxin (PTX)
$C_{129}H_{223}N_3O_{54}$, derived (with contraction) from the genus name *Palythoa* (corals), coined without any stated reason, and toxin

PAM
$C_7H_9IN_2O$, an abbreviation for 2-pyridinealdoxime methiodide

pamoic acid (embonic acid)
$C_{23}H_{16}O_6$, probably derived from pamaquine, the generic name of an antimalarial which was the first liquid basic drug to be formulated as its solid pamoate

[1]PAN
$C_2H_3NO_5$, an abbreviation for peroxyacetyl nitrate

[2]PAN
$C_{15}H_{11}N_3O$, an abbreviation for 1-(2-pyridylazo)-2-naphthol

panamine
$C_{20}H_{33}N_3$, derived from the specific epithet of the species name *Ormosia panamensis Benth.* (a tropical tree), from Panama, and -in(e)

panaxoside (ginsenoside)
derived from the genus name *Panax* (herbs), from *panax* (Greek: all-healing, panacea), -ose, and -ide

pancracine (hippagine)
$C_{16}H_{17}NO_4$, derived from the genus name *Pancratium* (daffodils), from *pankrates* (Greek: all-powerful), from pan(o)- and *kratos* (Greek: strength, power), and -in(e)

pancreastatin
derived from pancreas and -statin

pancreatin
derived from pancreas and -in(e)

pancreatopeptidase (elastase)
derived from pancreas and peptidase

pancreozymin (cholecystokinin)
derived (with contraction) from pancreas, enzyme, and -in(e)

pandermite (priceite)
$Ca_2B_5O_7(OH)_5\cdot H_2O$, named after the town of Panderma, Asia Minor, now Bandirma, Turkey, a port from which this mineral has been shipped

pangamic acid (vitamin B_{15})
an ill-defined mixture, derived from pan(o)- and gam(o)-

pankrin
derived from pancreas and -in(e)

pan(o)-
derived from *pas* (Greek: all, every)

panthenol (pantothenol)
$C_9H_{19}NO_4$, derived (with contraction) from pantothenic acid and -ol

pantetheine
$C_{11}H_{22}N_2O_4S$, derived from pant(o)-, thi(o), and -ein(e); patterned after cysteine

pantethine
$C_{22}H_{42}N_4O_8S_2$, derived from pant(o)-, thi(o), and -in(e); patterned after cystine

pant(o)-
derived from pantothenic acid

pantolactone
$C_6H_{10}O_3$, derived from pant(o)- and lactone

pantothenic acid
$C_9H_{17}NO_5$, derived from *pantothen* (Greek: from everywhere)

papain
derived from the specific epithet of the species name *Carica papaya* (green papaya, melon fruit), from papaya, from *ababai* (Carib: papaya), and -in(e)

papaver-
derived from the genus name *Papaver* (poppies), from *papaver* (Latin: poppy)

papaveraldine (xanthaline)
$C_{20}H_{19}NO_5$, coined by variation of papaverine

papaverine
$C_{20}H_{21}NO_4$, derived from papaver- and -in(e)

papuamine
$C_{25}H_{40}N_2$, derived from Papua, Papua-New Guinea, and amine

para-
derived from *para* (Greek: beside, similar, near, beyond, irregular, a modification of)

parabanic acid
$C_3H_2N_2O_3$, derived from *parabainein* (Greek: to pass over), from para- and *bainein* (Greek: to go)

paraben
$C_7H_6O_3$, coined by contraction of *p*-hydroxybenzoic acid

paracelsian
$Ba(AlSiO_4)_2$, derived from para- and celsian – referring to the dimorphism of celsian and paracelsian

paracelsin
a peptide antibiotic, named for the Swiss-German alchemist and physician Philippus Aureolus Paracelsus (Theophrastus Bombast von Hohenheim) (1493-1541), Paracelsus being the latinized form of the native family name Hohenheim, from para- and *celsus* (Latin: high)

parachalcogen
derived from paraelement and chalcogen

parachor
derived from para- and *choros* (Greek: space)

parachrysotile
$Mg_3Si_2O_5(OH)_4$, derived from para- and chrysotile – referring to this mineral's close relationship with chrysotile

paraelement
derived from para- and element

paraffin (alkane)
derived (with contraction) from *parum affinis* (Latin: little connected to, related to) – referring to the lack of reactivity of alkanes

paraformaldehyde
$(CH_2O)_n$, derived from para- and formaldehyde

parafuchsin (pararosaniline)
$C_{19}H_{18}ClN_3$, derived from para- and fuchsin

paragonite
$NaAl_2(Si_3Al)O_{10}(OH)_2$, derived from *paragein* (Greek: to mislead) and -ite – referring to the fact that this mineral could be mistaken for talc

parahalogen
derived from paraelement and halogen

parahelium
He, derived from para- and helium

paraherquamide
$C_{28}H_{35}N_3O_5$, derived from the specific epithet of the fungal species name *Penicillium paraherquei* and amide

parahydrogen
H_2, derived from para- and hydrogen

paraldehyde
$C_6H_{12}O_3$, coined by contraction of paraacetaldehyde, from para- and acetaldehyde

paramolybdate (heptamolybdate)
$Mo_7O_{24}^{6-}$, derived from para- and molybdate

parapepsin
derived from para- and pepsin

paraprotein
derived from para- and protein

pararosaniline (parafuchsin)
$C_{19}H_{18}ClN_3$, derived from para- and rosaniline

parasiticide
derived from parasite and -cide

parasorbic acid
$C_6H_8O_2$, derived from para- and sorbic acid – this compound is a lactone isomeric with sorbic acid, not an acid

parathyroid hormone (PTH)
derived from para- and thyroid, from *thyroides* (Greek: shaped like a shield), ultimately from *thyra* (Greek: door)

paratropic
coined by contraction of paramagnetism and -tropic

paratungstate
$H_2W_{12}O_{42}^{10-}$, derived from para- and tungstate

pargasite
$NaCa_2(Mg,Fe)_4Al(Si_6Al_2)O_{22}(OH)_2$, named after this mineral's locality Pargas, Finland

Parham cyclization
named for the US chemist William E. Parham (1922-1976)

Paris blue (Berlin blue, Milori blue, Prussian blue)
$C_6Fe_2KN_6$, named after the city of Paris, France

Paris green (Mitis green, Schweinfurt green)
named after the city of Paris, France

parisite
$Ca(Ce,La)_2(CO_3)_3F_2$, named for the 19th century Colombian mine proprietor J. J. Paris

Pariser-Parr-Pople method (PPP method)
named for the US chemists Rudolph Pariser (born 1923) and Robert Ghormley Parr (born 1921), and the British chemist Sir John Anthony Pople (1925-2004)

Parkes process
named for the British metallurgist Alexander Parkes (1813-1890)

Park nucleotide
named for the 20th century US biochemist James T. Park

paromomycin
$C_{23}H_{45}N_5O_{14}$, derived from the subspecific epithet of the bacterial species name *Streptomyces rimosus* forma *paromomycinus*, from para-, homo-, and -mycin

parotin
derived from parotid gland, from *parotis* (Greek: near the ear), from para- and *ous* (Greek: ear), and -in(e)

Parr bomb
named for the US chemist Samuel Wilson Parr (1857-1931)

Parri reaction
named for the 20th century Italian pharmacist Walter Parri

parthenin
$C_{15}H_{18}O_4$, derived from parthen(o)- and -in(e)

parthen(o)-
derived from the genus name *Parthenium* (herbs), from *parthenion* (Greek: feverfew), from *parthenos* (Greek: maiden, virgin)

parthenolide
$C_{15}H_{20}O_3$, derived from parthen(o)- and -olide

parylene
poly[*p*-xylylene], a trademark coined by contraction of poly[*p*-arylene]

PAS
$C_7H_7NO_3$, an abbreviation for *p*-aminosalicylic acid

pascoite
$Ca_3V_{10}O_{28}\cdot 17H_2O$, named after this mineral's locality Pasco province, Peru

Passerini reaction
named for the Italian chemist Mario Torquato Luigi Passerini (1891-1962)

Pasteur effect
named for the French biologist Louis Pasteur (1822-1895)

pasteurization
named for the French biologist Louis Pasteur (1822-1895)

Pasteur pipet
named for the French biologist Louis Pasteur (1822-1895)

patchouli alcohol
$C_{15}H_{26}O$, derived from patchouli oil, from patchouli (*Pogostemon cablin*), from *pacculi* (Tamil: woolly patchouli)

Paternò-Büchi reaction
named for the Italian chemist Emanuele Paternò, Marchese di Sessa (1847-1935) and the Swiss-US chemist Georg Hermann Büchi (1921-1998)

Pattinson process
named for the British chemist Hugh Lee Pattinson (1796-1858)

patulin
$C_7H_6O_4$, derived from the specific epithet of the fungal species name *Penicillium patulum*, from *patulus* (Latin: open), from *patere* (Latin: to be open), and -in(e)

Pauling complex theory
named for the US chemist Linus Pauling (1901-1994)

Pauling process
named for the 20th century German chemist Harry Pauling

Pauli principle
named for the Austrian-Swiss physicist Wolfgang Pauli (1900-1958)

Pauly reaction
named for the German chemist Hermann Pauly (1870-1950)

Pauson-Khand reaction
named for the German-British chemist Peter Ludwig Pauson (born 1925) and the Pakistani-British chemist Ihsan U. Khand

pavoninin
derived from the specific epithet of the species name *Pardachirus pavoninus* (Pacific sole), from *pavoninus* (Latin: of a peacock, resembling a peacock), from *pavo* (Latin: peacock), and -in(e)

Payne rearrangement
named for the 20th century US chemist George B. Payne

PBB
an abbreviation for polybrominated biphenyl(s)

PBN
$C_{11}H_{15}NO$, an abbreviation for *N-tert*-butyl-α-phenylnitrone

PC
an abbreviation for paper chromatography

PCB
an abbreviation for polychlorinated biphenyl(s)

PCDD
an abbreviation for polychlorodibenzo dioxin(s)

PCDF
an abbreviation for polychlorodibenzo-furan(s)

PCILO
an abbreviation for perturbative configuration interaction using localized orbitals

PCP
C_6HCl_5O, an abbreviation for pentachlorophenol

PCR
an abbreviation for polymerase chain reaction

pearlite
(Fe,Fe_3C), named for this alloy's mother-of-pearl-like appearance under the microscope

Pearlman's catalyst
$Pd(OH)_2$, named for the 20th century US chemist William M. Pearlman

Pearson concept (HSAB concept)
named for the US chemist Ralph Gottfrid Pearson (1919-1987)

Pearson symbol
named for the British chemist W. B. Pearson (born 1921)

Pechiney-Ugine-Kuhlmann process
named for the industrial company Pechiney-Ugine-Kuhlmann, Paris, France

Pechmann condensation
named for the German chemist Hans Freiherr von Pechmann (1850-1902)

Pechmann pyrazole synthesis
named for the German chemist Hans Freiherr von Pechmann (1850-1902)

pecilocin
$C_{17}H_{25}NO_3$, derived from the fungal genus name *Paecilomyces*, from *poikilos* (Greek: colorful), and -in(e)

pectin
derived from *pektos* (Greek: congealed, curdled), from *pegnynai* (Greek: to fasten together, to coagulate)

pectolinarigenin
$C_{17}H_{14}O_6$, derived from pectin, the genus name *Linaria* (herbs, undershrubs), from *linum* (Latin: flax), -gen, and -in(e)

PED
an abbreviation for photoelectric diffraction

pederin
$C_{25}H_{45}NO_9$, derived from the genus name *Paederus* (beetles), from *paideros* (Greek: an opalescent gem), and -in(e)

PEG
an abbreviation for poly[ethylene glycol]

pegmatite
a class of rocks, derived from *pegma* (Greek: something fastened together), from *pegnynai* (Greek: to fasten together, to coagulate), and -ite

pelargon-
derived from the genus name *Pelargonium* (herbs), from *pelargos* (Greek: stork)

pelargonic acid
$C_9H_{18}O_2$, derived from pelargon-

pelargonidin
$C_{15}H_{11}ClO_5$, derived from pelargon- and -idin(e)

pelargonin
$C_{27}H_{31}ClO_{15}$, derived from pelargon- and -in(e)

p-electron
named after the 'principal' line groups in atomic spectroscopy

π-electron
derived from π-orbital

pelletierine
$C_8H_{15}NO$, named for the French chemist Pierre Joseph Pelletier (1788-1842)

pellitorine
$C_{14}H_{25}NO$, derived from pellitory, ultimately from *pyrethron* (Greek: pyrethrum), from pyr(o)-, and -in(e)

Pellizzari reaction
named for the Italian chemist Guido Pellizzari (1858-1938)

pellotine
$C_{13}H_{19}NO_3$, derived from *pellote* (Spanish: peyote), from *peyotl* (Nahuatl: peyote), and -in(e)

peltatin
derived from the specific epithet of the species name *Podophyllum peltatum* L. (mayapple), from *peltatus* (New Latin: peltate), ultimately from *pelte* (Greek: small shield), and -in(e)

penaldic acid
$C_4H_5NO_4$, derived from penicillin and aldehyde

penam
$C_5H_7NO_S$, derived from penicillin and -am(e)

penem
C_5H_5NOS, derived from penam and -ene

penetratin
$C_{104}H_{168}N_{34}O_{20}S$, derived from penetration and -in(e)

penicillamine
$C_5H_{11}NO_2S$, derived from penicillin and amine

penicillin
derived from the fungal genus name *Penicillium*, ultimately from *penicillus* (Latin: brush), diminutive of *penis* (Latin: tail, male member)

penicillanic acid
$C_8H_{10}NO_3S$, derived from penicillin and -an(e)

penicillic acid
$C_8H_{10}O_4$, derived from penicillin

Penning effect
named for the Dutch physicist Frans Michel Penning (1894-1953)

pent(a)-
derived from *pente* (Greek: five)

pentacene
$C_{22}H_{14}$, derived from pent(a)- and -acene

pentaconta-
derived from *pentakonta* (Greek: fifty)

pentadeca-
derived from *pentadeka* (Greek: fifteen)

pentaerythritol
$C_5H_{12}O_4$, derived from penta- and erythritol – referring to this compound's five carbon atoms as well as to its chemical similarity with erythritol

pentagastrin
$C_{37}H_{49}N_7O_9S$, derived (with contraction) from pentapeptide and gastrin

pentahomoserine
$C_5H_{11}NO_3$, derived from pent(a)-, homo-, and serine – the number refers to the position of the hydroxyl group on C-5

pentakis-
derived from *pentakis* (Greek: five times)

pentalene
C_8H_6, derived from penta- and -alene

pentane
C_5H_{12}, derived from pent(a)- and -an(e)

pentaose
pentasaccharides, derived from pent(a)- and -ose

pentaprismo-
derived (with contraction) from pentagonal prism

pentel (pnicogen, pnictogen)
N, P, As, Sb, Bi, derived (with contraction) from pent(a)- and element – referring to the fifth main group of the periodic table

pentetic acid (DTPA)
$C_{14}H_{23}N_3O_{10}$, coined by contraction of diethylenetriaminepentaacetic acid

pentlandite
$(Ni,Fe)_9S_8$, named for the Irish natural historian Joseph Barclay Pentland (1797-1873)

1-pentol
C_6H_8O, coined by contraction of 3-methylpent-2-en-4-yn-1-ol

pentosan
derived from pentose and -an(e)

pentose
$C_5H_{10}O_5$, derived from pent(a)- and -ose

pentostatin
$C_{11}H_{16}N_4O_4$, derived from pentose and -statin

pentulose (ketopentose)
$C_5H_{10}O_5$, derived from pent(a)- and -ulose

peon-
derived from the genus name *Paeonia* (peonies), from *paeonia* (Greek: peony), named for its mythical discoverer Paeon, the physician of the Greek gods

peonidin
$C_{16}H_{13}ClO_6$, derived from peon- and -idin(e)

peonin
$C_{28}H_{33}ClO_{16}$, derived from peon- and -in(e)

PEP
$C_3H_4O_6P^-$, an abbreviation for phosphoenolpyruvate

pepsin
derived from *pepsis* (Greek: digestion), from *peptein* (Greek: to digest), and -in(e)

pepsinogen
derived from pepsin and -gen

pepstatin
$C_{34}H_{63}N_5O_9$, derived from pepsin and -statin – referring to this compound's inhibition of pepsin

peptaibol
coined by contraction of peptide, aminoisobutyric acid, and α-amino alcohol

peptide
derived from *pepsis* (Greek: digestion), from *peptein* (Greek: to digest), and [3]-id(e)

peptidase
derived from peptide and -ase

peptide T
$C_{35}H_{55}N_9O_{16}$, named after its high threonine content

peptization, to peptize
derived from from *pepsis* (Greek: digestion), from *peptein* (Greek: to digest)

peptoid
derived from peptide and -oid

peptolide (depsipeptide)
derived from peptide and -olide

peptone
derived from *pepsis* (Greek: digestion), from *peptein* (Greek: to digest), and -one

peptonization
derived from peptone

per-
derived from *per* (Latin: through, by means of, thoroughly)

peracetic acid
$C_2H_4O_3$, derived from per- and acetic acid

peracid
derived from per- and acid

peracidox process
coined by contraction of peracid oxidation process

peramine
$C_{12}H_{17}N_5O$, derived (with contraction) from the specific epithet of the species name *Lolium perenne* (rye grass), from *perennis* (Latin: perennial), from per- and *annus* (Latin: year), and amine

perbromate
BrO_4^-, derived from perbromic acid and-ate

perbromic acid
$HBrO_4$, derived from per- and bromic acid

perchlorate
ClO_4^-, derived from perchloric acid and -ate

perchloric acid
$HClO_4$, derived from per- and chloric acid

perchloryl
ClO_3-, derived from perchloric acid

perezone
$C_{15}H_{20}O_3$, derived from perezia (*Acourtia microcephala*), named for the 16th century Spanish pharmacist Lorenzo Perez, and -one

perforin
derived from perforation and -in(e)

performic acid
CH_2O_3, derived from per- and formic acid

peri-
derived from *peri* (Greek: around, about), from *peran* (Greek: to pass through)

periclase
MgO, derived from peri- and -clase – referring to this mineral's perfect cleavage

pericyclic
derived from peri- and cyclic

peridin-
derived from the genus name *Peridinium* (dinoflagellates), from *peridines* (Greek: whirled around), from peri- and *dinein* (Greek: to whirl, to rotate)

peridinine
$C_{39}H_{50}O_7$, derived from peridin- and -in(e)

peridinol
$C_{37}H_{48}O_6$, derived from peridin- and -ol

peridot (chrysolite, olivine)
$(Fe,Mg)_2SiO_4$, derived from *peritot* (Old French: peridot, literally unclear) or from *faridat* (Arabic: gem)

peridotite
a rock species, derived from peridot and -ite

perill(a)-
derived from the genus name *Perilla* (mints), perhaps from *perilla*, diminutive of *pera* (Latin: leather bag, wallet), of Greek origin

perilla ketone
$C_{10}H_{14}O_2$, derived from perill(a)-

perillaldehyde
$C_{10}H_{14}O$, derived from perill(a)- and aldehyde

perimidine
$C_{11}H_8N_2$, derived from peri- and amidine

perimycin
probably coined as a trademark, derived from peri- and -mycin

periodate
IO_4^-, derived from periodic acid and -ate

periodic acid (metaperiodic acid)
HIO_4, derived from per- and iodic acid

periodic system of the elements
derived from period

periplanone
derived from the genus name *Periplaneta* (cockroaches), from peri- and *planetes* (Greek: wanderer), and -one

periplo(c)-
derived from the genus name *Periploca* (woody vines), from peri- and *ploke* (Greek: action of twisting or turning), from *plekein* (Greek: to plait, to twine)

periplocin
$C_{36}H_{56}O_{13}$, derived from periplo(c)- and -in(e)

periplocymarin
$C_{30}H_{46}O_8$, derived from periplo(c)- and cymarin

periplogenin
$C_{23}H_{34}O_5$, derived from periplocin, -gen, and -in(e)

periselective
derived from pericyclic

peristaltic
derived from peristalsis, from peri- and *stalsis* (Greek: compression, constriction), from *stellein* (Greek: to place, to send)

peristerite
$NaAlSi_3O_8$, derived from *peristera* (Greek: dove, pigeon) and -ite – referring to this mineral's resemblance to the iridescent feathers of a pigeon's neck

peristylane
$C_{15}H_{20}$, derived from *peristylos* (Greek: surrounded by a colonnade), from peri- and *stylos* (Greek: pillar), and -an(e) – referring to this hydrocarbon's molecular shape

Perkin alicyclic synthesis
named for the British chemist Sir William Henry Perkin (1838-1907)

Perkin reaction
named for the British chemist Sir William Henry Perkin (1838-1907)

Perkin rearrangement
named for the British chemist Sir William Henry Perkin (1838-1907)

Perkow reaction
named for the German chemist Werner Perkow (1915-1994)

perlon
polyamides, a word invented as a trademark; patterned after nylon

permanganate
MnO_4^-, derived from per- and manganate

permeable
derived from *permeare* (Latin: to pass through, penetrate), from *meare* (Latin: to go, to pass)

perovskite
$CaTiO_3$, named for the Russian mineralogist Count L. A. Perovski (1794-1857)

peroxidase
derived from hydrogen peroxide and -ase

^{1}peroxide
O_2^{2-}, derived from per- and oxide

^{2}peroxide
R^1OOR^2, derived from per- and oxide

perrhenate
ReO_4^-, derived from per- and rhenate

perspex
originally coined as a trademark, derived from *perspicere* (Latin: to look through)

persulfate
derived from per- and sulfate

pertechnetate
TcO_4^-, derived from per- and technetate

perxenate (xenonate)
XeO_6^{4-}, derived from perxenic acid and -ate

perxenic acid (xenonic acid)
H_4XeO_6, derived from per- and xenic acid

perylene
$C_{20}H_{12}$, coined by contraction of the obsolete name *peri*-dinaphthylene

PES
an abbreviation for photoelectron spectroscopy

Pesez reactions
named for the French chemist Maurice Pesez (born 1914)

pesticide
derived from *pestis* (Latin: pest) and -cide

pestle
derived from *pistillum*, diminutive of *pilum* (Latin: pestle, javelin), from *pinsere* (Latin: to pound, to crush)

PET
an abbreviation for positron emission tomography

petalite (castorite)
$LiAlSi_4O_{10}$, derived from *petalon* (Greek: leaf) and -ite – referring to this mineral's perfect basal cleavage

Peterson reaction
named for the US chemist Donald J. Peterson (born 1935)

PETN
$C_5H_8N_4O_{12}$, an abbreviation for pentaerythritol tetranitrate

Petrenko-Kritschenko piperidone synthesis
named for the 20th century Ukrainian chemist P. Petrenko-Kritschenko

Petri dish
named for the German bacteriologist Julius Richard Petri (1852-1921)

petro-
derived from *petra* (Greek: rock) and/or *petros* (Greek: stone)

petroleum
derived from petro- and *oleum* (Latin: oil)

petroselinic acid
$C_{18}H_{34}O_2$, derived from the genus name *Petroselinum* (parsley), from *petroselinum* (Latin: parsley), from petro- and *selinon* (Greek: celery)

petunidin
$C_{16}H_{13}ClO_7$, derived from petunin and -idin(e)

petunin
$C_{28}H_{33}ClO_{17}$, derived from the genus name *Petunia* (petunia), from *petun* (Medieval French: tobacco), from *petyn* (Tupi: tobacco), and -in(e)

peucedanin
$C_{15}H_{14}O_4$, derived from the genus name *Peucedanum* (sulfurweed), from *peukedanon* (Greek: sulfurweed), from *peukedanos* (Greek: bitter tasting), from *peukos* (Greek: sharp), and -in(e)

peyonine
$C_{16}H_{19}NO_5$, derived from peyote (*Lophophera*), from *peyotl* (Nahuatl: peyote)

Peyrone's salt
$Cl_2H_6N_2Pt$, named for the Italian chemist Michele Peyrone (1813-1883)

Pfau-Plattner azulene synthesis
named for the Swiss chemists Alexandre Stanislav Pfau (1889-1938) and Placidus Andreas Plattner (1904-1975)

Pfeffer cell
named for the German botanist and chemist W. F. P. Pfeffer (1845-1920)

Pfeiffer effect
named for the German chemist Paul Pfeiffer (1875-1951)

Pfeiffer's substance (veramon)
$C_{21}H_{29}N_5O_4$, named for the German chemist Paul Pfeiffer (1875-1951)

Pfitzinger reaction
named for the 19th century German chemist W. Pfitzinger

Pfitzner-Moffat reaction
named for the US chemist K. E. Pfitzner and the Canadian-US chemist John Gilbert Moffat (born 1930)

Pfitzner-Moffat reagent
named for the US chemist K. E. Pfitzner and the Canadian-US chemist John Gilbert Moffat (born 1930)

PG
an abbreviation for prostaglandin

pH
an abbreviation of *potentia hydrogenii* (New Latin: strength of hydrogen), derived from the definition $[H^+] = 10^{-p}$

phae(o)-
derived from *phaios* (Greek: dusky, grayish brown)

phaeomelanin
derived from phae(o)- and melanin

phaeophorbide
derived from phae(o)- and phorbide

phaeophytin
derived from phae(o)- and phytin

phalloidin
$C_{35}H_{48}N_8O_{11}S$, derived from the specific epithet of the species name *Amanita phalloides* (Fr.) Seer. (death cap), from *phalloides* (Greek: resembling a phallus), and -idin(e)

phane
coined by contraction of phenylene and alkane

-phan(e)
derived from *phaneros* (Greek: apparent, visible), from *phainein* (Greek: to show)

pharmaco-
derived from *pharmakon* (Greek: drug, remedy)

phase
derived from *phasis* (Greek: appearance), from *phainein* (Greek: to show)

phaseolin
$C_{20}H_{18}O_4$, derived from the genus name *Phaseolus* (bean plants), from *phaseolus*, diminutive of *phaselus* (Latin: sword bean), of Greek origin, and -in(e)

phasin
derived from the genus name *Phaseolus* (bean plants), from *phaseolus*, diminutive of *phaselus* (Latin: sword bean), of Greek origin, and -in(e)

PHB
$C_7H_6O_3$, an abbreviation for *p*-hydroxybenzoic acid

phellandrene
$C_{10}H_{16}$, derived from the genus name *Phellandrium* (water fennel), from *phellandron* (Greek: plants with ivy-like leaves), and -ene

phell(o)-
derived from *phellos* (Greek: cork, bark), probably akin to *phloos* (Greek: bark)

phenacetin
$C_{10}H_{13}NO_2$, derived from phenol, acet(o)-, and -in(e)

phenacyl
C_8H_7O-, derived (anagrammatically and with contraction) from acetophenone and -yl

phenakite (phenacite)
Be_2SiO_4, derived from *phenax* (Greek: cheat, impostor) and -ite – referring to the fact that this mineral is readily mistaken for quartz

phenalene
$C_{13}H_{10}$, derived from phen(o)- and –alene

phenanthr(a)-
derived from phenanthrene

phenanthrene
$C_{14}H_{10}$, derived from phen(o)-, anthra-, and -ene

phenanthridine
$C_{13}H_9N$, coined by contraction of phenanthr(a)- and pyridine

phenanthroline
$C_{12}H_8N_2$, coined by contraction of phenanthrene and quinoline

phenazine
$C_{12}H_8N_2$, derived from phen(o)-, az(a)-, and -in(e)

phene
C_6H_6, an obsolete name for benzene, derived from *phainein* (Greek: to show) – referring to this compound's occurrence in illuminating gas

-phene
a suffix coined by contraction of phenanthrene

phenetidine
$C_8H_{11}NO$, derived from phenetole and -idin(e)

phenetole
$C_8H_{10}O$, coined by contraction of phenyl, ethyl, and -ol(e)

phenethyl
C_8H_9–, coined by contraction of 2-phenylethyl

phenicin
$C_{14}H_{10}O_6$, derived from the specific epithet of the fungal species name *Penicillium phoeniceum*, from *phoeniceus* (Latin: purple-red, literally from Phoenicia), and -icin(e)

phen(o)-
derived from phene

phenol
C_6H_6O, derived from phen(o)- and -ol

phenol oxidase
derived from phenol and oxidase

phenolphthalein
$C_{20}H_{14}O_4$, derived from phenol, phthal(o)-, and -ein(e)

phenolphthalin
$C_{20}H_{16}O_4$, derived from phenol, phthal(o)-, and -in(e)

phenolphthalol
$C_{20}H_{18}O_3$, derived from phenolphthalin and -ol

-phenone
a suffix coined by contraction of phenyl ketone

phenothiazine
$C_{12}H_9NS$, derived from phen(o)-, thi(a)-, az(a)-, and -in(e)

phenoxazine
$C_{12}H_9NO$, derived from phen(o)-, ox(a)-, az(a)-, and -in(e)

phentermine
$C_{10}H_{15}N$, coined by contraction of phenyl-*tert*-butylamine

phenyl
C_6H_5–, derived from phene and -yl

phenylalanine
$C_9H_{11}NO_2$, derived from phenyl and alanine

phenytoin
$C_{15}H_{12}N_2O_2$, coined by contraction of diphenylhydantoin

pheromone
derived (with contraction) from *pherein* (Greek: to bear) and hormone

-philic
derived from phil(o)-

phillipsite
$K(Ca,Na_2)(Si_5Al_3)O_{16}\cdot 6H_2O$, named for the British mineralogist William Phillips (1775-1829)

phillyrin
$C_{27}H_{34}O_{11}$, derived from the genus name *Phillyrea* (evergreen shrubs), from *philyrea* (Greek: mock privet), and -in(e)

phil(o)-
derived from *philos* (Greek: loving)

philodiene (dienophile)
derived from phil(o)- and diene

philosopher's stone
a concept created by the Arab alchemist Geber a.k.a. Jabir Ibn Hayyan (721-815) – the mythical philosopher's stone was believed to transmute base metals to gold

PhIP
$C_{13}H_{12}N_4$, a contracted abbreviation for 1-methyl-6-phenyl-1*H*-imidazo[4,5-*b*]pyridin-2-amine

phlegma
an alchemistic term for distillation residue, derived from *phlegma* (Greek: flame, inflammation), from *phlegein* (Greek: to burn)

phlobaphene
derived from phlo(ro)- and *baphe* (Greek: dye), from *baptein* (Greek: to dip, to dye)

phlogisticated air
N_2, an archaic name for nitrogen, derived from phlogiston

phlogisticated nitrous air
N_2O, an archaic name for dinitrogen oxide, derived from phlogiston and nitrous

phlogiston
a hypothetical substance erroneously believed to play a role in oxidations and reductions, thus a reduction was regarded as a phlogistication and an oxidation a dephlogistication, derived from *phlogistos* (Greek: burnt up, inflammable), from phlog(o)-, and [1]-on

phlog(o)-
derived from *phlox* (Greek: flame)

phlogopite
$K(Mg,Fe)_3(Si_3Al)O_{10}(F,OH)_2$, derived from *phlogopos* (Greek: fiery looking), from *phlox* (Greek: flame) and *ops* (Greek: eye, face), and -ite – referring to this mineral's fiery appearance

phloionic acid
$C_{18}H_{34}O_6$, derived from phlo(ro)- and -one

phloretin
$C_{15}H_{14}O_5$, derived from phlo(ro)- and -etin(e)

phloridzin (phlorhizin, phlorizin, phlorrhizin)
$C_{21}H_{24}O_{10}$, derived from phlo(ro)-, rhiz(o)-, and -in(e)

phlo(ro)-
derived from *phloios*, *phloos* (Greek: bark)

phloroglucinol
$C_6H_6O_3$, derived from phlo(ro)-, *glykys*

(Greek: sweet), and -ol

-phobic
derived from *phobos* (Greek: fear)

phon(o)-
derived from *phone* (Greek: sound, voice)

phonolite
a rock species, derived from phon(o)- and -lite – referring to a characteristic sound to be heard when a slab of this rock is struck

phorbin
$C_{22}H_{18}N_4$, derived from phorb(o)- and -in(e)

phorb(o)-
derived from *phorbe* (Greek: pasture, fodder)

phorbol
$C_{20}H_{28}O_6$, derived from the genus name *Euphorbia* (plants), named for the 1st century physician Euphorbus at the court of Juba, king of Mauretania, who discovered the use of these plants' latex as a remedy, from eu- and phorb(o)-, and -ol

-phore, -phoric
derived from *phoros* (Greek: carrying forward), from *pherein* (Greek: to bear)

-phoresis
derived from -phore

phorone
$C_9H_{14}O$, derived (with contraction) from camphorone – referring to this ketone's isomerism with camphorone

phos-
derived from from *phos* (Greek: light)

phosgene
$COCl_2$, derived from phos- and -gen(e) – referring to this compound's photochemical formation from carbon monoxide and chlorine

phosgenite
$Pb_2(CO_3)Cl_2$, derived from phosgene and -ite – referring to this mineral's formal composition $2PbO{\cdot}COCl_2$

phosphane (phosphine)
PH_3, derived from [1]phosph(o)- and -an(e)

phosphate
PO_4^{3-}, derived from [1]phosph(o)- and -ate

phosphide
P^{3-}, derived from [1]phosph(o)- and [2]-ide

phosphine (phosphane)
PH_3, derived from [1]phosph(o)- and -in(e)

phosphino-
H_2P-, derived from phosphine

phosphinothricin (PPT)
$C_5H_{12}NO_4P$, derived from phosphinic acid, an unexplained name fragment, and -icin(e)

phosphite
PO_3^{3-}, derived from [1]phosph(o)- and -ite

[1]**phosph(o)-**
derived from phosphorus

[2]**phosph(o)-**
H_2O_3P-, derived from phosphoric acid

phosphonium
PH_4^+, derived from [1]phosph(o)- and -onium

phosphor
derived from *phosphoros* (Greek: light bearer), from phos- and -phore

phosphorane
PH_5, derived from phosphorus and -an(e)

phosphorescence
derived from phosphor and -escence

phosphoric acid
H_3PO_4, derived from phosphorus

phosphorite
a rock species, derived from phosphorus and -ite

phosphorus
P, derived from *phosphoros* (Greek: light bearer), from phos- and -phore

phostamic acid
coined by contraction of phosphorus and lactam

phostonic acid
coined by contraction of phosphorus and lactone

phosvitin
derived from phosphorus, *vitellus* (Latin: egg yolk), and -in(e)

photo-
derived from *phos* (Greek: light)

photo-Fries rearrangement
named for the German chemist Karl Theophil Fries (1875-1962)

phren(o)-
derived from *phren* (Greek: diaphragm, mind)

phrenosin
$C_{48}H_{93}NO_9$, derived from phren(o)-, -ose, and -in(e)

phthalazine
$C_8H_6N_2$, derived from phthal(o)-, az(a)-, and -in(e)

phthalic acid
$C_8H_6O_4$, coined by contraction of naphthalic acid, from naphthalene

phthalide
$C_8H_6O_2$, derived from phthal(o)- and [3]-ide

phthal(o)-
derived from phthalic acid

phthalocyanin
$C_{32}H_{18}N_8$, derived from phthal(o)- and cyanin

phthaloyl
$-C_8H_4O_2-$, derived from phthal(o)- and -yl

phthalyl
$C_8H_5O_3-$, derived from phthal(o)- and -yl

phthiocol
$C_{11}H_5O_3$, derived from phthi(so)- and -ol – referring to this compound's formation in the microorganism *Mycobacterium tuberculosis*

phthioic acid
derived from phthi(so)- – referring to this compound's formation in the microorganism *Mycobacterium tuberculosis*

phthi(so)-
derived from *phthisis* (Greek: pulmonary tuberculosis), from *phthiein*, *phthinein* (Greek: to decay, to wane)

phyc(o)-
derived from *phykos* (Greek: seaweed)

phycobilin
derived from phyc(o)- and bilin

phyll(o)-
derived from *phyllon* (Greek: leaf), from *phyein* (Greek: to bring forth, to grow, to generate)

phylloquinone (phytomenadione, phytonadione, vitamin K_1)
$C_{31}H_{46}O_2$, derived from phyll(o)- and quinone

phyllosilicate
derived from phyll(o)- and silicate

phyl(o)-
derived from *phylon* (Greek: tribe, clan)

physalaemin
$C_{58}H_{84}N_{14}O_{16}S$, derived from the genus name *Physalaemus* (amphibians), from phys(o)- and *laima* (Greek: throat, gorge), and -in(e)

physics
derived from *physis* (Greek: nature), from *phyein* (Greek: to bring forth, to grow, to generate)

phys(o)-
derived from *physa* (Greek: bellows)

physodic acid
$C_{26}H_{30}O_8$, derived from the specific epithet of the species name *Parmelia physodes* (L.) Ach. (a lichen), from *physodes* (Greek: full of wind), from phys(o)-

physostigmine (eserine)
$C_{15}H_{21}N_3O_2$, derived from the genus name *Physostigma* (vines), from phys(o)- and *stigma* (Greek: mark), from *stizein* (Greek: to tattoo), and -in(e)

physovenine
$C_{14}H_{18}N_2O_3$, derived (with contraction) from the species name *Physostigma venenosum* Balf. (Calabar bean), from phys(o)- and *venenosus* (Late Latin: poisonous), from *venenum* (Latin: poison, love potion), after the Roman love goddess Venus, and -in(e)

phytic acid
$C_6H_{18}O_{24}P_6$, derived from phyt(o)-

phyt(o)-
derived from *phyton* (Greek: plant), from *phyein* (Greek: to bring forth, to grow, to generate)

phytoalexin
derived from phyt(o)-, *alexein* (Greek: to defend), and -in(e)

phytochlorin
$C_{34}H_{36}N_4O_6$, derived from phyt(o)- and chlorin

phytochrome
derived from phyt(o)- and -chrome

phytoene
$C_{40}H_{64}$, derived from phyt(o)- and -ene

phytofluene
$C_{40}H_{62}$, derived from phyt(o)-, fluorescence, and -ene

phytol
$C_{20}H_{40}O$, derived from phyt(o)- and -ol

phytosterol
derived from phyt(o)- and sterol

phytyl
$C_{20}H_{39}-$, derived from phytol and -yl

picein
$C_{14}H_{18}O_7$, derived from the genus name *Picea* (evergreen trees), from *picea* (Latin: pitch pine), from pico-, and -in(e)

picene
$C_{22}H_{14}$, derived from pic(o)- and -ene

Pickering emulgator
named for the British chemist P. S. U. Pickering (1858-1920)

picloram (Agent White)
$C_6H_3Cl_3N_2O_2$, coined by anagrammatical contraction of 4-amino-3,5,6-trichloro-picolinic acid

pic(o)-
derived from *pix* (Latin: pitch)

picoline
C_6H_7N, derived from pic(o)-, -ol(e), and -in(e)

picolinic acid
$C_6H_5NO_2$, derived from picoline

picramic acid
$C_6H_5N_3O_5$, derived from picr(o)- and -amic acid

picramide
$C_6H_4N_3O_4$, derived from picr(o)- and amide

picrasane
$C_{20}H_{34}O$, derived from picras(m)- and -an(e)

picras(m)-
derived from the genus name *Picrasma* (bitterwood), from *pikrasmos* (Greek: bitterness), from picr(o)-

picrasmin (isoquassin)
$C_{22}H_{28}O_6$, derived from picras(m)- and -in(e)

picric acid
$C_6H_3N_3O_7$, derived from picr(o)- – referring to this phenol's bitter taste

picr(o)-
derived from *pikros* (Greek: sharp, bitter)

picrocrocin
$C_{16}H_{26}O_7$, derived from picr(o)- and the genus name *Crocus* (perennial herbs), from *krokos* (Greek: saffron, crocus), of Semitic origin, and -in(e)

picrolonic acid
$C_{10}H_8N_4O_5$, coined by contraction of picr(o)- and pyrazolone

picromerite (schoenite)
$K_2Mg(SO_4)_2 \cdot 6H_2O$, derived from picr(o)-, -mer, and -ite – referring to this mineral's bitter taste

picromycin
$C_{28}H_{47}NO_8$, derived from picr(o)- and -mycin – referring to this antibiotic's bitter taste

picrotin
$C_{15}H_{18}O_7$, derived (with contraction) from picrotoxin and -in(e)

picrotoxin
$C_{30}H_{34}O_{13}$, derived from picr(o)- and toxin

picrotoxinin
$C_{15}H_{16}O_6$, derived from picrotoxin and -in(e)

picryl
$C_6H_2N_3O_6-$, derived from picric acid and -yl

Pictet-Gams isoquinoline synthesis
named for the Swiss chemists Amé Pictet (1857-1937) and A. Gams

Pictet-Hubert reaction (Morgan-Walls reaction)
named for the Swiss chemists Amé Pictet (1857-1937) and A. Hubert

Pictet-Spengler isoquinoline synthesis
named for the Swiss chemists Amé Pictet (1857-1937) and T. Spengler

Pictet-Trouton rule
named for the Swiss physicist R. P. Pictet (1846-1929) and the Irish physicist F. T. Trouton (1863-1922)

piemontite
$Ca_2MnAl_2(Si_2O_7)(SiO_4)(O,OH)_2$, named after this mineral's locality, the province of Piemont, Italy

PID
an abbreviation for photoionization detector

PIES
an abbreviation for Penning ionization electron spectroscopy

piezo-
derived from *piezein* (Greek: to squeeze, to press)

pigeonite
$(Mg,Fe,Ca)_2Si_2O_6$, named after this mineral's locality Pigeon Point, MN, USA

pigment
derived from *pigmentum* (Latin: coloring matter), from *pingere* (Latin: to paint)

pil(o)-
derived from *pilos* (Greek: felt)

pilocarpine
$C_{11}H_{16}N_2O_2$, derived from the genus name *Pilocarpus* (jaborandi), from pil(o)- and carp(o)-, and -in(e)

pilocereine
$C_{45}H_{65}N_3O_6$, derived from the genus name *Pilocereus* (cacti), from pil(o)- and cere(o)-, and -in(e)

pilosine
$C_{16}H_{18}N_2O_3$, derived from the genus name *Pilocarpus* (jaborandi), from pil(o)-, and -in(e)

Piloty-Robinson synthesis
named for the German chemist Oskar Piloty (1866-1915) and the British chemist Sir Robert Robinson (1885-1975)

Piloty's acid
$C_6H_7NO_3S$, named for the German chemist Oskar Piloty (1866-1915)

pimarane
$C_{20}H_{36}$, derived from pimar(o)- and -an(e)

pimaric acid
$C_{20}H_{30}O_2$, derived from pimar(o)-

pimaricin (natamycin)
$C_{33}H_{47}NO_{13}$, derived (with contraction) from Pietermaritzburg, Natal, South Africa and -icin – referring to the habitat of the pimaricin-producing bacterial species *Streptomyces natalensis*

pimar(o)-
derived (with contraction) from the species name *Pinus maritima* (cluster pine), from *pinus* (Latin: pine) and *maritimus* (Latin: of the sea), from *mare* (Latin: sea)

pimelic acid
$C_7H_{12}O_4$, derived from *pimele* (Greek: lard)

pimer (π-mer)
coined by contraction of π-dimer

pimpinellin
$C_{13}H_{10}O_5$, derived from the genus name *Pimpinella* (herbs), from *pimpinella* (Late Latin: a medicinal plant), from *pipinella*, diminutive of *piper* (Latin: pepper), and -in(e)

pinacol (pinacone)
$C_6H_{14}O_2$, derived from *pinax* (Greek: board, tablet, picture) and -ol – referring to this compound's characteristic crystal shape

pinacolin (pinacolone)
$C_6H_{12}O$, derived from pinacol and -in(e)

pinacolone (pinacolin)
$C_6H_{12}O$, derived from pinacol and –one

pinacone (pinacol)
$C_6H_{14}O_2$, derived (with contraction) from pinacol and -one

pinane
$C_{10}H_{18}$, derived from pin(o)- and -an(e)

pinchbeck
(Cu,Zn), named for the British watchmaker Christopher Pinchbeck (1670-1732)

pinene
$C_{10}H_{16}$, derived from pin(o)- and -ene

pinguinain
derived from the specific epithet of *Bromelia pinguin* Plum. ex L. (pinguin), from *pinguin*, a native West Indian word, and -ain(e)

pINN
an abbreviation for proposed international nonproprietary (drug) name

Pinner reaction
named for the German chemist Adolf Pinner (1842-1909)

Pinner's salt
named for the German chemist Adolf Pinner (1842-1909)

Pinner triazine synthesis
named for the German chemist Adolf Pinner (1842-1909)

pinnoite
$MgB_2O_4{\cdot}3H_2O$, named for the 19th century German mining official M. Pinno

pin(o)-
derived from the genus name *Pinus* (pines), rom *pinus* (Latin: pine)

pinonic acid
$C_{10}H_{16}O_3$, derived from pin(o)- and -one

pinosylvin
$C_{14}H_{12}O_2$, derived from the species name *Pinus sylvestris* L. (Scotch pine), from pin(o)- and *silvestris* (Latin: of a forest), from *silva* (Latin: forest), and -in(e)

pionium
$\pi^+\mu^-$, $\pi^-\mu^+$, derived from pion and -ium

pipecolic acid (dihydrobaikiane, homoproline)
$C_6H_{11}NO_2$, derived from pipecoline

pipecoline
$C_6H_{13}N$, coined by contraction of piperidine and picoline

piperazine
$C_4H_{10}N_2$, derived from piper(o)-, az(a)-, and -in(e); patterned after pyrazine

piperic acid
$C_{12}H_{10}O_4$, derived from piper(o)-

piperidine
$C_5H_{11}N$, derived from piperine and -idin(e)

piperine
$C_{17}H_{19}NO_3$, derived from piper(o)- and -in(e)

piperitone
$C_{10}H_{16}O$, derived from the specific epithet of the species name *Eucalyptus piperita* (Sydney peppermint), from *piperitus* (Latin: seasoned with pepper), from piper(o)-, and -one

piper(o)-
derived from the genus name *Piper* (peppers), from *piper* (Latin: pepper), ultimately from *pippali* (Sanskrit: long pepper)

piperonal (heliotropin)
$C_8H_6O_3$, derived from piperine and -al –

referring to the formation of piperonal by oxidation of piperine

piperonylic acid
$C_8H_6O_4$, derived from piperonal

PIPES
$C_8H_{18}N_2O_6S_2$, an abbreviation for piperazine-1,4-di(ethanesulfonic acid)

pipet
derived from *pipette* (French: little pipe)

pipsyl
$C_6H_4IO_2S-$, coined by contraction of *p*-iodophenylsulfonyl

Piria reaction
named for the Italian chemist Raffaele Piria (1815-1865)

Piria's acid
$C_{10}H_9NO_3S$, named for the Italian chemist Raffaele Piria (1815-1865)

Pirkle alcohol (Pirkle's alcohol)
$C_{16}H_{11}F_3O$, named for the US chemist William Howard Pirkle (born 1934)

pirssonite
$Na_2Ca(CO_3)_2{\cdot}2H_2O$, named for the US petrographer L. V. Pirsson (1860-1919)

PIS
an abbreviation for Penning ionization spectroscopy

pisatin
$C_{17}H_{14}O_6$, derived (with contraction) from he species name *Pisum sativum* (field and garden pea), from *pisum* (Latin: pea) and *sativus* (Latin: cultivated), and -in(e)

pitchblende
UO_2, derived from *Pechblende* (German: pitchblende)

Pitzer strain
named for the US chemist Kenneth Sanborn Pitzer (1914-1997)

pivalic acid
$C_5H_{10}O_2$, coined by contraction of pinacolin and valeric acid – referring to the facts that pivalic acid can be obtained by oxidation of pinacolin and is isomeric with valeric acid

pivaloyl (pivalyl)
C_5H_9O-, derived from pivalic acid and -oyl

pivalyl (pivaloyl)
C_5H_9O-, derived from pivalic acid and -yl

PIXE
an abbreviation for particle-induced (proton-induced) X-ray emission

pixyl
$C_{19}H_{13}O-$, coined by contraction of 9-phenyl-9*H*-xanthen-9-yl

pK
derived from K, from *Konstante* (German: a constant); patterned after pH

placebo
derived from *placebo* (Latin: I shall please), from *placere* (Latin: to please)

plagioclase
$(Ca,Na)(Al,Si)_4O_8$, derived from *plagios* (Greek: oblique, slanting) and -clase – referring to this mineral's obtuse cleavage angles

Planck's constant
named for the German physicist Max Karl Ernst Ludwig Planck (1858-1947)

[1]**plasma**
derived from *plasma* (Greek: something molded), from *plassein* (Greek: to mold)

[2]plasma (green chalcedony)
SiO_2, derived from *plasma* (Greek: something molded), from *plassein* (Greek: to mold)

plasmalogen
coined by contraction of plasma, alkali, and -gen

plasmin
derived from plasma and -in(e)

plasminogen
derived from plasmin and -gen

plaster of Paris
$2CaSO_4 \cdot H_2O$, named after deposits at Montmartre, Paris, France

plastic
derived from plast(o)-

plast(o)-
derived from *plastikos* (Greek: pliable), from *plassein* (Greek: to form, to mold)

plastocyanin
derived (with contraction) from chloroplast and cyanin

plastoquinone
derived (with contraction) from chloroplast and quinone

platforming
derived (with contraction) from platinum and reforming

platinocene
$C_{10}H_{10}Pt$, derived from platinum and -ocene

platinum
Pt, derived from *platina* (Spanish: platinum), diminutive of *plata* (Spanish: silver) – referring to the perceived uselessness of the silver-like platinum because of its high melting point

plattnerite
α-PbO_2, named for the German metallurgist K. F. Plattner (1800-1858)

platy-
derived from *platys* (Greek: flat, broad)

platyphylline
$C_{18}H_{27}NO_5$, derived from the specific epithet of the species name *Senecio platyphyllus* DC. (a groundsel), from platy-, phyll(o)-, and -in(e)

plaunotol
$C_{20}H_{34}O_2$, coined as a trademark

pleiadene
$C_{18}H_{12}$, derived from the Pleiads, a constellation of seven stars, according to Greek mythology seven daughters of Atlas turned into stars, and -ene – referring to the seven-membered central ring of pleiadene

ple(o)-
derived from *pleion* (Greek: more)

pleonaste (ceylonite)
$Al_2(Fe,Mg)O_4$, derived from *pleonastos* (Greek: abundant) – referring to the many faces of the crystals of this mineral

plesi(o)-
derived from *plesios* (Greek: near), from *pelas* (Greek: near)

plessite
(Fe,Ni), derived from plesi(o)- and -ite – referring to the intergrowth of kamacite and taenite in this mineral

pleur(o)-
derived from *pleura* (Greek: side, rib)

pleuromutilin
$C_{22}H_{34}O_5$, derived from the species name *Pleurotus mutilus* (Fr.) Sacc. (a mushroom), from pleur(o)- and *mutilus* (Latin: mutilated, maimed), and -in(e)

pleurotin
$C_{21}H_{22}O_5$, derived from the genus name *Pleurotus* (mushrooms), from pleur(o)- and -in(e)

plicatic acid
$C_{20}H_{22}O_{10}$, derived from the specific epithet of the species name *Thuja plicata* (western red cedar), from *plicatus* (Latin: folded), from *plicare* (Latin: to fold)

PLNM
an abbreviation for principle of least nuclear motion

plumb(a)-
derived from plumbum

plumbagin
$C_{11}H_8O_3$, derived from the genus name *Plumbago* (herbs, shrubs), from *plumbago* (Latin: leadwort), from *plumbum* (Latin: lead), and -in(e)

plumbago
C_n, archaic name for graphite, derived from *plumbago* (Latin: lead ore), from *plumbum* (Latin: lead)

plumbane
PbH_4, derived from plumb(o)- and -an(e)

plumbi-
Pb^{4+}, derived from plumbum

plumbic
derived from plumb(o)- and -ic

plumb(o)-
derived from plumbum

plumbo-
Pb^{2+}, derived from plumbum

plumbous
derived from *plumbosus* (Latin: full of lead), from *plumbum* (Latin: lead)

plumbum
Pb, New Latin name for lead, from *plumbum* (Latin: lead)

plumericin
$C_{15}H_{14}O_6$, derived from the genus name *Plumeria* (shrubs, trees), named for the French botanist Charles Plumier (1646-1704), and -icin(e)

plumieride
$C_{21}H_{26}O_{12}$, derived from the genus name *Plumeria* (shrubs, trees), after the French botanist Charles Plumier (1646-1704), and ^{3}-ide

plutonium
Pu, named for the planet Pluto, ultimately derived from Pluton, the Greek god of the underworld

PLZT
(Pb,La,Zr,Ti), an abbreviation of the element symbols of the components of this alloy

PMR
an abbreviation for proton magnetic resonance

PNDO
an abbreviation for partial neglect of differential overlap

pnicogen (pentel)
N, P, As, Sb, Bi, derived from *pniktos* (Greek: suffocated) and -gen

pnictide
derived from pnicogen and -ide

POC
an abbreviation for particulate organic carbon

podand
derived from pod(o)- and -and

podaplex
derived (with contraction) from podand and complex

podate
derived from podand and -ate

-podium
derived from pod(o)- and -ium

pod(o)-
derived from *pous* (Greek: foot)

podocarpane
$C_{17}H_{30}$, derived from podocarp(o)- and -an(e)

podocarpic acid
$C_{17}H_{22}O_3$, derived from podocarp(o)-

podocarp(o)-
derived from the genus name *Podocarpus* (yellowwood), from podo- and carp(o)-

pododacric acid
$C_{20}H_{28}O_5$, derived (with contraction) from the species name *Podocarpus dacrydioides* (Kahikatea white pine), from podocarp(o)- and *dakrydion* (Greek: scammony), diminutive of *dakry* (Greek: drop, tear)

podophyllic acid
$C_{22}H_{24}O_9$, derived from podophyll(o)-

podophyll(o)-
derived from the genus name *Podophyllum* (herbs), from pod(o)- and phyll(o)-

podophyllotoxin
$C_{22}H_{22}O_8$, derived from podophyll(o)- and toxin

poiet(o)-
derived from *poietikos* (Greek: productive, formative), from *poiein* (Greek: to make, to do, to create, to compose)

polane
PoH_2, derived (with contraction) from polonium and -an(e)

polarimetry
derived (with contraction) from optical polarization and -metry

polarography
derived (with contraction) from electrochemical polarization and -graphy

polianite (pyrolusite)
MnO_2, derived from *poliainesthai* (Greek: to become white with foam), from poli(o)- – referring to this mineral's gray color

poli(o)-
derived from *polios* (Greek: gray, pale)

pollucite
$(Cs,Na)_2Al_2Si_4O_{12}\cdot H_2O$, named for the ancient mythical figure Pollux, latinized from of *Polydeukes* (Greek: very thoughtful), from poly- and *endukeos* (Greek: steadfast, thoughtful) – referring to this mineral's common association with castorite

polonium
Po, named for Poland, derived from *Polonia* (New Latin: Poland)

Polonovski reaction
named for the Russian-French chemist Max

Polonovski (1861-1939) and the French chemist Michel Polonovski (1889-1954)

poly-
derived from *polys* (Greek: much, many)

polybasite
$Ag_{12}Cu_4Sb_2S_{11}$, derived from poly-, basic, and -ite – referring to the basic nature of this mineral

polycrase
$Y(Ti,Nb)_2(O,OH)_6$, derived from poly- and -crase – referring to the large number of elements occurring in this mineral

polydymite
Ni_3S_4, derived from poly-, didymos (Greek: twin), and -ite – referring to this mineral's numerous twinned crystals

polygodial
$C_{15}H_{22}O_2$, derived from the genus name *Polygonum* (herbs), from *polygonon* (Greek: knot grass), from poly- and -gon, di-, and -al

polyhalite
$K_2Ca_2(SO_4)_4{\cdot}2H_2O$, derived from poly-, hal(o)-, and -ite – referring to the many salt components in this mineral

polylithionite
$KLi_2AlSi_4O_{10}(F,OH)_2$, derived from poly-, lithium, and -ite – referring to this mineral's high lithium content

polymer
derived from poly- and -mer

polymerase
derived from polymerization and -ase

polymorphic
derived from poly- and -morphic

polymyxin
derived from the specific epithet of the bacterial species name *Bacillus polymyxa*, from poly- and myx(o)-, and -in(e)

polynosic
derived (with contraction) from *polymère non synthétique* (French: non-synthetic polymer)

polypeptide
derived from poly- and peptide

polyprenol
derived from poly- and [1]prenol

polyquinane (cyclopentanoid, quinane)
derived from poly-, *quini* (Latin: five apiece), and -an(e) – referring to these hydrocarbons' skeleton of condensed cyclopentane rings

polysaccharide (glycan)
derived from poly- and saccharide

polyterpene
$(C_5H_8)_n$, derived from poly- and terpene

Polzeniusz-Krauss process
named for the 20th century German chemists Ferdinand Eduard Polzeniusz and Krauss

POM
an abbreviation for polyoxometalate

Pomeranz-Fritsch reaction
named for the 19th century Austrian chemist C. Pomeranz and the 19th century German chemist P. Fritsch

ponasterone
derived (with contraction) from the species name *Podocarpus nakaii* Hay. (Nakai podocarp), sterol, and -one

ponceau
derived from *pouncel* (Old French: red poppy), utlimately from *pavo* (Latin: peacock)

Ponzio reaction
named for the 19th century Italian chemist Giacomo Ponzio

POP
an abbreviation for persistent organic pesticide(s)

populin
$C_{20}H_{22}O_8$, derived from the genus name *Populus* (poplars), from *populus* (Latin: poplar), and -in(e)

p-orbital
named after the spectroscopic 'principal' series

π-orbital
derived from p-orbital – referring to the common symmetry of π- and p-orbitals

porfiromycin
$C_{16}H_{20}N_4O_5$, derived from porphyr(o)- and -mycin – referring to this antibiotic's deep purple color

poriferastane
$C_{29}H_{52}$, derived from the phylum name *Porifera* (invertebrate animals), from *porus* (Latin: pore) and -fer, sterol, and -an(e)

porphobilinogen
$C_{10}H_{14}N_2O_4$, derived from porphyrin, bil(i)-, and -gen

porphyrin (porphine)
$C_{20}H_{14}N_4$, derived from porphyr(o)- and -in(e) – referring to this heterocycle's dark red color

porphyr(o)-
derived from *porphyra* (Greek: purple)

porphyropsin
derived from porphyr(o)- and opsin

porphyry
a rock species, ultimately derived from *porphyrites lithos* (Greek: purple colored stone)

Porter-Silber reaction
named for the US chemists C. C. Porter (born 1914) and R. H. Silber (born 1915)

Portland cement
a name reflecting the resemblance to Portland stone, quarried on the Portland Peninsula, Dorsetshire, UK

positronium
e^+e^-, derived from positron and -ium

post-
derived from *post* (Latin: after)

potash
K_2CO_3, derived from pot ashes, ultimately of Dutch origin

potassa
K_2CO_3, New Latin word for potash, derived from potash

potassium
K, derived from potassa and -ium

potentiometry
derived from potential and -metry

Potier-Polonovski reaction
named for the French chemist P. Potier and the Russian-French chemist Max Polonovski (1861-1939)

Pott-Broche process
named for the German chemists Alfred Pott (1882-1951) and H. Broche (1896-1963)

povidone
coined by contraction of polyvinylpyrrolidone

powellite
$CaMoO_4$, named for the US geologist John Wesley Powell (1834-1902)

PP pathway
an abbreviation for pentose phosphate pathway

PPP method
an abbreviation for Pariser-Parr-Pople method

PQ
an abbreviation for plastoquinone

PQQ (methoxatin)
$C_{14}H_6N_2O_8$, an abbreviation for pyrroloquinolinequinone

prase(o)-
derived from *prasios* (Greek: onion green), from *prason* (Greek: onion)

praseodidymium
Pr, now obsolete, derived from prase(o)- and *didymos* (Greek: twin)

praseodymia
Pr_2O_3, derived from praseodymium and -a

praseodymium
Pr, coined by alteration of obsolete praseodidymium

prasterone (DHEA)
$C_{19}H_{28}O_2$, coined by contraction and alteration of dehydroepiandrosterone

pratensein
$C_{16}H_{12}O_6$, derived from the specific epithet of the species name *Trifolium pratense* L. (red clover), from *pratensis* (Latin: of the meadow), from *pratum* (Latin: meadow), and -in(e)

pre-
derived from *prae* (Latin: before)

precipitate
derived from *praecipitare* (Latin: to throw down headlong), from *praeceps* (Latin: headlong)

precocene
derived (with contraction) from precocious metamorphosis and -ene

prednisolone
$C_{21}H_{28}O_5$, derived from pregn(a)-, di-, -ene, and [1]-olone

prednisone
$C_{21}H_{26}O_5$, derived from pregn(a)-, di-, and -one

pregn(a)-
derived from pregnant

pregnane
$C_{21}H_{36}$, derived from pregn(a)- and -an(e)

prehnite
$Ca_2Al(Si,Al)_4O_{10}(OH)_2$, named for the Dutch colonel H. van Prehn (1733-1785)

prehnitene
$C_{10}H_{14}$, derived from prehnitic acid and -ene

prehnitic acid
$C_{10}H_6O_8$, derived from prehnite – probably referring to similarities of the crystal shapes of prehnite and prehnitic acid

Prelog-Djerassi lactone
$C_{10}H_{16}O_4$, named for the Yugoslav-Swiss chemist Vladimir Prelog (1906-1998) and the Austrian-US chemist Carl Djerassi (born 1923)

Prelog strain
named for the Bosnian-Swiss chemist Vladimir Prelog (1906-1998)

[1]**prenol**
coined by contraction of isoprenoid alcohol

[2]**prenol (prenyl alcohol)**
$C_5H_{10}O$, derived from [1]prenol

prenyl
C_5H_9-, derived from [2]prenol and -yl

prephenic acid
$C_{10}H_{10}O_6$, derived from pre- and phene – referring to this compound's role as an intermediate in the biosynthesis of aromatic compounds

Prévost reaction
named for the French chemist C. Prévost (born 1899)

priceite (pandermite)
$Ca_2B_5O_7(OH)_5{\cdot}H_2O$, named for the US mineralogist Thomas Price (born 1837)

Prilezhaev reaction
named for the Russian chemist Nikolai Aleksandrovich Prilezhaev (1872-1944)

primapterin
$C_9H_{11}N_5O_3$, derived from prima, from *primus* (Latin: the first), and pterin

primever-
derived from *primevère* (French: cowslip), from *flour de primevoire* (Old French: spring flower)

primeverin
$C_{20}H_{28}O_{13}$, derived from primever- and -in(e)

primeverose
$C_{11}H_{20}O_{10}$, derived from primever- and -ose

prim(o)-
derived from *primus* (Latin: the first)

primocarcin
$C_8H_{12}N_2O_3$, derived from prim(o)-, carcin(o)-, and -in(e)

primulaverin
$C_{20}H_{28}O_{11}$, coined by variation of primeverin, derived from the genus name *Primula* (perennial herbs), from *primula veris* (Latin: firstling of spring), from *primulus* (Latin: firstling) and *ver* (Latin: spring)

Prins reaction
named for the Dutch chemist Hendrik Jacobus Prins (1889-1958)

prion
coined by contraction of proteinaceous infectious particle

priorite (eschynite, aeschynite, blomstrandine)
$(Ce,Ca,Fe)(Ti,Nb)_2(O,OH)_6$, named for the British mineralogist Granville T. Prior (deceased 1936)

prismane (Ladenburg benzene)
C_6H_6, derived from prism and -an(e) – referring to the molecular shape of this hydrocarbon

prismatine
$(\square,Fe,Mg)(Mg,Al,Fe)_5Al_4Si_2(Si,Al)_2(B,Si,Al)(O,OH,F)_{22}$, derived from prismatic and -in(e) – referring to this mineral's prismatic crystals

pristane
$C_{19}H_{40}$, derived from *pristis* (Latin: shark, sawfish) and -an(e)

pristinamycin
derived from the specific epithet of the bacterial species name *Streptomyces pristinaespiralis*, from *pristinus* (Latin: earlier, pristine), and -mycin

pro-
derived from *pro* (Latin: for)

proazulene
derived from pro- and azulene

probertite
$NaCaB_5O_7(OH)_4{\cdot}3H_2O$, named for the British-US scientist Frank H. Probert (1876-1940)

prochiral
derived from pro- and chiral

prodigiosin
$C_{30}H_{25}N_3O$, derived from the specific epithet of the bacterial species name *Chromobacterium prodigiosum*, from *prodigiosus* (Latin: prodigious), from *prodigium* (Latin: omen)

prodlure
$C_{16}H_{28}O_2$, probably derived from to prod and lure

proenzyme (zymogen)
derived from pro- and enzyme

progesterone
$C_{21}H_{30}O_2$, derived from pro-, *gestatio* (Latin: pregnancy), from *gerere* (Latin: to bear), sterol, and -one

proinsulin
derived from pro- and insulin

prolactin (mammatropin)
derived from pro, lactogenic, and -in(e)

prolamin
coined by contraction of proline and *Glutaminsäure* (German: glutamic acid)

prolan
derived from *proles* (Latin: offspring, progeny) and -an(e)

proline
$C_5H_9NO_2$, coined by contraction of pyrrolidine

promethium
Pm, named for Prometheus (Greek: the one who thinks ahead), a Greek mythical figure, from *promethes* (Greek: foresightful) – referring to the notion that nuclear power would benefit mankind in modern time as fire did in antiquity

promine
a name referring to the cancer growth-promoting acitvity of this protein

prop(a)-
derived from propane

propane
C_3H_8, derived from propionic acid and -ane

propargyl
C_3H_3-, derived from prop(a)- and *argyros* (Greek: silver) – referring to the possible replacement of one acidic proton of the propargyl group by silver ion

propellane
derived from propeller and -an(e) – referring to the molecular shape of these hydrocarbons

properdin (complement factor P)
derived from pro-, *perdere* (Latin: to

destroy), and -in(e) – referring to this serum protein's destructive action on bacteria, neutralization of viruses, and lysis of red blood cells in the presence of complement and magnesium ions

propiolactone
$C_3H_4O_2$, derived from propio(n)- and lactone

propiolic acid
$C_3H_2O_2$, coined by modification of propionic acid; patterned after stearolic acid

propio(n)-
derived from propionic acid

propionic acid
$C_3H_6O_2$, derived from *proteon pion* (Greek: the first fat) – referring to the fact that propionic acid is to be regarded as the first member of the homologous series of fatty acids judged by the soapy feel of its salts

propiophenone
$C_9H_{10}O$, derived from propio(n)- and -phenone

propyl
C_3H_7–, derived from prop(a)- and -yl

propylene
C_3H_6, derived from propyl- and -ene

propylure
$C_{18}H_{32}O_2$, derived (with contraction) from (*E*)-10-propyltrideca-5,9-dien-1-ol acetate and lure

proscillaridin
$C_{30}H_{42}O_8$, derived from pro-, scillaren A, and -idin(e)

prost(a)-
derived from prostate gland, ultimately from *proistanai* (Greek: to put in front) and *glans* (Latin: acorn)

prostacyclin
$C_{20}H_{32}O_5$, derived from prost(a)-, cycl(o)-, and -in(e)

prostaglandin
derived from prostate gland, ultimately from *proistanai* (Greek: to put in front) and *glans* (Latin: acorn), and -in(e)

prostane
$C_{20}H_{40}$, derived from prost(a)- and -an(e)

prostanoid
derived (with contraction) from prostaglandin and -oid

prosthetic group
derived from *prosthetikos* (Greek: adding, furthering), from *prostithenai* (Greek: to add)

protactinium
Pa, derived from prot(o)- and actinium – referring to the fact that this element forms actinium in the course of its radioactive decay

protamine
derived (with contraction) from protein and amine – referring to the high content of arginine residues in these strongly basic proteins

protease (proteinase)
derived from protein and -ase

proteide
derived from protein and [1]-ide

protein
derived from *proteion* (Greek: the first place, the chief rank) – referring to the proteins' key role in living organisms

proteinase (protease)
derived from protein and -ase

proteinogenic
derived from protein and -gen – referring to amino acids obtainable by hydrolysis of proteins

proteo-
derived from protein

prothrombin (thrombogen)
derived from pro- and thrombin

protic
derived from proton

protide
$^{1}H^{-}$, derived from protium and ^{2}-ide

protio-
derived from protium

protirelin (thyrotropin-releasing hormone)
$C_{16}H_{22}N_6O_4$, coined by contraction of prolactin, thyrotropin, release, and -in(e)

protium
^{1}H, derived from prot(o)- and -ium

prot(o)-
derived from *protos* (Greek: the first)

protoanemonin
$C_5H_4O_2$, derived from prot(o)- and anemonin

protoberberine (berbine)
$C_{17}H_{17}N$, derived from prot(o)- and berberine

protocatechu-
derived from prot(o)- and catechu-

protocatechuic acid
$C_7H_6O_4$, derived from protocatechu-

protocatechuic aldehyde (protocatechu aldehyde)
$C_7H_6O_3$, derived from protocatechu-

protofluorine
a hypothetical element inferred from certain lines in the spectra of nebulae, derived from prot(o)- and fluorine

protohydrogen
a hypothetical element inferred from certain lines in the spectra of nebulae, derived from prot(o)- and hydrogen

protokosin
derived from prot(o)- and kosin

protolysis
derived from proton and lysis

protolyte
derived from protolysis

proton
derived from prot(o)- and ^{3}-on

protopine
$C_{20}H_{19}NO_5$, derived from prot(o)-, opium, and -in(e)

protoporphyrin IX (ooporphyrin, Kammerer's porphyrin)
$C_{34}H_{34}N_4O_4$, derived from prot(o)- and porphyrin

protostane
$C_{30}H_{54}$, derived from prot(o)-, sterol, and -ane

protostephanine
$C_{21}H_{27}NO_4$, derived from prot(o)- and stephanine

protoveratrine
derived from prot(o)- and veratrine

protoverine
$C_{27}H_{43}NO_9$, coined by variation of protoveratrine

protyl
H, an obsolete name for hydrogen, derived from prot(o)- and -yl

proustide (daltonide)
stoichiometric compound, named for the French chemist Louis Joseph Proust (1754-1826)

proustite (light red silver ore)
Ag_3AsS_3, named for the French chemist Louis Joseph Proust (1754-1826)

Proust's law
named for the French chemist Joseph Louis Proust (1754-1826)

Prout's hypothesis
named for the British physician and chemist William Prout (1785-1850)

provitamin
derived from pro- and vitamin

prun-
derived from the genus name *Prunus* (fruit trees), from *prunus* (Latin: plum tree)

prunase
derived from prun- and -ase

prunetin
$C_{16}H_{12}O_5$, derived from prun- and -etin(e)

Prussian blue (Berlin blue, Milori blue, Paris blue)
$C_6Fe_2KN_6$, named after Prussia, Germany

prussic acid
HCN, derived from Prussian blue – referring to the formation of hydrogen cyanide by acid treatment of Prussian blue

[1]prusside (prussiate)
$Fe(CN)_6^{4-}$, derived (with contraction) from Prussian blue and [2]-ide

[2]prusside (prussiate)
CN^-, derived from Prussic acid

PS
an abbreviation for photoelectron spectroscopy

Pschorr reaction
named for the German chemist Robert Pschorr (1868-1930)

PSE
an abbreviation for the periodic system of the elements

pseud(o)-
ultimately derived from *pseudein* (Greek: to lie, to cheat, to falsify)

pseudoacid
derived from pseudo- and acid – referring to the required tautomeric reorganization before these compounds' acidity can manifest itself

pseudobaptigenin
$C_{16}H_{10}O_5$, derived from pseudobaptisin, -gen, and -in(e)

pseudobaptisin
$C_{28}H_{30}O_{14}$, derived from pseudo- and baptisin

pseudocodeine (neoisocodeine)
$C_{18}H_{21}NO_3$, derived from pseudo- and codeine

pseudoconhydrine
$C_8H_{17}NO$, derived from pseud(o)- and conhydrine

pseudocumene
C_9H_{12}, derived from pseudo- and cumene

pseudoelement
derived from pseudo- and element

pseudoephedrine
$C_{10}H_{15}NO$, derived from pseudo- and ephedrine

pseudofructose (psicose)
$C_6H_{12}O_6$, derived from pseudo- and fructose

pseudoguaiane (ambrosane)
$C_{15}H_{28}$, derived from pseud(o)- and guaiane

pseudohalogen
derived from pseudo- and halogen

pseudoionone
$C_{13}H_{20}O$, derived from pseudo- and ionone

pseudomalachite (dihydrite)
$Cu_5(PO_4)_2(OH)_4$, derived from pseud(o)- and malachite – referring to this mineral's similarity with malachite

pseudomerism (tautomerism)
derived from pseudo- and -mer

pseudomonic acid (mupirocin, turixin)
$C_{26}H_{44}O_9$, derived from the bacterial species name *Pseudomonas*, from pseudo- and -monas

pseudomorph
derived from pseud(o)- and -morph

pseudomorphine
$C_{34}H_{36}N_2O_6$, derived from pseudo- and morphine

pseudonitrol
$R_2C(NO_2)(NO)$, derived from pseudo-, nitr(o)-, and -ol

pseudopelletierine
$C_9H_{15}NO$, derived from pseudo- and pelletierine

pseudoracemate
derived from pseudo- and racemate

pseudorotation
derived from pseudo- and rotation

pseudotropine
$C_8H_{15}NO$, derived from pseudo- and tropine

pseudourea
CH_4N_2O, derived from pseudo- and urea – referring to this tautomer's isomerism with urea

pseudouridine
$C_9H_{12}N_2O_6$, derived from pseudo- and uridine – referring to this ribonucleoside's isomerism with uridine

pseudoyohimbine
$C_{21}H_{25}N_2O_3$, derived from pseudo- and yohimbine

psicofuranine
$C_{11}H_{15}N_5O_5$, derived from psicose, furan, and -in(e)

psicose
$C_6H_{12}O_6$, coined by contraction and alteration of pseudofructose

psil(o)-
derived from *psilos* (Greek: bare), related to *psen* (Greek: to rub)

psilocin
$C_{12}H_{16}N_2O$, derived from psiloc(yb)- and -in(e)

psiloc(yb)-
derived from the genus name *Psilocybe* (mushrooms), from psil(o)- and *kybe*

(Greek: head)

psilocybin
$C_{12}H_{17}N_2O_4P$, derived from psiloc(yb)- and -in(e)

psilomelane
MnO_2, derived from psil(o)- and -melane – referring to the smooth and black appearance of this mineral

psoralen
$C_{11}H_6O_3$, derived from the genus name *Psoralea* (breadroot), from *psoraleos* Greek: scabby, itchy), from *psora* (Greek: itch)

psychotrine
$C_{28}H_{36}N_2O_4$, derived from the genus name *Psychotria* (shrubs, trees, herbs), from *psychotrios* (Middle Greek: vivifying), and -in(e)

psychro-
derived from *psychros* (Greek: cold), from *psychein* (Greek: to make cold)

ptaquiloside
$C_{20}H_{30}O_8$, derived (with contraction) from the species name *Pteridium aquilininum* var *latiusculum* (a bracken fern), from pter(o)- and *aquilinus* (Latin: eagle-like), from *aquila* (Latın: eagle), -ose, and [3]-ide

PTC
an abbreviation for phase transfer catalysis

-pten(e)
derived from *ptenos* (Greek: winged), from *petesthai* (Greek: to fly)

pteridine
$C_6H_4N_4$, derived from pter(o)- and -idine

pterin
$C_6H_5N_5O$, derived from pter(o)- and -in(e)

pter(o)-
derived from *pteron* (Greek: feather, wing)

pterocarpin
$C_{17}H_{14}O_5$, derived from the genus name *Pterocarpus* (tropical trees), from pter(o)- and carp(o)-, and -in(e)

pteroic acid
$C_{14}H_{12}N_6O_3$, derived from pterin

pteropterin (PTGA, teropterin)
$C_{29}H_{33}N_9O_{12}$, derived from pteroyl and pterin

pteroyl
$C_{14}H_{11}N_6O_2-$, derived from pteroic acid and -yl

ptil(o)-
derived from *ptilon* (Greek: down, feather)

ptilolite (clinoptilolite, mordenite)
$(K,Ca,Na)_6(Al_9Si_{39})O_{96}{\cdot}29H_2O$, derived from ptil(o)- and -lite – referring to this mineral's fibrous appearance

ptoma-
derived from *ptoma* (Greek: corpse), from *piptein* (Greek: to fall)

ptomaine (ptomatine)
derived from ptoma- and -in(e)

ptyalin
derived from ptyal(o)- and -in(e)

ptyal(o)-
derived from *ptyalon* (Greek: saliva), from *ptyein* (Greek: to spit)

ptych(o)-
derived from *ptyssein* (Greek: to fold)

pukateine
$C_{18}H_{17}NO_3$, derived from pukatea tree

(*Laurelia novae-zealandiae*), from pukatea (Maori: pukatea tree), and -in(e)

pulegone
$C_{10}H_{16}O$, derived from the specific epithet of the species name *Mentha pulegium* L. (pennyroyal), from *puleium*, *pulegium* (Latin: pennyroyal), and -one

pullulan
derived from the specific epithet of the fungal species name *Aureobasidium pullulans*, from *pullulans* (Latin: sprouting), from *pullulare* (Latin: to sprout), from *pullulus*, diminutive of *pullus* (Latin: young of an animal), from *putus* (Latin: boy), and -an

pullulanase
derived from pullulan and -ase

pulvinic acid (pulvic acid)
$C_{18}H_{12}O_5$, derived anagrammatically from vulpinic acid

pumice
derived from *pumex* (Latin: pumice), from *spuma* (Latin: foam)

Pummerer rearrangement
named for the Austrian-German chemist Rudolf Pummerer (1882-1973)

pumpellyite
$Ca_2(Al,Mg,Fe)_3Al_2(SiO_4)(Si_2O_7)(OH)_2 \cdot H_2O$, named for the US geologist Raphael Pumpelly (1837-1923)

Purdie methylation
named for the 20th century Scottish chemist Thomas Purdie

Purex process
coined by contraction of plutonium-uranium recovery by extraction

purine
$C_5H_4N_4$, derived from *purus* (Latin: pure), (*acidum*) *uricum* (New Latin: uric acid), and -in(e)

puro-
derived from purine

puromycin
$C_{22}H_{29}N_7O_5$, probably coined as a trademark, derived from puro- and -mycin

purpurin
$C_{14}H_8O_5$, derived from purpur(o)- and -in(e)

purpurite
$MnPO_4$, derived from purpur(o)- and -ite – referring to this mineral's purple red color

purpur(o)-
derived from *purpureus* (Latin: purple red), from *purpura* (Latin: purple color), ultimately from *porphyra* (Greek: purple stone)

purpurogallin
$C_{11}H_8O_5$, derived from purpur(o)-, gall, and -in(e)

putrescine
$C_4H_{12}N_2$, derived from *putrescere* (Latin: to rot), from *puter* (Latin: rotten), and -in(e)

PVC
an abbreviation for poly[vinyl chloride]

PVD
an abbreviation for physical vapor deposition

pycnometer (pyknometer)
derived from *pyknos* (Greek: thick, dense) and meter

pyo-
derived from *pyon* (Greek: pus)

pyocyanine
$C_{13}H_{19}N_2O$, derived from the specific epithet of the bacterial species name *Bacillus pyocyaneus*, from py(o)- and cyan(o)-, and -in(e)

Py-Phe
$C_{29}H_{25}NO_3$, coined by contraction of *N*-[4-(pyren-1-yl)butyryl]-L-phenylalanine

pyralspite
a group name for garnets coined by contraction of pyrope, almandine, spessartine, and -ite

pyran
C_5H_6O, derived from pyrone and -an(e)

pyranose
derived from pyran and -ose – referring to the six-membered ring in these sugars

pyranoside
derived from pyranose and [3]-ide

pyranthrene
$C_{30}H_{16}$, derived from pyr(o)-, anthr(a)-, and -ene

pyrargyrite (dark red silver ore)
Ag_3SbS_3, derived from pyr(o)-, argyr(o)-, and -ite – referring to this mineral's red color and silver content

pyrazine
$C_4H_4N_2$, derived from pyr(o)-, az(a)-, and -in(e)

pyrazole
$C_3H_4N_2$, derived from pyr(o)-, az(a)-, and -ol(e)

pyrene
$C_{16}H_{10}$, derived from pyr(o)- and -ene – referring to pyrene's formation upon dry distillation of coal

pyrethrin
derived from pyrethrum and -in(e)

pyrethroid
derived from pyrethrum and -oid

pyrethrosin
$C_{17}H_{22}O_5$, derived from pyrethrum, -ose, and -in(e)

pyrethrum
derived from the genus name *Pyrethrum* (chrysanthemum), ultimately from *pyrethron* (Greek: pellitory), from pyr(o)- – referring to the spicy taste of the root

pyridazine
$C_4H_4N_2$, derived from pyridine and hydrazine

pyridine
C_5H_5N, derived from pyr(o)-, [2]-ide, and -in(e)

pyridomycin
$C_{27}H_{32}N_4O_8$, derived from pyridine and -mycin

pyridoxal
$C_8H_9NO_3$, derived from pyridoxine and -al

pyridoxine (adermine, vitamin B_6)
$C_8H_{11}NO_3$, derived from pyridine, ox(o)-, and -in(e)

pyrimidine
$C_4H_4N_2$, coined by contraction of pyridine and amidine

pyrite
FeS_2, derived from *pyrites* (Greek: of fire) – referring to the sparks observed when iron is struck with a lump of this mineral

pyr(o)-
derived from *pyr* (Greek: fire)

pyrocatechol
$C_6H_6O_2$, derived from pyr(o)- and catechol

pyrochlore
$(Na,Ca)_2Nb_2(O,OH,F)_7$, derived from pyr(o)- and [1]chlor(o)- – referring to this mineral's fiery colors

pyrochroite
$Mn(OH)_2$, derived from pyr(o)- *chroma* (Greek: color), and -ite – referring to this mineral's color change upon ignition

pyrocomane (γ-pyrone)
$C_5H_4O_2$, derived from pyr(o)- and comanic acid

pyrogallol
$C_6H_6O_3$, derived from pyr(o)-, gallic acid, and -ol

pyroglutamic acid
$C_5H_7NO_3$, derived from pyr(o)- and glutamic acid – referring to this acid's formation upon heating of glutamic acid

pyroligneous acid
derived from pyr(o)- and *lignum* (Latin: wood)

pyrolusite (polianite)
MnO_2, derived from pyr(o)-, *loyein* (Greek: to wash), and -ite – referring to the use of this mineral to remove the greenish color of glass caused by iron impurities

pyrolysis
derived from pyr(o)- and lysis

pyromellitic acid
$C_{10}H_6O_8$, derived from pyr(o)- and mellitic acid – referring to the formation of this acid by pyrolysis of mellitic acid

pyromorphite
$Pb_5Cl(PO_4)_3$, derived from pyr(o)-, morph(o)-, and -ite – referring to the recrystallization reaction of the molten mineral

pyrone
$C_5H_4O_2$, coined by contraction of pyrocomane

pyrope
$Mg_3Al_2(SiO_4)_3$, derived from *pyropos* (Greek: fiery-eyed) – referring to this mineral's red hue

pyrophoric
derived from pyr(o)- and -phoric

pyrophosphatase
derived from pyrophosphate and -ase

pyrophosphoric acid (diphosphoric acid)
$H_4P_2O_7$, derived from pyr(o)- and phosphoric acid

pyrophyllite
$Al_2Si_4O_{10}(OH)_2$, derived from pyr(o)-, phyll(o)-, and -ite – referring to this mineral's voluminous exfoliation upon heating with the blowpipe

pyrosmalite
$(Fe,Mn)_8Si_6O_{15}(OH,Cl)_{10}$, derived from pyr(o)-, *osme* (Greek: smell), from *ozein* (Greek: to smell), and -lite – referring to the unpleasant odor this mineral gives off in the flame of the blowpipe

pyrostilpnite
Ag_3SbS_3, derived from pyr(o)- and *stilpnos* (Greek: shining) – referring to this mineral's fiery luster

pyrosulfuric acid (disulfuric acid)
$H_2S_2O_7$, derived from pyr(o)- and sulfuric acid

pyroxene
a rock species, derived from pyr(o)- and *xenos* (Greek: stranger) – referring to this rock species' appearance as perceived impurities in lava glass

pyroxenite
a rock species, derived from pyroxene and -ite

pyroxmangite
$(Mn,Fe)SiO_3$, derived from pyroxene, manganese, and -ite

pyrrhotite (magnetopyrite)
FeS, derived from *pyrrhotes* (Greek: redness) and -ite – referring to this mineral's red color

pyrr(o)-
derived from *pyrros* (Greek: reddish, literally flame colored)

pyrrocoline (indolizine)
C_8H_7N, derived from pyrrole, an unidentified name fragment, and -in(e)

pyrrole
C_4H_5N, derived from pyrr(o)- – referring to the red color developing when pyrrole vapor acts upon a pine splinter moistened with hydrochloric acid

pyrrolidine
C_4H_9N, derived from pyrrole and -idin(e)

pyrroline
C_4H_7N, derived from pyrrole and -in(e)

pyrrolizine
C_7H_7N, derived from pyrrole and -izin(e)

pyrrolnitrin
$C_{10}H_6Cl_2N_2O_2$, derived from pyrrole, nitr(o)-, and -in(e)

pyrrolysine
$C_{12}H_{21}N_3O_3$, derived from pyrrole and lysine

pyrromycin
$C_{30}H_{35}NO_{11}$, derived from pyrr(o)- and -mycin – referring to this antibiotic's red color

pyruvic acid
$C_3H_4O_3$, derived from pyr(o)- and uv(a)- – referring to this compound's formation upon dry distillation of, *inter alia*, tartaric acid

pzc
an abbreviation for point of zero charge

PZT
(Pb,Zr,Ti), (Pb,La,Zr,Ti), an abbreviation of the element symbols of the components of this alloy

Q

QSAR
an abbreviation for quantitative structure-activity relationship(s)

quadr(i)-
derived from *quadr(i)-* (Latin: four-)

quadricyclane
C_7H_8, derived from quadr(i)-, cycl(o)-, and -an(e)

quadro-
derived from *quadratum* (Latin: quadrangle)

quantum dye
$C_{38}H_{36}EuN_8S_2$, a name which refers to this compound's use as quantitative fluorescent label

quartz
SiO_2, derived from *Quarz* (German: quartz), from *kwardy* (West Slavic: hard)

quartzite
a rock species, derived from quartz and -ite

quasi-
derived from *quasi* (Latin: as if)

quassin
$C_{22}H_{28}O_6$, derived from Surinam quassia (*Picrasma amara* L.), named for its discoverer Quassi, an 18th century Surinamese slave, and -in(e)

quater-
derived from *quater* (Latin: four times)

quaternary
derived from *quaternarius* (Latin: consisting of four each, containing four)

quebrachamine
$C_{19}H_{26}N_2$, derived from quebrach(o)- and amine

quebrach(o)-
derived from quebracho, from *quiebrahacha* (American Spanish: quebracho), from *quiebra* (Spanish: it breaks) and *hacha* (Spanish: ax)

Queen's metal
(Sn,Sb,Zn,Pb,Cu), so named because this alloy first was used to manufacture goods for the British royal household

queen substance
$C_{10}H_{16}O_3$, named after this compound's secretion by queen honey bees (*Apis mellifera*)

Quelet reaction
named for the 20th century French chemist R. Quelet

quercetagetin
$C_{15}H_{10}O_8$, derived from querci-, the genus name *Tagetes* (tropical herbs), probably after the Etruscan god Tages, and -in(e)

quercetin (quercitin)
$C_{15}H_{10}O_7$, coined by modification of quercitrin

querci-
derived from *quercus* (Latin: oak)

quercitol
$C_6H_{12}O_5$, derived from querci- and -itol

quercitrin
$C_{21}H_{20}O_{11}$, derived from querci- and *citrus*

(Latin: citrus tree)

quicklime
CaO, derived from *calx viva* (Latin: quicklime, literally living lime)

quicksilver
Hg, archaic name for mercury, derived from *argentum vivum* (Latin: mercury, literally living silver)

quinaldic acid
$C_{10}H_7NO_2$, derived from quinaldine

quinaldine
$C_{10}H_9N$, derived from quin(o)-, aldehyde, and -in(e) – referring to this compound's formation from acetaldehyde and aniline

quinalizarin
$C_{14}H_8O_6$, derived from quin(o)- and alizarin

-quinane
derived from *quini* (Latin: five apiece) and -an(e)

quinarene
coined by contraction of quinonoid arene

quinazoline
$C_8H_6N_2$, derived from quin(o)-, az(a)-, -ol(e), and -in(e)

quinhydrone
$C_{12}H_{10}O_4$, coined by contraction of quinone and hydroquinone

quinic acid
$C_{11}H_9NO_3$, derived from quin(o)-

quinidine
$C_{20}H_{24}N_2O_2$, derived from quin(o)- and -idin(e)

quinine
$C_{20}H_{24}N_2O_2$, derived from *quinquina* (Spanish: cinchona bark), ultimately from *quinaquina* (Quechua: cinchona bark)

quininic acid
$C_{11}H_9NO_3$, derived from quinine

quininone
$C_{20}H_{22}N_2O_2$, derived from quinine and -one

quinizarin
$C_{14}H_8O_4$, derived (with contraction) from quin(o)- and alizarin

quin(o)-
derived from quinine

quinoid
derived from quinone and -oid

quinol
derived from quinone and -ol

quinoline
C_9H_7N, derived from quin(o)-, -ol(e), and -in(e)

quinolinic acid
$C_7H_5NO_4$, derived from quinoline

quinolizine
C_9H_9N, derived from quinoline and -izin(e)

quinomethane
derived from quin(o)- and methane

quinone
derived from quin(o)- and -one

quinovic acid
$C_{30}H_{46}O_5$, derived from quinovin

quinovin
derived from *quina nova* (New Latin: cinchona tree, *Cinchona*) and -in(e)

quinovose
$C_6H_{12}O_5$, derived from quinovin and -ose

quinoxaline
$C_8H_6N_2$, coined by contraction of quin(o)-, glyoxal, and -in(e)

quinqu(e)-
derived from *quinquies* (Latin: five times), from *quinque* (Latin: five)

quintessence
derived from *quinta essentia* (Latin: the fifth (and finest) essence or element (after fire, earth, air, and water), from *pempte oysia* (Greek: fifth substance)

quinuclidine
$C_8H_{13}N$, coined by contraction of quin(o)-, nucle(o)-, and -idin(e)

quisqualic acid
$C_5H_7N_3O_5$, derived from the genus name *Quisqualis* (woody vines), from *quis* (Latin: who?) and *qualis* (Latin: of what kind?) – referring to these plants' changing appearance which makes their identification difficult

R

R
derived from *rectus* (Latin: right)

Racah parameter
named for the Italian-Israeli physicist Giulio (Yoel) Racah (1909-1965)

racemic
derived from racemic acid

racemic acid
$C_4H_6O_6$, derived from *racemus* (Latin: cluster of grapes)

radialene
C_nH_n, derived from *radialis* (Latin: radial) and -ene

radical
derived from *radix* (Latin: root)

radicinin
$C_{12}H_{12}O_5$, derived from the specific epithet of the fungal species name *Stemphylium radicinum* and -in(e)

radio-
derived from *radius* (Latin: ray)

radioactive
derived from radio- and active

radio-iodine
^{131}I, derived from radio- and iodine

radio-phosphorus
^{32}P, derived from radio- and phosphorus

radio-sodium
^{24}Na, derived from radio- and sodium

radium
Ra, derived from *radius* (Latin: ray) and -ium – referring to the radioactivity of this element

radium emanation
Rn, an obsolete name for radon, derived from radium and emanation

radon
Rn, coined by contraction of the obsolete name radium emanation

Raecke process (Henkel process)
named for the 20th century German chemist Bernhard Raecke

raffinose (melitose, melitriose)
$C_{18}H_{32}O_{16}$, derived from *raffiner* (French: to refine) and -ose

Raman effect
named for the Indian physicist Chandrasekhara Venkata Raman (1888-1970)

Ramberg-Bäcklund rearrangement
named for the Swedish chemists Ludvig Ramberg (1874-1940) and Birger Bäcklund (1908-1997)

rammelsbergite
$NiAs_2$, named for the German chemist and mineralogist Karl Friedrich Rammelsberg (1813-1899)

Ramsay grease
named for the British chemist Sir William Ramsay (1852-1916)

ramsdellite
MnO_2, named for the US mineralogist Lewis Stephen Ramsdell (1895-1975)

Raney nickel
named for the US mechanical engineer Murray Raney (1885-1966)

Rankine temperature scale
named for the British physicist William John Macquorn Rankine (1820-1872)

Raoult's law
named for the French chemist François Marie Raoult (1830-1901)

rapamycin
$C_{51}H_{79}NO_{13}$, derived from *Rapa Nui* (Rapanui: Easter Island, Chile) and -mycin – referring to the habitat of the rapamycin-producing bacterium *Streptomyces hydroscopicus*

rare earth
derived from earth in its alchemistic meaning, metal oxide

rare earth metal
derived from rare earth

Raschig phenol process
named for the German chemist Friedrich August Fritz Raschig (1863-1928)

Raschig rings
named for the German chemist Friedrich August Fritz Raschig (1863-1928)

raubasine
$C_{21}H_{24}N_2O_3$, derived from the genus name *Rauwolfia* (trees, shrubs), after the German botanist Leonhard Rauwolf (1535-1596), base, and -in(e)

Raymond reaction
named for the British chemist W. D. Raymond (1901-1988)

rayon
derived from *rayon* (French: ray) – referring to this material's glossy appearance

RBS
an abbreviation for Rutherford backscattering spectroscopy

rDNA
an abbreviation for ribosomal deoxy-ribonucleic acid

RDX (cyclonite, hexogen)
$C_3H_6N_6O_6$, most likely an abbreviation for research department explosive

RE
an abbreviation for rare earth

re-
derived from *re-* (Latin: back, again)

REACH
an EU abbreviation for Registration, Evaluation, and Authorization of Chemicals

realgar (sandarac)
As_4S_4, derived from *rahj al-ghar* (Arabic: powder of the mine)

Réaumur temperature scale
named for the French naturalist René Antoine Ferchault de Réaumur (1683-1757)

rebaudioside
derived from the specific epithet of the genus name *Stevia rebaudiana* (candyleaf), -ose, and [3]-ide

rectification
derived from *rectus* (Latin: right) and, ultimately, *facere* (Latin: to make) – referring to purification by repeated distillation

red beryl
$Be_3Al_2Si_6O_{18}$, named after its color

redox
coined by contraction of reduction and oxidation

redoxase (oxidoreductase)
derived from redox and -ase

redoxin
derived from redox and -in(e)

reductase
derived from reduction and -ase

reductic acid
$C_5H_6O_3$, derived from reduction – referring to this compound's antioxidant action

reduction
derived from *reductio* (Latin: the act of bringing back), from *reducere* (Latin: to bring back)

reductone
coined by contraction of reduction and ketone

Reed reaction
named for the US chemist Cortes F. Reed (1891-1972)

REELS
an abbreviation for reflection electron energy loss spectroscopy

R-effect (M-effect)
coined by contraction of resonance effect

Reformatsky reaction
named for the Russian chemist Sergei Nikolaevich Reformatsky (1860-1934)

refractory
derived from *refractarius* (Latin: stubborn, obstinate), from *refringere* (Latin: to break)

regioselectivity
derived from *regio* (Latin: region)

Regitz diazo transfer reaction
named for the German chemist Manfred Regitz (born 1935)

regulus
derived from *regulus*, diminutive of *rex* (Latin: king)

Reichardt's dye
$C_{41}H_{29}NO$, named for the German chemist Christian Reichardt (born 1934)

Reichert-Meissl number
named for the German chemist E. Reichert (1838-1894) and the Austrian chemist E. Meißl (1855-1905)

Reimer-Tiemann reaction
named for the German chemists Karl Ludwig Reimer (1845-1883) and Johann Karl Wilhelm Ferdinand Tiemann (1848-1899)

Reinecke's salt
$C_4H_{10}CrN_7S_4$, named for the German chemist Albert Reinecke (born 1843)

Reinhardt-Zimmermann titration
named for the German chemists C. Reinhardt (1859-1905) and J. C. Zimmermann (1856-1885)

Reiset salt
named for the French chemist Jules Reiset (1818-1896)

Reissert compound
named for the German chemist C. Arnold Reissert (1860-1945)

Reissert indole synthesis
named for the German chemist C. Arnold Reissert (1860-1945)

Reissert reaction
named for the German chemist C. Arnold Reissert (1860-1945)

relaxin (RLX)
derived from *relaxare* (Latin: to relax) and -in(e) – referring to this hormone's relaxing action on the pubic symphysis and the cervix

REM
an abbreviation for reflection electron microscopy

REMPI
an abbreviation for resonance enhanced multiphoton ionization

renin
derived from ren(o)- and -in(e)

rennet
derived from *rennen* (Middle English: to cause to coagulate)

rennin (chymosin)
derived from rennet and -in(e)

ren(o)-
derived from *ren* (Latin: kidney)

replicase
derived from *replicare* (Latin: to repeat, to fold back) and -ase

Reppe chemistry
named for the German chemist Walter Reppe (1892-1969)

resacetophenone
$C_8H_8O_3$, derived (with contraction) from resorcinol and acetophenone

resazurin
$C_{12}H_7NO_4$, derived from resorcinol, azur, and -in(e)

rescinnamine
$C_{35}H_{42}N_2O_9$, derived from reserpine, cinnamic acid, and amine

reserpic acid
$C_{22}H_{28}N_2O_5$, derived from reserpine

reserpine
$C_{33}H_{40}N_2O_9$, anagrammatically derived from the specific epithet of the species name *Rauwolfia serpentina* (Indian snakeroot), from *serpentinus* (Latin: of the snake), from *serpens* (Latin: snake), from *serpere* (Latin: to creep), and -in(e)

resibufogenin
$C_{24}H_{32}O_4$, derived from *resina* (Latin: resin), buf(o)-, -gen, and -in(e)

resilin
derived from resilience and -in(e) – referring to the rubber-like elacticity of this protein

resin
ultimately derived from *rhetine* (Greek: pine resin)

resistomycin
$C_{22}H_{16}O_6$, derived from *resistere* (Latin: to resist) and -mycin

resorcinol
$C_6H_6O_2$, derived from resin and, ultimately, *orcello* (Old Italian: archil)

resorcylic acid
$C_7H_6O_4$, derived from resorcinol

resveratrol
$C_{14}H_{12}O_3$, derived (with contraction) from resorcinol, veratr(o)-, and -ol

retamine
$C_{15}H_{26}N_2O$, derived from the genus name *Retama* (tropical shrubs), from *retama*

(American Spanish: canary broom, *Genista canariensis*), and amine

retene
$C_{18}H_{18}$, derived from *rhetine* (Greek: pine resin) and -ene

Retgers' solution
(I_2,CHI_3,CH_2I_2), named for the Dutch mineralogist and crystallographer J. W. Retgers (1856-1896)

reticulin
derived from *reticulum*, diminutive of *rete* (Latin: net), and -in(e)

reticuline
$C_{19}H_{23}NO_4$, derived from the specific epithet of the genus name *Anona reticulata* Linn. (custard apple), from *reticulatus* (Latin: consisting of or provided with a small net), and -in(e)

retinal (axerophthal, retinene, vitamin A aldehyde)
$C_{20}H_{28}O$, derived from retin(o)- and -al

retine
derived from *retardare* (Latin: to retard), from *tardus* (Latin: slow), and -in(e) – referring to the malignant growth-inhibiting action of this protein

retinoid
derived from retinol and –oid

retin(o)-
derived from retina, from *rete* (Latin: net)

retinoic acid (tretinoin)
$C_{20}H_{28}O_2$, derived from retin(o)-

retinol (axerophthol, vitamin A$_1$)
$C_{20}H_{30}O$, derived from retin(o)- and -ol

retort
derived from re- and *tortus* (Latin: twisted), from *torquere* (Latin: to twist)

retro-
derived from *retro* (Latin: behind, backward)

retron
coined by contraction of retro- and synthon

retronecine
$C_8H_{13}NO_2$, derived from retro-, necine, and -in(e)

retrorsine
$C_{18}H_{25}NO_6$, derived from the specific epithet of the species name *Senecio retrorsus* DC (bushweed), from *retrorsus* (Latin: bent backwards), from *retroversus* (Latin: turned backwards), and -in(e)

Reverdin reaction
named for the Swiss chemist Frédéric Reverdin (1849-1931)

revertase
coined by contraction of reverse transcriptase

reyerite
$(Na,K)_4Ca_4Si_{22}Al_2O_{58}(OH)_8 \cdot 6H_2O$, named for the Austrian geologist Eduard Reyer (1849-1907)

rhamnetin
$C_{16}H_{12}O_7$, derived from rhamn(o)-, -ete, and -in(e)

rhamn(o)-
derived from the genus name *Rhamnus* (buckthorns), from *rhamnos* (Greek: buckthorn)

rhamnose
$C_6H_{12}O_5$, derived from rhamn(o)- and -ose

rhapontin
$C_{21}H_{24}O_9$, derived from the specific epithet of the species name *Rheum rhaponticum* L. (rhubarb), from *rha ponticum* (Late Latin: Pontic rhubarb), from *Pontus* (Latin: Black Sea), and -in(e)

rhead-
derived from the specific epithet of the species name *Papaver rhoeas* (corn poppy), from *rhoias* (Greek: corn poppy)

rheadan
$C_{17}H_{17}NO$, derived from rhead- and -an(e)

rheadine
$C_{21}H_{21}NO_6$, derived from rhead- and -in(e)

RHEED
an abbreviation for reflection high energy electron diffraction

rhein (cassic acid)
$C_{15}H_8O_6$, derived from the genus name *Rheum* (rhubarb), from *rheon* (Greek: rhubarb), and -in(e)

rhenium
Re, named for the Rhine, derived from *Rhenus fluvius* (Latin: Rhine) – referring to Rheinland, Germany, the birth region of the codiscoverer of this element, the German chemist Ida Eva Tacke (1896-1978)

rhenocene
$C_{10}H_{10}Re$, derived from rhenium and -ocene

rheo-
derived from *rheos* (Greek: stream), from *rhein* (Greek: to flow)

rhiz(o)-
derived from *rhiza* (Greek: root)

rhodanide (thiocyanate)
SCN^-, derived from [1]rhod(o)-, -an(e), and [2]-ide – referring to the red color of this ion's 1:1 complex with Fe^{3+}

rhodanine (rhodanic acid)
$C_3H_3NOS_2$, derived from [1]rhod(o)-, -an(e), and -in(e)

rhodinol
$C_{10}H_{20}O$, derived from [1]rhod(o)-, -in(e), and -ol – referring to this alcohol's rose odor

rhodium
Rh, derived from rhod(o)- and -ium – referring to the characteristic red color of rhodium salts

rhodizite
$(K,Cs)Be_4Al_4(B,Be)_{12}O_{28}$, derived from *rhodizein* (Greek: to be rose-red) and -ite – referring to the fact that this mineral tinges the blowpipe flame red

rhodizonic acid
$C_6H_2O_6$, derived from *rhodizein* (Greek: to color rose red) and -one – referring to this compound's dark orange color

[1]rhod(o)-
derived from *rhodon* (Greek: rose)

[2]rhod(o)-
derived from rhodium

rhodocene
$C_{10}H_{10}Rh$, derived from [2]rhod(o)- and -ocene

rhodochrosite
$MnCO_3$, derived from *rhodochroos* (Greek: rose colored) and -ite – referring to this mineral's pink or red color

rhododendrin
$C_{16}H_{24}O_7$, derived from the genus name

Rhododendron (shrubs, trees), from [1]rhod(o)- and dendr(o)-, and -in(e)

rhodomycin
derived from [1]rhod(o)- and -mycin – referring to the purplish color of the rhodomycin-producing bacterium *Streptomyces pupurascens*

rhodonite
$MnSiO_3$, derived from [1]rhod(o)- and -ite – referring to this mineral's often pink to red color

rhodopin
$C_{40}H_{58}O$, coined by contraction of [1]rhod(o)-, lycopene, and -in(e)

rhodopsin
derived from [1]rhod(o)-, ops-, and -in(e)

rhodoquinone
$C_{58}H_{89}NO_3$, derived from the bacterial genus name *Rhodospirillum*, from [1]rhod(o)-, and quinone

rhodoviolascin
$C_{42}H_{60}O_2$, derived from the bacterial genus name *Rhodovibrio violascens*, from [1]rhod(o)- and *vibrare* (Latin: to shake), and *violascens* (New Latin: purple), and -in(e)

rhodoxanthin
$C_{40}H_{50}O_2$, derived from [1]rhod(o)-, xanth(o)-, and -in(e)

rhynch(o)-
derived from *rhynchos* (Greek. snout, bill, beak)

rhynchophylline
$C_{22}H_{28}N_2O_4$, derived from the specific epithet of the species name *Uncaria rhynchophylla* Miq. (the Chinese medicinal herb *gou teng*), from rhynch(o)- and phyll(o)-, and -in(e)

rhyolite
a rock species, derived from *rhyax* (Greek: stream of lava) and -lite

RIA
an abbreviation for radioimmunoassay

ribazole
$C_{14}H_{18}N_2O_4$, derived from ribose, az(a)-, and -ol(e)

ribitol (adonitol)
$C_5H_{12}O_5$, derived from ribose and -itol

riboflavin
$C_{17}H_{20}N_4O_6$, derived from ribitol and flavin

ribonic acid
$C_5H_{10}O_5$, derived anagramatically (with contraction) from arabinonic acid

ribonuclease (RNase)
derived from ribonucleic acid and -ase

ribonucleic acid
derived from ribose and (cell) nucleus

ribonucleoside
derived from ribose and nucleoside

ribonucleotide
derived from ribose and nucleotide

ribose
$C_5H_{10}O_5$, derived from ribonic acid

riboside
derived from ribose and [3]-ide

ribostamycin
$C_{17}H_{34}N_4O_{10}$, derived from the specific epithet of the bacterial species name *Streptomyces ribosidificus*, from *ribosidificus* (New Latin: ribosidifying), from riboside and *facere* (Latin: to make), and –mycin

ribozyme
coined by contraction of ribonucleic acid and enzyme

RIBS
an abbreviation for Rutherford ion backscattering

ribulose
$C_5H_{10}O_5$, derived from ribose and -ulose

Rice-Herzfeld mechanism
named for the British chemist Francis Owen Rice (1890-1989) and the Austrian-US physicist Karl F. Herzfeld (1892-1978)

Rice-Ramsperger-Kassel-Marcus theory
named for the US chemists Oscar Knefler Rice (1903-1978), Herman Carl Ramsperger (1896-1932), and Louis S. Kassel (1905-1973), and the Canadian chemist Rudolph Arthur Marcus (born 1923)

richterite
$Na_2CaMg_3Fe_2(Si_8O_{22})(OH)_2$, named for the German mineralogist Theodor Richter (1824-1898)

von Richter cinnoline synthesis
named for the German chemist Victor von Richter (1841-1891)

von Richter rearrangement
named for the German chemist Victor von Richter (1841-1891)

Richter's law
named for the German chemist Jeremias Benjamin Richter (1762-1807)

Richter system
named for the German chemist Max M. Richter (1861-1936)

ricin
derived from ricin- and -in(e)

ricin-
derived from the genus name *Ricinus* (castor oil plant), from *ricinus* (Latin: castor oil plant), from *ricinus* (Latin: dog tick) – referring to the tick-shaped seeds of this plant

ricinine
$C_8H_8N_2O_2$, derived from ricin- and -in(e)

ricinoleic acid
$C_{18}H_{34}O_3$, derived from ricin- and -ol

riebeckite
$Na_2(Fe^{II},Mg)_3(Fe^{III})_2Si_8O_{22}(OH,F)_2$, named for the German explorer Ernst Riebeck (1853-1885)

Riehm quinoline synthesis
named for the 19th century German chemist P. Riehm

Riemschneider thiocarbamate synthesis
named for the German chemist Randolph Riemschneider (born 1920)

rifa-
derived from rifamycin

rifamycin
derived by alteration of rifomycin, an antibiotic arbitrarily named for the classical French gangster movie *Rififi*, from *rififi* (French slang: heist), and -mycin

Riley oxidation
named for the British chemist Harry Lister Riley (1899-1986)

rimocidin
$C_{39}H_{61}NO_{14}$, derived from the specific epithet of the bacterial species name *Streptomyces rimosus*, from *rimosus* (Latin: fissured), from *rima* (Latin: slit, crack, fissure), and -idin(e)

RIMS
an abbreviation for resonance ionization mass spectroascopy

Ringer's solution
named for the British biochemist Sydney Ringer (1835-1910)

Rinman's green
$ZnCo_2O_4$, named for the Swedish mineralogist Sven Rinman (1720-1792)

rINN
an abbreviation for recommended international nonproprietary (drug) name

RIS
an abbreviation for resonance ionization spectroscopy

RIST
an abbreviation for radioimmunosorbent test

Ritter reaction
named for the US chemist John Joseph Ritter (1895-1975)

RLCC
an abbreviation for rotating locular counter-current chromatography

RNA
an abbreviation for ribonucleic acid

RNase
an abbreviation for ribonuclease

robinin
$C_{33}H_{40}O_{19}$, derived from the genus name *Robinia* (trees, shrubs), after the French botanist Jean Robin (1550-1629), and -in(e)

Robinson annulation
named for the British chemist Sir Robert Robinson (1885-1975)

Robinson-Schöpf reaction
named for the British chemist Sir Robert Robinson (1885-1975) and the German chemist Clemens Schöpf (1899-1970)

Robison ester
$C_6H_{13}O_9P$, named for the British chemist Robert Robison (1883-1941)

roccellic acid
$C_{17}H_{32}O_4$, derived from the genus name *Roccella* (lichens), from *roccella* (Italian: archil), an alteration of *oricello* (Italian: archil)

Rochelle salt (Seignette salt)
$KNa(C_4H_4O_6)$, named for the town of La Rochelle, France

Rochow-Müller process
named for the US chemist Eugene G. Rochow (1909-2002) and the German chemist R. Müller (1903-1999)

rock crystal
SiO_2, derived from rock and crystal

Rockwell hardness
named for the US metallurgist S. P. Rockwell (1886-1940)

rodenticide
derived from rodent and -cide

rodiasine
$C_{38}H_{42}N_2O_6$, derived from the specific epithet of the species name *Ocotea rodiaei* (bebeeru, Demerara greenheart) and -in(e)

Roelen reaction (oxo synthesis)
named for the German chemist Otto Roelen (1897-1993)

roentgenium
Rg, named in honor of the German physicist Wilhelm Conrad Röntgen (1845-1923), also

in order to commemorate the centenary of Röntgen's discovery of X-rays which coincided with the preparation of this element

ROESY
an abbreviation for rotational frame nuclear Overhauser effect spectroscopy

Rohrbach's solution
$BaHgI_4$, named for the German naturalist C. E. M. G. Rohrbach (1861-1932)

Rohrschneider constant
named for the German chemist Lutz Rohrschneider (born 1927)

romanèchite
$(Ba,H_2O)_2(Mn^{IV},Mn^{III})_5O_{10}$, named after its locality Romanèche, Département Sâone-et-Loire, France

ROMP
an abbreviation for ring opening metathesis polymerization

rongalite
CH_3NaO_3S, derived from *rongeage* (French: an etching) and -ite

rosalic acid (aurin, pararosolic acid)
$C_{19}H_{14}O_3$, named after this dye's pink color

rosane
$C_{20}H_{36}$, derived from the genus name *Rosa* (roses), from *rosa* (latin: rose) and -an(e)

rosaniline (fuchsin)
$C_{20}H_{19}N_3$, derived from *rosa* (latin: rose) and aniline – referring to this dye's formation from aniline and toluidine

rosaramicin (rosamicin)
$C_{31}H_{51}NO_9$, derived from the specific epithet of the bacterial species name *Micromonospora rosaria*, from *rosarius* (Latin: of roses), from *rosa* (Latin: rose), and -micin

roscoelite
$K(V,Al,Mg)_2AlSi_3O_{10}(OH)_2$, named for the British chemist Sir Henry Enfield Roscoe (1833-1915) who was the first to prepare pure vanadium

rose bengal
$C_{20}H_2Cl_4I_4K_2O_5$, derived from rose and Bengal, India and Bangladesh

rose beryl (morganite)
$Be_3Al_2Si_6O_{18}$, named after its pink color

rosenbuschite
$(Ca,Na)_3(Zr,Ti)(Si_2O_8)F$, named for the German geologist and mineralogist Carl Harry Ferdinand Rosenbusch (1836-1914)

Rosenmund reduction
named for the German chemist Karl Wilhelm Rosenmund (1884-1965)

Rosenmund-von Braun synthesis
named for the German chemists Karl Wilhelm Rosenmund (1884-1965) and Julius von Braun (1875-1939)

rose quartz
SiO_2, named after its pink color

Rose's metal
(Bi,Pb,Sn), named for the German chemist and pharmacist Valentin Rose, Sr. (1736-1771)

Rose crucible
named for the German chemist H. Rose (1795-1864)

Rosiwal hardness
named after the Austrian mineralogist August Karl Rosiwal (1860-1923)

rosolic acid
$C_{20}H_{16}O_3$, coined by contraction of rose and carbolic acid

rotamer
derived from rotation and -mer

rotane
derived from *rota* (Latin: wheel) and -an(e)

rotaxane
derived from *rota* (Latin: wheel), *axis* (Latin: axis), and -an(e)

rotenoid
derived from rotenone and -oid

rotenone
$C_{23}H_{22}O_6$, derived from *roten* (Japanese: derris plant, *Derris elliptica*) and -one

Rothemund reaction
named for the 20th century US chemist Paul Rothemund

rottlerin
$C_{30}H_{28}O_8$, derived from the genus name *Rottlera* (East Indian trees), after the German-Danish missionary Johann Peter Rottler (1749-1836), and -in(e)

Roussin's black salt
$Fe_4KN_7O_7S_3$, named for the French pharmacist F.-Z. Roussin (1827-1894)

Roussin's red salt
$Fe_2K_2N_4O_4S_2$, named for the French pharmacist F.-Z. Roussin (1827-1894)

RPC
an abbreviation for reversed phase chromatography

RRKM theory
an abbreviation for Rice-Ramsperger-Kassel-Marcus theory

rRNA
an abbreviation for ribosomal ribonucleic acid

rubber
derived from to rub – referring to the use of rubber in erasers

rubeanic acid
$C_2H_4N_2S_2$, derived from rubi- and -an(e) – referring to this compound's red color

ruberythric acid
$C_{25}H_{26}O_{13}$, derived from the genus name *Rubia* (Old World herbs), from *rubia* (Latin: madder), and erythr(o)-

rubi-
derived from *rubeus* (Latin: red, reddish)

rubiadin
$C_{15}H_{10}O_4$, derived from the genus name *Rubia* (Old World herbs), from *rubia* (Latin: madder), and -in(e)

rubicene
$C_{26}H_{14}$, derived from rubi- and -acene – referring to this compound's red color

rubidium
Rb, derived from rubid(o)- and -ium – referring to the red color of this element's characteristic spectral lines

rubid(o)-
from *rubidus* (Latin: dark red)

rubidomycin (daunomycin, daunorubicin)
$C_{27}H_{29}NO_{10}$, derived from rubid(o)- and -mycin – referring to this antibiotic's dark red color

rubijervine
$C_{27}H_{43}NO_2$, derived from rubi- and jervine

rubisco
coined by contraction of ribulose bisphosphate carboxylase

rubixanthin
$C_{40}H_{56}O$, derived from rubi-, xanth(o)-, and -in(e)

Rubottom oxidation
named for the 20th century US chemist George M. Rubottom

rubredoxin
derived from rubri-, redox, and -in(e)

rubri-
derived from *ruber* (Latin: dark red)

ruby
Al_2O_3, ultimately derived from *ruber* (Latin: dark red)

Ruff-Fenton degradation
named for the German chemist Otto Ruff (1871-1939) and the British chemist Henry John Horstman Fenton (1854-1929)

rufigallol
$C_{14}H_8O_8$, derived from *rufus* (Latin: orange, yellow), gallic acid, and -ol

Ruggli dilution principle
named for the Swiss chemist Paul Ruggli (1884-1945)

rugulovasine
$C_{16}H_{16}N_2O_2$, derived from the specific epithet of the fungal species name *Penicillium concavo-rugulosum*, from *rugulosus* (Latin: wrinkled), from *ruga* (Latin: wrinkle), cardiovascular activity, and -in(e)

Rupe rearrangement
named for the Swiss chemist Hans Rupe (1866-1951)

russellite
Bi_2WO_6, named for the British mineralogist Sir Arthur Edward Ian Montagu Russell (1878-1964)

Russell-Saunders coupling
named for the US astronomer Henry Norris Russell (1877-1957) and the Canadian physicist F. A. Saunders (1875-1963)

rutecarpine
$C_{18}H_{13}N_3O$, derived from the specific epithet of the species name *Evodia rutaecarpa* Hook & Thoms. (the Chinese medicinal herb *wu zhu yu*), from *ruta* (Latin: rue) and carp(o)-, and -in(e)

ruthenium
Ru, named for Russia, derived from *Ruthenia* (New Latin: Russia)

ruthen(o)-
derived from ruthenium

ruthenocene
$C_{10}H_{10}Ru$, derived from ruthen(o)- and -ocene

rutherfordium
Rf, named for the New Zealand-British physicist Sir Ernest Rutherford (1871-1937)

rutile
TiO_2, derived from *rutilis* (Latin: red, shining), from *ruber* (Latin: dark red) – referring to the deep red color observed in some specimens when viewed in transmitted light

rutin (rutoside)
$C_{27}H_{30}O_{16}$, derived from the genus name *Ruta* (rue), from *ruta* (Latin: rue), and -in(e)

rutinose
$C_{12}H_{22}O_{10}$, derived from rutin and -ose

Ruzicka large-ring synthesis
named for the Croatian-Swiss chemist Leopold Ruzicka (1887-1976)

ryanodine
$C_{25}H_{35}NO_9$, derived from the genus name *Ryania* (South American shrubs), after the 18th century British physician John Ryan, and -in(e)

Rydberg's constant
named for the Swedish physisict Johannes Robert Rydberg (1854-1919)

S

S
derived from *sinister* (Latin: left)

sabadine
$C_{29}H_{51}NO_8$, derived from sabadilla, from *cebadilla* (Spanish: sabadilla, *Schoenocaulon officinale*), diminutive of *cebada* (Spanish: barley), ultimately from *cibus* (Latin: food, meal), and -in(e)

Sabatier-Senderens reduction
named for the French chemists Paul Sabatier (1854-1941) and J. B. Senderens

saccharase
derived from saccharose and -ase

saccharide
derived from sacchar(o)- and ^{3}-ide

saccharin
$C_7H_5NO_3S$, derived from sacchar(o)- and -in(e) – referring to the sweet taste of this compound

sacchar(o)-
derived from *saccharum* (Latin: sugar), ultimately from *sarkara* (Sanskrit: sugar)

saccharopine
$C_{11}H_{20}N_2O_6$, derived (with contraction) from the microbiological genus name *Saccharomyces* – referring to this compound's role as a lysine precursor in yeast metabolism – and octopine – also referring to this compound's structure, two amino acids (here glutamic acid and lysine) with a shared amino group, as first observed in octopine

saccharose
$C_{12}H_{22}O_{11}$, derived from sacchar(o)- and -ose

saccharum
New Latin name for sucrose, derived from sacchar(o)-

Sachse-Mohr theory
named for the German chemists Hermann Sachse (1862-1893) and Ernst Mohr (1873-1926)

Saegusa oxidation
named for the 20th century Japanese chemist Takeo Saegusa

safflorite
$CoAs_2$, derived from *Safflor* (German: dyer's saffron), ultimately from *zafaran* (Arabic: saffron), and -ite – referring to the use of this mineral in the manufacture of *Saflor* (German: zaffer, cobalt blue)

safranal
$C_{10}H_{14}O$, derived from *Safran* (German: saffron), ultimately from *zafaran* (Arabic: saffron), and -al

safrole
$C_{10}H_{10}O_2$, derived (with contraction) from sassafras, ultimately from *saxifraga herba* (Late Latin: rock-breaking herb), and -ol(e)

SAICAR
$C_{13}H_{19}N_4O_{12}P$, an abbreviation for *N*-succino-5-aminoimidazole-4-carboxamide-ribonucleotide

sakuranetin
$C_{16}H_{14}O_5$, derived from *sakura* (Japanese: Japanese flowering cherry, *Prunus yedoensis* Matsum.) and -etin(e)

sal ammoniac
NH_4Cl, derived from *sal ammoniacum* (Latin: salt of the Old Egyptian god Ammon), from *Amon* (Old Egyptian: literally the hidden one) – referring to the oasis of Siwa, Egypt, in antiquity dedicated to the Egyptian god Ammon, where this salt was supposedly found

salatrim
an abbreviation for short- and long-acyl triglyceride molecule

salen
$C_{16}H_{16}N_2O_2$, coined by contraction of *N,N'*-ethylenebis(salicylideneimine)

salic-
derived from the genus name *Salix* (willows), from *salix* (Latin: willow)

salicin
$C_{13}H_{18}O_7$, derived from salic- and -in(e)

salicylic acid
$C_7H_6O_3$, derived from salic- and -yl

salin(o)-
derived from *salinus* (Latin: salty), from *sal* (Latin: salt)

salinomycin
$C_{42}H_{70}O_{11}$, derived from salin(o)- and -mycin – referring to this antibiotic's ionophoric properties

Salkowski reaction
named for the German biochemist Ernst Leopold Salkowski (1844-1923)

salmiac
NH_4Cl, derived from *sal ammoniacum* (Latin: salt of Ammon) – referring to the oasis of Siwa, Egypt, in antiquity dedicated to the Egyptian god Ammon, where this salt was supposedly found

salsoline
$C_{11}H_{15}NO_2$, derived from the genus name *Salsola* (Old World herbs), from *salso* (Italian: salty), from *salsus* (Latin: salted, salty), from *sal* (Latin: salt), and -in(e)

salt
an indigenous English word, akin to *sal* (Latin: salt)

saltpeter
KNO_3, derived from *sal petrae* (Latin: rock salt)

salutaridine (floripavine)
$C_{19}H_{21}NO_4$, derived from the specific epithet of the species name *Croton salutaris* (a tropical herb), from *salutaris* (Latin: wholesome), from *salus* (Latin: health, safety), and -idin(e)

[1]SAM
$C_{15}H_{22}N_6O_5S$, an abbreviation for *S*-adenosyl-L-methionine

[2]SAM
an abbreviation for scanning Auger microscopy

samaderin
derived from the genus name *Samadera* (trees), from *samadara* (Sinhalese: name of a tree), and -in(e)

samandarine
$C_{19}H_{31}NO_2$, derived (with contraction) from the genus name *Salamander* (amphibians), ultimately from *salamandra* (Greek: lizard), and -in(e)

samaria
Sm_2O_3, derived from samarskite and -a

samarium
Sm, derived from samaria and -ium

samarskite
$(Y,Ce,U,Fe^{III},Nb)(Nb,Ta,Ti)O_4$, named for Colonel Vasilii Yefrafovich Samarski-Bykhovets (1803-1870), chief of staff of the Russian corps of mining engineers

sanbornite
$BaSi_2O_5$, named for the US mineralogist Frank B. Sanborn (1862-1936)

sandarac (realgar)
As_4S_4, derived from *sandarake* (Greek: realgar), probably akin to *candana* (Sanskrit: sandalwood)

Sandmeyer diphenylurea isatin synthesis
named for the Swiss chemist Traugott Sandmeyer (1854-1922)

Sandmeyer isonitrosoacetanilide isatin synthesis
named for the Swiss chemist Traugott Sandmeyer (1854-1922)

Sandmeyer reaction
named for the Swiss chemist Traugott Sandmeyer (1854-1922)

Sanger's reagent
$C_6H_3FN_2O_4$, named for the British molecular biologist Frederick Sanger (born 1918)

sanguinarine
$C_{20}H_{14}NO_4^+$, derived from the genus name *Sanguinaria* (herbs), from *sanguinarius* (Latin: blood-thirsty), from *sanguis* (Latin: blood), and -in(e)

sanidine
$KAlSi_3O_8$, derived from *sanis* (Greek: board) and -in(e) – referring to this mineral's flat crystals

santalol
$C_{15}H_{24}O$, derived from the genus name *Santalum* (sandalwood), from *sandalum*, *santalum* (Medieval Latin: sandalwood), ultimately from *zandal* (Arabic: sandalwood), and -ol

santonic acid
$C_{15}H_{20}O_4$, derived from santonin

santonin
$C_{15}H_{18}O_3$, derived from *herba santonica* (Latin: herb of the Santoni, a Celtic people of Aquitania; *Artemisia pauciflora*, European wormwood), and -in(e)

sapogenin
derived from sapo(n)-, -gen, and -in(e)

sapo(n)-
derived from *sapo* (Latin: soap)

saponarin
$C_{27}H_{30}O_{15}$, derived from the genus name *Saponaria* (Old World herbs), from *saponarius* (Latin: of soap), from sapo(n)-, and -in(e)

saponification
derived from sapo(n)- and *facere* (Latin: to make)

saponin
derived from sapo(n)- and -in(e) – referring to these compounds' emulsifying properties

saponite
$(Ca,Na)_{0.3}(Mg,Fe^{II})_3(Si,Al)_4O_{10}\cdot 4H_2O$, derived from sapo(n)- and -ite – referring to this mineral's soapy consistency

saporin
derived from the genus name *Saponaria* (Old World herbs), from *saponarius* (Latin: of soap), from sapo(n)-, and -in(e)

sapphire
Al_2O_3, ultimately derived from *sappir*

(Hebrew: sapphire, lapis lazuli)

sapphirine
$Mg_7Al_{18}Si_3O_{40}$, derived from sapphire and -in(e) – referring to this mineral's sapphire-blue color

sapr(o)-
derived from *sapros* (Greek: rotten)

sapropterin
$C_9H_{15}N_5O_3$, derived from sapr(o)- and pterin

sarafotoxin (SRTX)
derived from *saraf* (Hebrew: a biblical poisonous snake, possibly *Atractaspis engaddensis*) and toxin

sarcosine
$C_3H_7NO_2$, derived from *sarx* (Greek: flesh) and -in(e)

sard
SiO_2, probably named after the town of Sardis, the capital of the ancient kingdom of Lydia, Asia Minor

sardonyx
SiO_2, derived from sard and onyx

Sarett oxidation
named for the US chemist Lewis Hastings Sarett (1917-2000)

sarin (GB, trilon 46)
$C_4H_{10}FO_2P$, probably coined as an abbreviation of the names of the German chemists Gerhard Schrader (1903-1990), O. Ambros (1901-1990), F. Ritter, and H. J. von der Linde

sarkomycin
$C_7H_8O_3$, derived from *sarx* (Greek: flesh) and -mycin

sarmentose
$C_7H_{14}O_4$, derived from the specific epithet of the species name *Strophanthus sarmentosus* DC. (arrow poison strophantus), from *sarmentosus* (Latin: producing twigs), from *sarmentum* (Latin: twig), from *sarpere* (Latin: to prune), and -ose

sarpag-
derived from *sarpagandha* (Hindi: Indian snakeroot, *Rauwolfia serpentina*)

sarpagan
$C_{19}H_{22}N_2$, derived from sarpag- and -an(e)

sarpagine
$C_{19}H_{22}N_2O_2$, derived from sarpag- and -in(e)

sarsasapogenin
$C_{27}H_{44}O_3$, derived (with contraction) from sarsaparilla (*Smilax ornata* Hooker), from *zarzaparrilla* (Spanish: sarsaparilla), from *zarza* (Spanish: bush) and *parrilla*, diminutive of *parra* (Spanish: vine), saponin, -gen, and -in(e)

SAS
an abbreviation for secondary alkane-sulfonate

sassolite (sassoline)
$B(OH)_3$, named after this mineral's locality Mount Sasso, Italy

saxi-
derived from *saxum* (Latin: rock)

saxitoxin (STX)
$C_{10}H_{17}N_7O_4^+$, derived from the genus name *Saxidomus* (clams), from saxi-, and toxin

SBR
an abbreviation for styrene-butadiene rubber

scabiolide
$C_{21}H_{28}O_8$, derived from the specific epithet of the species name *Centaurea scabiosa* (L.) Presl. (greater knapweed), from *scabiosus* (Latin: mangy), from *scabies* (Latin: mange), and -olide

scandium
Sc, named for Scandinavia - also referring to the fact that the scandium-containing minerals euxenite and gadolinite are found only in Scandinavia

scapolite (wernerite)
derived from *scapus* (Latin: shaft) and -lite – referring to the prismatic crystals of this group of minerals

SCF method
an abbreviation for self-consistent field method

Schadt-Helfrich effect
named for the Swiss physicist M. Schadt (born 1938) and the German physicist W. Helfrich (born 1932)

Schäffer's acid
$C_{10}H_8O_4S$, named for the German chemist Ernst Schäffer (1889-1915)

schapbachite
$AgSbS_2$, named after this mineral's (now discredited) locality Schapbach, Germany

Schardinger dextrin (α-cyclodextrin)
$C_{36}H_{60}O_{30}$, named for the German biochemist Franz Schardinger (1853-1917)

Schardinger enzyme (xanthine oxidase)
named for the German biochemist Franz Schardinger (1853-1917)

Schardinger test
named for the German biochemist Franz Schardinger (1853-1917)

Scheele's green
$CuHAsO_3$, named for the Swedish chemist Karl Wilhelm Scheele (1742-1786)

scheelite
$CaWO_4$, named for the Swedish chemist Karl Wilhelm Scheele (1742-1786)

scheelium
W, a name unsuccessfully suggested for tungsten, in honor of the Swedish chemist Karl Wilhelm Scheele (1742-1786)

schefferite
$(Ca,Mn)(Mg,Fe,Mn)Si_2O_6$, named for the Swedish chemist Henrik Theophil Scheffer (1710-1759)

Scheibler's reagent
named for the German chemist Karl Scheibler (1827-1899)

Schellbach burette
named for the German mathematician and physicist K. H. Schellbach (1804-1892)

Schenck mechanism
named for the German chemist Günther Otto Schenck (1913-2003)

Schiemann reaction (Balz-Schiemann reaction)
named for the German chemist Günther Robert Arthur Schiemann (1899-1969)

Schiff base
named for the German-Italian chemist Hugo (Ugo) Joseph Schiff (1834-1915)

schiz(o)-
derived from *schizein* (Greek: to split)

Schlegel diagram
named for the German mathematician S. F. V. Schlegel (1843-1905)

Schlenk equilibrium
named for the German chemist Wilhelm Schlenk (1879-1943)

[1]**Schlenk's hydrocarbon**
$C_{37}H_{27}$, named for the German chemist Wilhelm Schlenk (1879-1943)

[2]**Schlenk's hydrocarbon**
$C_{38}H_{28}$, named for the German chemist Wilhelm Schlenk (1879-1943)

Schlenk tube
named for the German chemist Wilhelm Schlenk (1879-1943)

Schlesinger process
named for the US chemist Hermann Irving Schlesinger (1882-1960)

Schlippe's salt
Na_3SbS_4, named for the German-Russian pharmacist and chemist C. F. von Schlippe (1799-1874)

Schlittler-Müller modification
(of the Pomeranz-Fritsch reaction), named for the 20th century Swiss chemists E. Schlittler and J. Müller

Schmidt reaction
named for the German chemist Karl Friedrich Schmidt (1887-1971)

Schmidt's double bond rule
named for the German chemist Otto Schmidt (1874-1943)

Schoenflies symbols
named for the German mathematician Arthur Moritz Schoenflies (1853-1928)

schoenite (picromerite)
$K_2Mg(SO_4)_2 \cdot 6H_2O$, possibly derived from *schoinos* (Greek: rush, reed) and -ite

schoepite
$UO_3 \cdot 2H_2O$, named for the Belgian mineralogist Alfred Schoep (1881-1966)

Scholl reaction
named for the German chemist Roland Scholl (1865-1945)

Schöllkopf bis-lactim amino acid synthesis
named for the German chemist Ulrich Schöllkopf (born 1927)

scholzite
$CaZn_2(PO_4)_2 \cdot 2H_2O$, named for the German chemist and mineral collector Adolf Scholz (1894-1950)

Schomaker-Stevenson equation
named for the US chemists Verner Schomaker (1914-1997) and D. P. Stevenson

Schönberg reaction
named for the German chemist Alexander Schönberg (1892-1985)

Schönherr process
named for the German chemist Otto Schönherr (1861-1926)

schorl
$NaFe_3Al_6(BO_3)_3Si_6O_{18}(OH)_4$, derived from *Schörl* (German: waste)

schorlomite
$Ca_3(Ti,Fe)_2[(Si,Fe)O_4]_3$, derived from schorl, hom(o)-, and -ite – referring to this mineral's resemblance to schorl

Schotten-Baumann reaction
named for the German chemists Carl Ludwig Johannes Schotten (1853-1910) and Eugen Baumann (1846-1896)

Schottky defect
named for the German physicist Walter Schottky (1886-1976)

schradan
$C_8H_{24}N_4O_3P_2$, named for the German chemist Gerhard Schrader (1903-1990)

Schrock's molybdenum catalyst
$C_{30}H_{35}F_{12}MoNO_2$, named for the US chemist Richard Royce Schrock (born 1945)

Schrödinger equation
named for the Austrian physicist Erwin Schrödinger (1887-1961)

Schulze-Hardy rule
named for the German chemist H. O. Schulze (1853-1892) and the British chemist Sir W. B. Hardy (1864-1934)

Schwartz-Negishi reagent
$C_{10}H_{11}ClZr$, named for the US chemist Jeffrey Schwartz (born 1945) and the Japanese-US chemist Ei-Ichi Negishi (born 1935)

Schweinfurt green (Mitis green, Paris green)
named for the city of Schweinfurt, Germany where this dye was first manufactured

Schweizer's reagent (cuoxam)
$CuH_{14}N_4O_2$, named for the Swiss chemist Mathias Eduard Schweizer (1818-1860); commonly misspelled Schweitzer's reagent

scill(a)-
derived from the genus name *Scilla* (herbs), from *scilla* (Latin: squill)

scillaren
derived from scill(a)- and -en(e)

scillarenin
$C_{24}H_{32}O_4$, derived from scillaren and -in(e)

scilliroside
$C_{32}H_{44}O_{12}$, derived from scill(a)-, ros(a)-, and ^{3}-ide

scintillation
derived from *scintilla* (Latin: spark)

scler(o)-
derived from *skleros* (Greek: hard)

scolecite
$Ca(Si_3Al_2)O_{10}{\cdot}3H_2O$, derived from scolec(o)- and -ite – referring to this mineral's worm-like behavior in the flame of the blowpipe

scolec(o)-
derived from *skolex* (Greek: worm)

scoparin
$C_{22}H_{22}O_{11}$, derived from the specific epithet of the species name *Spartium scoparium* L. (the herb broom), from *spartum* (Latin: the herb broom) and *scopae* (Latin: the tool broom), and -in(e)

scoparone
$C_{11}H_{10}O_4$, derived from scoparin and -one

scopol-
derived from the genus name *Scopola* (herbs), after the Italian-Austrian naturalist Giovanni Antonio Scopoli (1723-1788)

scopolamine
$C_{17}H_{21}NO_4$, derived from scopol- and amine

scopoletin
$C_{10}H_8O_4$, derived from scopol- and -etin(e)

scopolin
$C_{16}H_{18}O_9$, derived from scopol- and -in(e)

scopoline
$C_8H_{13}NO_2$, derived from scopol- and -in(e)

-scopy
ultimately derived from *skopein* (Greek: to observe, to examine)

scorodite
$FeAsO_4 \cdot 2H_2O$, derived from *skorodon* (Greek: garlic) and -ite – referring to the unpleasant odor this mineral gives off when heated

scorzalite
$(Fe,Mg)Al_2(PO_4)_2(OH)_2$, named for the Brazilian mineralogist Evaristo Penna Scorza (1899-1969)

scotophobin
$C_{62}H_{97}N_{23}O_{26}$, derived from *skotos* (Greek: dark), *phobos* (Greek: fear), and -in(e) – referring to this peptide's occurrence in the brain of rats with an acquired fear of the dark

SCP
an abbreviation for single cell protein

scutellarein
$C_{15}H_{10}O_6$, coined by variation of scutellarin

scutellarin
$C_{21}H_{18}O_{12}$, derived from the genus name *Scutellaria* (skullcap), from *scutella* (Latin: drinking bowl), diminutive of *scutra* (Latin: shallow bowl), and -in(e)

scyllitol
$C_6H_{12}O_6$, derived from the genus name *Scyllium* (fish), ultimately from *skylion* (Greek: dogfish), and -ol

scyllo-
derived from scyllitol

SDS
$C_{12}H_{25}NaO_4S$, an abbreviation for sodium dodecyl sulfate

se-
derived from *se-* (Latin: apart, without)

seaborgium
Sg, named for the US chemist Glenn Theodore Seaborg (1912-1999)

sebacate
$C_{10}H_{16}O_4^{2-}$, derived from sebacic acid and -ate

sebacic acid
$C_{10}H_{16}O_4^{2-}$, derived from *sebum* (Latin: tallow)

sebacil
$C_{10}H_{18}O_4$, derived from sebacic acid and -il(e); patterned after benzil

sebacoin
$C_{10}H_{18}O_4$, derived from sebacic acid and -oin; patterned after benzoin

sebacoyl
$-C_{10}H_{16}O_2-$, derived from sebacic acid and -oyl

SEC
an abbreviation for size exclusion chromatography

secalin
derived from secal(o)- and -in(e)

secalinin
derived from secal(o)- and -inin(e)

secal(o)-
derived from the genus name *Secale* (rye), from *secale* (Latin: rye)

secalonic acid
$C_{32}H_{30}O_{14}$, derived from secal(o)-

secalose (trifructosan)
$C_{18}H_{30}O_{15}$, derived from secal(o)- and –ose

seco-
derived from *secare* (Latin: to cut)

secretin
derived from secretion and -in(e)

securinine
$C_{13}H_{15}NO_2$, derived from the genus name *Securinega* (tropical trees), from *securis* (Latin: hatchet, ax) and *negare* (Latin: to refuse, to negate) – referring to the hardness of the wood of these trees – and -in(e)

sedoheptulose
$C_7H_{14}O_7$, derived from the genus name *Sedum* (herbs), from *sedum* (Latin: houseleek), and heptulose

Seger cone
named for the German chemist Hermann August Seger (1839-1893)

Seignette salt (Rochelle salt)
$KNa(C_4H_4O_6)$, named for the French naturalist Elie Seignette (1632-1698)

selachyl alcohol
$C_{21}H_{42}O_3$, derived from the genus name *Selachius* (fish), from *selachos* (Greek: cartilagenous fish), akin to *selas* (Greek: light, brightness)

selagine (huperzine A)
$C_{15}H_{18}N_2O$, derived from the specific epithet of the species name *Lycopodium selago* L. (fir club moss), from *selago* (Latin: a plant resembling the savin), and -in(e)

selane
H_2Se, derived from selen(o)- and -an(e)

selanyl
HSe–, derived from selane and -yl

selectin
derived from selective and -in(e)

s-electron
named after the 'sharp' line groups in atomic spectroscopy

σ-electron
derived from σ-orbital

selenate
SeO_4^{2-}, derived from selen(o)- and -ate

selenic acid
H_2SeO_4, derived from selen(o)-

selenide
Se^{2-}, derived from selen(o)- and [2]-ide

[1]selenite
$CaSO_4 \cdot 2H_2O$, ultimately derived from *selenites lithos* (Greek: moon stone) – referring to an ancient belief that this mineral waxed and waned with the moon

[2] selenite
SeO_3^{2-}, derived from selen(o)- and -ite

selenium
Se, named after Selene, the ancient Greek moon goddess, from *selene* (Greek: moon), and -ium – referring to the close relationship between this element and tellurium

selen(o)-
derived from selenium

selenocysteine
$C_3H_7NO_2Se$, derived from selen(o)- and cysteine – in this unusual case selen(o)- is not used to indicate the replacement of oxygen by selenium, but that of sulfur

selenomethionine
$C_5H_{11}NO_2Se$, derived from selen(o)- and methionine – in this unusual case selen(o)-

is not used to indicate the replacement of oxygen by selenium, but that of sulfur

selenone
R_2SeO_2, derived from selen(o)- and -one – patterned after sulfone

selenophen
C_4H_4Se, derived from selen(o)- and phene

selenous acid
H_2SeO_3, derived from selen(o)- and -ous

selinene
$C_{15}H_{24}$, derived from selin(o)- and -ene

selin(o)-
ultimately derived from *selinon* (Greek: celery)

Seliwanoff reaction
named for the Russian chemist T. Seliwanoff (1859-1938)

sellaite
MgF_2, named for the Italian mining engineer and mineralogist Quintino Sella (1827-1884)

selone
$R_2C{=}Se$, derived from selen(o)- and -one; patterned after ketone

SEM
an abbreviation for scanning electron microscopy

semi-
derived from *sem(i)-* (Latin: half-), from *semis* (Latin: half)

semibullvalene
C_8H_8, derived from semi- and bullvalene

semicarbazide
CH_5N_3O, derived (with contraction) from semi-, carbamide, and hydrazide

semidine
$C_{12}H_{12}N_2$, coined by contraction of semi- and benzidine

semiochemical
derived from *semeion* (Greek: sign) and chemical

semipermeable
derived from semi- and permeable

semiquinone
derived from semi- and quinone

Semmler-Wolff reaction
named for the German chemists Friedrich Wilhelm Semmler (1860-1931) and Ludwig Wolff (1857-1919)

sempervirine
$C_{19}H_{16}N_2$, derived from the specific epithet of the species name *Gelsemium sempervirens* (L.) (Carolina jessamine), from *sempervirens* (Latin: evergreen), from *semper* (Latin: always) and *virens* (Latin: greening), from *virere* (Latin: to green), and -in(e)

semseyite
$PbSb_8S_{21}$, named for the Hungarian amateur mineralogist Andor von Semsey (1833-1923)

senarmontite
Sb_2O_3, named for the French mineralogist Henri Hureau de Sénarmont (1808-1862)

Sendzimir process
named for the French industrialist Tadeusz Sendzimir (1894-1989)

senecic acid
$C_{10}H_{16}O_5$, derived from seneci(o)-

seneci(o)-
derived from the genus name *Senecio* (herbs, shrubs, trees), from *senecio* (Latin: old man, groundsel), from *senex* (Latin: old man) – referring to the gray hair on the seeds of these plants

senecioic acid
$C_5H_8O_2$, derived from seneci(o)-

senecionan
$C_{18}H_{29}NO_2$, derived from seneci(o)- and -an(e)

senecionine
$C_{18}H_{25}NO_5$, derived from seneci(o)- and -in(e)

seneciphylline
$C_{18}H_{23}NO_5$, derived (with contraction) from the species name *Senecio platyphyllus* DC. (ragwort), from seneci(o)- and phyll(o)-, and -in(e)

sennoside
$C_{42}H_{38}O_{20}$, derived from senna (*Cassia*), from *sana* (Arabic: senna), -ose, and [3]-ide

sepio-
derived from the genus name *Sepia* (cuttlefish), ultimately from *sepia* (Greek: cuttlefish), from *sapros* (Greek: rotten)

sepiolite (meerschaum)
$Mg_4(OH)_2(Si_2O_5)_3 \cdot 6H_2O$, derived from *sepion* (Greek: cuttlebone) and -lite – referring to a perceived resemblance of this mineral to the porous bones of cuttlefish

sepiomelanin
derived from sepio- and melanin

septi-
derived from *septies* (Latin: seven times)

septum
derived from *saeptum* (Latin: wall, enclosure), from *saepire* (Latin: to hedge in)

sérandite
$Na(Mn,Ca)_2Si_3O_8(OH)$, named for the 20th century West African mineral collector J. M. Sérand

seric-
derived from *sericum* (Latin: silk), from *Seres* (Latin: the Chinese)

sericin
derived from seric- and -in(e)

sericite
$KAl_2(Si_3Al)O_{10}(OH,F)_2$, derived from seric- and -ite – referring to the silky sheen of sericite-containing rocks

serine
$C_3H_7NO_3$, derived from sericin and -in(e)

Serini reaction
named for the German chemist Arthur Serini (1897-1945)

serotonin
$C_{10}H_{12}N_2O$, derived from serum and -tonin

Serpek process
named for the 20th century Swiss-French chemist Ottokar Serpek

[1]serpentine
$C_{21}H_{20}N_2O_3$, derived from the specific epithet of the species name *Rauwolfia serpentina* (Indian snakeroot), from *serpentinus* (Latin: of the snake), from *serpens* (Latin: snake), from *serpere* (Latin: to creep), and -in(e)

[2]serpentine
$Mg_3Si_2O_5(OH)_4$, derived from *serpens* (Latin: serpent) and -in(e) – referring either

to this mineral's serpent-like colors and patterns or to an ancient belief that this mineral protects its bearer from snake bites

serpentinite
a rock species, derived from serpentine and -ite

serpin
coined by contraction of serine proteinase inhibitor

SERRS (SERS)
an abbreviation for surface-enhanced resonance Raman scattering/spectroscopy

serum
derived from *serum* (Latin: whey, serum)

ses-
derived (with contraction) from *sestertius* (Latin: two and a half)

sesamin
$C_{20}H_{18}O_6$, derived from sesame (*Fagara*), ultimately from *simsim* (Arabic: sesame), and -in(e)

sesamolin
$C_{20}H_{18}O_7$, coined by variation of sesamin

sesqui-
derived from *sesqui* (Latin: one and a half), contracted form of *semisque* (Latin: one half more, one and a half)

sesquiterpene
$(C_5H_8)_3$, derived from sesqui- and terpene

sester-
derived from *sestertius* (Latin: a coin worth two and a half as, literally two and a half)

sesterterpene
$(C_5H_8)_5$, derived from sester- and terpene

SET
an abbreviation for single electron transfer

SEXAFS
an abbreviation for surface extended X-ray absorption fine structure

sexi-
derived from *sexies* (Latin: six times)

Seyferth's reagent
$C_7H_5BrCl_2Hg$, named for the German-US chemist Dietmar Seyferth (born 1929)

SFC
an abbreviation for supercritical fluid chromatography

SFE
an abbreviation for supercritical fluid extraction

Shannon-Prewitt ion radius
named for the 20th century US chemists R. D. Shannon and C. T. Prewitt

Shapiro reaction
named for the US chemist Robert Howard Shapiro (born 1935)

Sharpless epoxidation
named for the US chemist Karl Barry Sharpless (born 1941)

Sharpless-Jacobsen dihydroxylation
named for the US chemists Karl Barry Sharpless (born 1941) and Eric Niels Jacobsen (born 1960)

Sharpless oxyamination
named for the US chemist Karl Barry Sharpless (born 1941)

shellac
derived from *laque en écailles* (French: scaly lacquer)

Shell-Deacon process
named for the energy and petrochemical company Royal Dutch Shell plc, Den Haag, The Netherlands, and the Deacon process

shellolic acid
$C_{15}H_{20}O_6$, derived from shellac and -ol

Shell process
named for the energy and petrochemical company Royal Dutch Shell plc, Den Haag, The Netherlands

sherardization
named for the British chemist and inventor Sherard Osborn Cowper-Coles (1866-1936)

sherwoodite
$Ca_9Al_2V^{IV}{}_4V^{V}{}_{24}O_{80}\cdot 56H_20$, named for the US chemist A. M. Sherwood (born 1888)

shikimic acid
$C_7H_{10}O_5$, derived from *shikimi* (Japanese: star anise, *Illicium verum*)

shionone
$C_{30}H_{50}O$, derived from *shi-on* (Japanese: a drug prepared from the roots of the Tatarian aster, *Aster tataricus* L.) and -one

Shoolery NMR rule
named for the US chemist James Nelson Shoolery (born 1925)

Shore hardness
named for the 20th century US manufacturer Albert F. Shore

showdomycin
$C_9H_{11}NO_4$, derived from the specific epithet of the bacterial species name *Streptomyces showdoensis* and -mycin

SI
an abbreviation for *Système International d'Unités* (French: international system of units)

sialic acid
derived from sial(o)-

sialidase (neuraminidase)
coined by contraction of sialic acid and glycosidase

sial(o)-
derived from *sialon* (Greek: saliva)

sialon
(Al,N,O,Si), coined by contraction of silicon aluminum oxynitride

siccanin
$C_{22}H_{30}O_3$, derived from the specific epithet of the fungal species name *Helminthosporium siccans*, from *siccans* (Latin: drying), from *siccare* (Latin: to dry), from *siccus* (Latin: dry), and -in(e)

sideramine
derived from sider(o)- and amine

siderite
$FeCO_3$, derived from sider(o)- and -ite

sider(o)-
derived from *sideros* (Greek: iron)

siderochrome
derived from sider(o)- and -chrome

sideromycin
derived from sider(o)- and -mycin – referring to this compound's siderophoric properties

siderophilin (transferrin)
derived from sider(o)-, -philic, and -in(e)

siderophore
derived from sider(o)- and -phore

Sidot blende
ZnS, named for the 19th century French chemist Theodor Sidot

Siemens-Martin process
named for the German-British industrialists and inventors Friedrich Siemens (1826-1904) and Karl Wilhelm Siemens (1823-1883), and the French industrialists F. M. E. Martin (1794-1871) and Pierre-Emile Martin (1824-1915)

Siemens ozonizer
named for the German-British engineer Sir Charles William Siemens (1823-1883)

sigmatropic
derived from σ-electrons and -tropic

sil(a)-
derived from silicon

silane
SiH_4, derived from sil(a)- and -an(e)

silhydrite
$3SiO_2 \cdot H_2O$, derived (with contraction) from silica hydrate and -ite

sil(ic)-
derived from silicon

silica
SiO_2, derived from *silex* (Latin: pebble, flint)

silicalite
SiO_2, derived from silica and -lite

silicate
derived from sil(ic)- and -ate

silicic acid
H_4SiO_4, derived from sil(ic)-

silicide
Si^{4-}, derived from sil(ic)- and -ide

silicium
Si, New Latin name for silicon, derived from *silex* (Latin: flint) and -ium

silicocarnotite
$Ca_5SiP_2O_{12}$, derived from silicon and carnotite

silicon
Si, derived from silica and ^{2}-on

silicone
coined by contraction of silicon ketone

sillimanite (fibrolite)
$Al(AlSiO_5)$, named for the US chemist and mineralogist Benjamin Silliman (1779-1864)

siloxane
derived from sil(a)-, ox(a)-, and -an(e)

silver
Ag, a prehistoric Germanic word, ultimately derived from *sarpu* (Akkadian: silver, literally purified by melting), from *sarapu* (Akkadian: to melt)

silybin
$C_{25}H_{22}O_{10}$, derived from the genus name *Silybum* (thistles), from *silybos* (Greek: thistle), and -in(e)

silymarin
coined by contraction of the species name *Silybum marianum* (L.) Gaertn. (milk thistle), from *silybos* (Greek: thistle) and *marianus* (Latin: of Maria), and -in(e)

similor
(Cu,Zn,Sn), derived from *simili* (French: fake), from *similis* (Latin: similar), and *or* (French: gold), from *aurum* (Latin: gold)

Simmons-Smith reaction
Zn/CH_2I_2, named for the US chemists Howard Ensign Simmons, Jr. (1929-1997) and Ronald Dean Smith (born 1930)

Simmons-Smith reagent
Zn/CH_2I_2, named for the US chemists Howard Ensign Simmons, Jr. (1929-1997) and Ronald Dean Smith (born 1930)

Simonini reaction
named for the Austrian chemist A. Simonini (born 1867)

Simonis chromone cyclization
named for the German chemist Hugo Simonis (1874-1949)

SIMS
an abbreviation for secondary ion mass spectrometry

sin(a)-
derived from *sinapis* (Latin: mustard), of Greek origin

sinalbin
$C_{30}H_{42}N_2O_{15}S_2$, derived (with contraction) from the species name *Sinapis alba* (white mustard), from sin(a)- and alb(o)-, and -in(e)

sincalide
$C_{49}H_{62}N_{10}O_{16}S_3$, a name probably coined as a trademark without any stated reason

SINDO
an abbreviation for symmetrically orthogonalized intermediate neglect of differential overlap

sinefungin
$C_{15}H_{23}N_7O_5$, probably derived from *sine* (Latin: without), fungus, and -in(e) – referring to the fungicidal activity of this antibiotic

sinigrin (potassium myronate)
$C_{10}H_{16}KNO_9S_2$, derived (with contraction) from the species name *Sinapis nigra* (black mustard), from sin(a)- and *niger* (Latin: black), and -in(e)

sinomenine
$C_{19}H_{23}NO_4$, derived from the genus name *Sinomenium* (orient vine), from *sinai* (Greek: the Chinese) and *mene* (Greek: moon), and -in(e)

sinter
an indigenous Germanic word

sito-
derived from *sitos* (Greek: grain, food)

sitosterol
derived from sito- and sterol

sizofiran
coined by variation of schizophyllane, from schiz(o)-, phyll(o)-, and -an(e)

skarn
a rock species, derived from *skarn* (Old Norse: dirt)

skatole (scatole)
C_9H_9N, derived from *skor* (Greek: dung) and -ol(e) – uncharacteristically the suffix here neither refers to an oil nor to an alcohol

skimmi(a)-
derived from the genus name *Skimmia* (evergreen shrubs), from *shikimi* (Japanese: star anise, *Illicium verum*)

skimmianine
$C_{14}H_{13}NO_4$, derived from skimmi(a)- and -in(e)

skimmin
$C_{15}H_{16}O_8$, derived from skimmi(a)- and -in(e)

Skita's rule
named for the Austrian-German chemist Aladar Skita (1876-1953)

sklodowskite
$(H_3O)_2Mg(UO_2)_2(SiO_4)_2 \cdot 4H_2O$, named for the Polish-French chemist Marie Curie-Sklodowska (1867-1934)

Skraup reaction
named for the Austrian chemist Zdenko Hans Skraup (1850-1910)

skutterudite
$(Co,Ni)As_{3-x}$, derived from this mineral's locality Skutterud, Norway and -ite

slaked lime
$Ca(OH)_2$, an indigenous English term

Slater orbital
named for the US physicist John Clarke Slater (1900-1976)

smaltite
$(Co,Fe,Ni)As_2$, derived from smalt and -ite – referring to the fact that this mineral was useful in the production of smalt for coloring porcelain and glass blue

SMB chromatography
an abbreviation for simulated moving bed chromatography

SMEAH (Red-Al)
$C_6H_{16}AlNaO_4$, an abbreviation for sodium bis(2-methoxyethoxy)aluminum hydride

smectic
ultimately derived from *smechein* (Greek: to cleanse)

smectite
ultimately derived from *smechein* (Greek: to cleanse) and -ite

Smekal-Raman effect
named for the Austrian physicist Adolf Gustav Stephan Smekal (1895-1959) and the Indian physicist Sir Chandrasekhara Venkata Raman (1888-1970)

smilagenin
$C_{27}H_{44}O_3$, derived from the genus name *Smilax* (plants), from *smilax* (Latin: bindweed), ultimately perhaps from *smile* (Greek: wood-carving knife), -gen, and -in(e)

Smiles rearrangement
named for the British chemist Samuel Smiles (1877-1953)

smithite
$AgAsS_2$, named for the British mineralogist George Frederick Herbert Smith (1872-1953)

smithsonite
$ZnCO_3$, named for the British chemist and mineralogist James Smithson (1765-1829)

smoky quartz
SiO_2, named after its appearance

SNG
an abbreviation for substitute natural gas and/or synthetic natural gas

snRNA
an abbreviation for small nuclear ribonucleic acid

sobrerol
$C_{10}H_{18}O_2$, named for the Italian chemist Ascanio Sobrero (1812-1888) who determined this terpenoid's structure

SOD
an abbreviation for superoxide dismutase

soda
$Na_2CO_3 \cdot 10H_2O$, derived from *soda* (Medieval Latin: barilla), ultimately maybe from *suwwad* (Arabic: barilla)

sodalite
$Na_4Al_3Cl(SiO_4)_3$, derived from soda and -lite

Söderberg electrode
named for the Swedish-Norwegian engineer C. W. Söderberg (1876-1955)

sodium
Na, derived from soda and -ium

sol
derived from *solutio* (Latin: solution)

solanesol
$C_{45}H_{74}O$, derived from solan(o)- and -ol

solanidine
$C_{27}H_{43}NO$, derived from solan(o)- and -idin(e)

solanine
$C_{45}H_{73}NO_{15}$, derived from solan(o)- and -in(e)

solan(o)-
derived from the genus name *Solanum* (nightshades), from *solanum* (Latin: nightshade), of dubious etymology

solanocapsine
$C_{27}H_{46}N_2O_2$, coined by contraction of the species name *Solanum pseudocapsicum* L. (Jerusalem cherry) and/or *Solanum capsicastrum* Link. (false Jerusalem cherry), from solan(o)- and caps(a)-, and -in(e)

solanone
$C_{13}H_{22}O$, derived from solan(o)- and -one

solasodine
$C_{27}H_{43}NO_2$, coined by contraction of solasonine and solanidine

solasonine
$C_{45}H_{73}NO_{16}$, derived (with contraction) from the species name *Solanum sodomaeum* (apple of Sodom), from solan(o)- and *sodomaeus* (Late Latin: of Sodom), and -in(e)

solution
ultimately derived from *solvere* (Latin: to loosen, to dissolve)

solvation
ultimately derived from *solvere* (Latin: to loosen, to dissolve)

Solvay process
named for the Belgian chemist Ernest Solvay (1838-1922)

solvent
ultimately derived from *solvere* (Latin: to loosen, to dissolve)

soman (GD, trilon 300)
$C_7H_{16}FO_2P$, an invented name of undisclosed etymology

somat(o)-
derived from *soma* (Greek: body)

somatoliberin (GRF, somatocrinin)
derived from somatotropin and liberin

somatomedin
coined by contraction of somatotropin, intermediary, and -in(e)

somatostatin (GH-RIF)
$C_{76}H_{104}N_{18}O_{19}S_2$, derived from somatotropin and -statin

somatotropin (GH, phyone)
derived from somat(o)- and -tropin

Sommelet-Hauser rearrangement
named for the French chemist Marcel Sommelet (1877-1952) and the US chemist Charles Roy Hauser (1900-1970)

Sommelet reaction
named for the French chemist Marcel Sommelet (1877-1952)

songorine
$C_{22}H_{31}NO_3$, derived from the specific epithet of the species name *Aconitum songoricum* Popov (a monkshood), from *songoricus* (New Latin: of the Songo River, Manchuria), and -in(e)

Sonnenschein's reagent
$Na_3PMo_{12}O_{40}$, named for the German forensic chemist F. L. Sonnenschein (1819-1879)

Sonn-Müller method
named for the 20th century German chemists Adolf Sonn and Ernst Müller

Sonogashira-Hagihara cross-coupling reaction
named for the 20th century Japanese chemists Kenkichi Sonogashira and Nobue Hagihara

soot
C_n, an indigenous English word

sophor(a)-
derived from the genus name *Sophora* (trees, shrubs), ultimately from *sufayra* (Arabic: *Sophora*)

sophorabioside
$C_{27}H_{30}O_{14}$, derived from sophor(a)-, bi-, -ose, and -ide

sophorose
$C_{12}H_{22}O_{11}$, derived from sophor(a)- and -ose

sorbic acid
$C_6H_8O_2$, derived from sorb(o)-

sorbide
$C_6H_{10}O_4$, derived from sorb(o)- and ^{3}-ide

s-orbital
named after the 'sharp' line group in atomic spectroscopy

σ-orbital
derived from s-orbital – referring to the common symmetry of σ- and s-orbitals

sorbitan
derived from sorbitol and -an

sorbite
iron alloy, named for the British geologist and metallurgist Henry Clifton Sorby (1826-1908)

sorbitol
$C_6H_{14}O_6$, derived from sorb(o)- and -itol

sorb(o)-
derived from the genus name *Sorbus* (trees, shrubs), from *sorbus* (Latin: service tree)

sorbose
$C_6H_{12}O_6$, derived from sorbitol and -ose

Sorel cement
$(MgCl_2,MgO)$, named for the French inventor S. Sorel (1802-1871)

Sørensen buffer
(KH_2PO_4,Na_2HPO_4), named for the Danish chemist Søren Peder Lauritz Sørensen (1868-1939)

sorensenite
$Na_4Be_2Sn(Si_3O_9)_2{\cdot}2H_2O$, named for the

Danish mineralogist and petrographer Henning Sørensen (born 1926)

sorosilicate
derived from *soros* (Greek: heap) and silicate

sorption
derived from *sorbere* (Latin: to suck in, to swallow up)

Southern blot (Southern blotting)
named for the British molecular biologist E. M. Southern (born 1938)

Soxhlet extraction
named for the German chemist Franz Ritter von Soxhlet (1848-1926)

sozoiodolic acid
$C_6H_4I_2O_4S$, derived from sozolic acid and ^{2}iod(o)-

sozolic acid
$C_6H_6O_4S$, derived from *sozein* (Greek: to save) and -ol – referring to the life-saving antiseptic properties of this compound

sparassol
$C_{10}H_{12}O_4$, derived from the genus name *Sparassis* (mushrooms), from *sparassein* (Greek: to tear, to rend), and -ol

sparsomycin
$C_{13}H_{19}N_3O_5S_2$, derived from the specific epithet of the bacterial species name *Streptomyces sparsogenes* var *sparsogenes*, from *spargere* (Latin: to scatter, to sprinkle) and -gen, and -mycin

sparteine
$C_{15}H_{26}N_2$, derived from the genus name *Spartium* (shrubs), from *spartum* (Latin: the herb broom), and -in(e)

spatula
derived from *spatula*, diminutive of *spatha* (Latin: broad tool or weapon), from *spathe* (Greek: broad blade of wood or metal)

SPE
an abbreviation for solid-phase extraction

spectinomycin (actinospectacine)
$C_{14}H_{24}N_2O_7$, coined by variation of actinospectacine and -mycin

spectrin
named for the isolation of these proteins from erythrocyte 'ghosts'

spectro-
derived from spectrum

spectrum
derived from *spectrum* (Latin: appearance, image), from *spectare* (Latin: to watch, to look at), from uncommon *specere* (Latin: to look)

speleand
derived from *speleion* (Greek: cave) and -and

speleate
derived from speleand and -ate

spermidine
$C_7H_{19}N_3$, derived from sperm(o)- and -idin(e)

spermine
$C_{10}H_{26}N_4$, derived from sperm(o)- and -in(e)

sperm(o)-
derived from *sperma* (Greek: seed, sperm)

sperrylite
$PtAs_2$, named for the US-Canadian chemist Francis Louis Sperry (deceased 1906)

spessartine
$Mn_3Al_2(SiO_4)_3$, named after this mineral's type locality Spessart Mountains, Germany

sphalerite
(Zn,Fe)S, derived from *sphaleros* (Greek: treacherous, deceitful) and -ite – referring to this mineral's resemblance to galena

sphene (titanite)
$CaTiO(SiO_4)$, derived from *sphen* (Greek: wedge) – referring to this mineral's wedge-shaped crystal habit

spherand
derived from *sphaira* (Greek: sphere) and -and

spheraplex
coined by contraction of spherand and complex

sphing(o)-
derived from *sphingein* (Greek: to bind fast)

sphingofungin
derived from sphing(o)-, fungus, and -in(e)

sphingomyelin
derived from sphing(o)- and myelin

sphingosine (sphingenine)
$C_{18}H_{37}NO_2$, derived from sphing(o)- and -in(e)

spinasterol
$C_{29}H_{48}O$, derived from the genus name *Spinacia* (spinach), ultimately from *isfanakh* (Persian: spinach), and sterol

spinel
$MgAl_2O_4$, ultimately derived from *spinella*, diminutive of *spina* (Latin: thorn, spine) – referring to this mineral's sharp-pointed crystals

spinosyn
derived from the specific epithet of the microbiological species name *Saccharopoly spora spinosa*, from *spinosus* (Latin: thorny), from *spina* (Latin: thorn, spine), probably systemic (action), and -in(e)

spiraeic acid (salicylic acid)
$C_7H_6O_3$, an obsolete name for salicylic acid, derived from the genus name *Spiraea* (shrubs), from *speira* (Greek: coil)

spirane
derived from spir(o)- and -an(e)

spirein (helicin)
$C_{13}H_{16}O_7$, derived from the genus name *Spiraea* (spirea), from *speira* (Greek: coil), and -in(e)

spiritus fumans Libavii
$SnCl_4$, (New Latin: fuming liquor of Libavius), named for the German alchemist Andreas Libavius (Libau or Liebau) (1540-1616 or 1560-1616) and after this compound's hydrolysis by water vapor

spiritus vini
C_2H_6O, alchemistic name for ethanol, from *spiritus* (Latin: spirit) and *vinum* (Latin: wine)

spir(o)-
ultimately derived from *speira* (Greek: coil, twist)

spironolactone
$C_{24}H_{32}O_4S$, derived from spir(o)- and lactone

spirosolane
$C_{27}H_{45}NO$, derived from spir(o)-, solan(o)-, and an(e)

spirostan
$C_{27}H_{44}O_2$, derived from spir(o)-, sterol, and

-an(e)

splendipherin
$C_{105}H_{185}N_{29}O_{32}$, derived (with contraction) from the specific epithet of the species name *Litoria splendida* (magnificent tree frog), from *splendidus* (Latin: magnificent), from *splendere* (Latin: to shine), pheromone, and -in(e)

SPME
an abbreviation for solid-phase micro-extraction

spodumene
$LiAlSi_2O_6$, derived from *spodoumenos* (Greek: burnt to ashes), ultimately from *spodos* (Greek: wood ash) – referring to this mineral's ashen color

sporidesmin
$C_{18}H_{20}ClN_3O_6S_2$, derived from sporidesm-(o)- and -in(e)

sporidesm(o)-
derived from the fungal genus name *Sporidesmium*, from spor(o)- and *desme* (Greek: bundle), from *dein* (Greek: to bind, to tie)

sporidesmolide
derived from sprodesm(o)- and -olide

spor(o)-
derived from *spora* (New Latin: seed, spore)

spuma lupi (lupi spuma)
$(Fe,Mn)WO_4$, an alchemistic name for wolframite, derived from *lupi spuma* (Latin: wolf's foam)

squalamine
$C_{34}H_{67}N_3O_5S$, derived from squal(o)- and amine

squalane
$C_{30}H_{62}$, derived from squal(o)- and -an(e)

squalene
$C_{30}H_{50}$, derived from squal(o)- and -ene

squal(o)-
derived from the genus name *Squalus* (shark), from *squalus* (Latin: a sea fish)

squaric acid
$C_4H_2O_4$, derived from square – referring to this compound's square molecular shape

sRNA
an abbreviation for soluble ribonucleic acid

SRPES
an abbreviation for synchroton radiation photoelectron spectroscopy

SRS-A
an abbreviation for slow reacting substances of anaphylaxis

SRXRF
an abbreviation for synchroton radiation X-ray fluorescence spectroscopy

stachy-
derived from the genus name *Stachys* (herbs), from *stachys* (Greek: ear of grain)

stachydrine
$C_7H_{13}NO_2$, derived from stachy-, hydr(o)-, and -in(e)

stachyose
$C_{24}H_{42}O_{21}$, derived from stachy- and -ose

stalactite
derived from *stalaktos* (Greek: dripping), from *stalassein* (Greek: to drop) and -ite

stalagmite
derived from *stalagma* (Greek: a drop) and

-ite

stallimycin
$C_{22}H_{27}N_9O_4$, derived from the specific epithet of the bacterial species name *Streptomyces distallicus* and -mycin

stannane
SnH_4, derived from stannum and -an(e)

stanni-
Sn^{4+}, derived from stannum

stannic
derived from stannum

stannite
Cu_2FeSnS_4, derived from stannum and -ite

stanno-
Sn^{2+}, derived from stannum

stann(o)-
derived from stannum

stannocene
$C_{10}H_{10}Sn$, derived from stann(o)- and -ocene

stannous
derived from stannum

stannum
Sn, New Latin name for tin, from *stagnum* (Latin: tin), probably of Celtic origin

staphyl(o)-
derived from *staphyle* (Greek: bunch of grapes)

starch
derived from *stercan* (Old English: to stiffen)

Stas-Otto separation
named for the Belgian chemist Jean Servais Stas (1813-1891) and the German chemist Friedrich Julius Otto (1809-1870)

-statin
derived from *-states* (Greek: one that causes to stand), from *histanai* (Greek: to cause to stand), and -in(e)

statine
$C_8H_{17}NO_3$, derived from pepstatin

statolon
derived from the specific epithet of the fungal species name *Penicillium stoloniferum*, from stolon, from *stolo* (Latin: shoot, branch), and -on(e)

Staudinger reaction
named for the German chemist Hermann Staudinger (1881-1965)

Staudinger's ketene
$C_{14}H_{10}O$, named for the German chemist Hermann Staudinger (1881-1965)

Stauffer grease
named for the 19th century US chemist and industrialist J. Stauffer

staur(o)-
derived from *stauros* (Greek: pale, stake, cross)

staurolite
$(Fe,Mg)_4Al_{17}(Si,Al)_8O_{45}(OH)_3$, derived from staur(o)- and -lite – referring to this mineral's commonly cross-shaped twin crystals

staurosporine
$C_{28}H_{26}N_4O_3$, derived from the specific epithet of the bacterial species name *Streptomyces staurospores*, from staur(o)- and spor(o)-, and -in(e)

stearic acid
$C_{18}H_{36}O_2$, derived from stear(o)-

stearin
$C_{57}H_{110}O_6$, derived from stear(o)- and -in(e)

stear(o)-
derived from *stear* (Greek: fat, tallow)

stearoptene
the solid part of a partially solidified natural oil, derived from stear(o)- and -pten(e)

steatite (talc, talcum, soapstone)
$Mg_3Si_4O_{10}(OH)_2$, derived from steat(o)- and -ite – referrring to this mineral's soapy consistency

steat(o)-
derived from *stear* (Greek: fat, tallow)

steel
ultimately derived from *stakati* (Sanskrit: he resists)

steenstrupine
$Na_{14}Ce_6Mn_2Fe_2Zr(PO_4)_7Si_{12}O_{36}(OH)_2{\cdot}3H_2O$ named for the Danish geologist Knud Johannes Vogelius Steenstrup (1842-1913)

Stefan-Boltzmann constant
named for the Austrian physicists Josef Stefan (1835-1893) and Ludwig Eduard Boltzmann (1844-1906)

stellite
(C,Co,Cr,Fe,Mn,Mo,Ni,Si,W), coined as a trademark of the metallurgical company Deloro Stellite Inc., Belleville, Ontario, Canada

Stelzner nomenclature
named for the German chemist R. Stelzner (1869-1943)

STEM
an abbreviation for scanning transmission electron microscopy

stenhuggarite
$CaFe^{III}(As^{III}O_2)(As^{III}Sb^{III}O_5)$, named for the US scientist Brian Harold Mason (born 1917), jocularly from *stenhuggare* (Swedish: mason)

stephanine
$C_{19}H_{19}NO_3$, derived from the genus name *Stephania* (vines), from stephan(o)-, and -in(e)

stephanite
Ag_5SbS_4, named for the Austrian mining director and engineer Archduke Victor Stephan (1817-1867)

stephan(o)-
derived from *stephanos* (Greek: crown), from *stephein* (Greek: to encircle, to crown)

Stephen aldehyde synthesis
named for the British chemist Henry Stephen (1889-1965)

sterc(o)-
derived from Sterculius, Roman god of cultivation and manuring, from *stercus* (Latin: manure)

stercobilin
derived from sterc(o)- and bilin

stercorin (coprosterol)
$C_{27}H_{48}O$, coined by contraction of the name fragments sterc(o)- and sterin

stere(o)-
derived from *stereos* (Greek: solid)

stereogenic
derived from stere(o)- and -gen

stereoheterotopic
derived from stere(o)-, heter(o)-, and -topic

sterin
an obsolete name for sterol, derived from stere(o)- and -in(e)

Stern double layer
named for the German-US physicist Otto Stern (1888-1969)

steroid
derived from sterol and -oid

sterol
derived from sterin and -ol

Stetter reaction
named for the German chemist Hermann Stetter (1917-1993)

Stevens rearrangement
named for the Scottish chemist Thomas Stevens Stevens (1900-2000)

stevioside
$C_{38}H_{60}O_{18}$, derived from the genus name *Stevia* (shrubs, herbs), after the Spanish botanist Pedro Jaime Esteve (latinized Stevius) (1550-1556), -ose, and ^{3}-ide

-sthene
derived from *sthenos* (Greek: force)

stibane (stibine)
SbH_3, derived from stibium and -an(e)

stibarsen (allemontite)
(As,Sb), derived from stibium and *arsène* (French: arsenic)

stibiconite
$Sb_3O_6(OH)$, derived from stibium, *konia* (Greek: ashes, dust, powder), and -ite – referring to this mineral's composition and habit

stibine (stibane)
SbH_3, derived from stibium and -in(e)

stibium
Sb, New Latin name for antimony, derived from *stibium* (Latin: antimony), ultimately from *stm* (Old Egyptian: antimony)

stibnite (antimonite)
Sb_2S_3, derived from stib(o)- and -ite

stib(o)-
derived from stibium

Stieglitz rearrangement
named for the US chemist Julius Stieglitz (1867-1937)

stigma-
derived from the genus name *Physostigma* (vines), from phys(o)- and *stigma* (Greek: tattoo, mark), from *stizein* (Greek: to tattoo)

stigmastane
$C_{29}H_{52}$, derived from stigma-, sterol, and -an(e)

stigmastanol (fucostanol)
$C_{29}H_{52}O$, derived from stigma-, sterol, -ane, and -ol

stigmasterol
$C_{29}H_{48}O$, derived from stigma- and sterol

stigmatellin
$C_{30}H_{42}O_7$, derived from the bacterial genus name *Stigmatella*, from *stigmatella*, diminutive of *stigma* (Greek: tattoo, mark), from *stizein* (Greek: to tattoo), and -in(e)

stilbene
$C_{14}H_{12}$, derived from stilb(o)- and -ene

stilbite (desmine)
$NaCa_2Al_5Si_{13}O_{36}\cdot 14H_2O$, derived from stilb(o)- and -ite – referring to this mineral's

pearly to vitreous luster

stilb(o)-
derived from *stilbein* (Greek: to glitter)

Stille coupling reaction
named for the US chemist John Kenneth Stille (1930-1989)

Still-Wittig rearrangement
named for the US chemist W. Clark Still, Jr. (born 1946) and the German chemist Georg Wittig (1897-1987)

stilpnomelane
$(K,Ca,Na)(Fe,Mg,Al)_8(Si,Al)_{12}(O,OH)_{36}\cdot nH_2O$, derived from *stilpnos* (Greek: glistening) and -melane – referring to this mineral's luster and black color

stipitatic acid
$C_8H_6O_5$, derived from the specific epithet of the fungal genus name *Penicillium stipitatum*, from *stipitatus* (Latin: having or borne on a stipe), from *stipes* (Latin: stipe), akin to *stipare* (Latin: to press together)

stishovite
SiO_2, named for the Russian crystallographer Sergei Mikhailovich Stishov (born 1937) who synthesized this mineral

[1]STM
an abbreviation for scanning transmission electron microscopy

[2]STM
an abbreviation for scanning tunneling microscopy

STO
an abbreviation for Slater-type orbital

Stobbe condensation
named for the German chemist Johann Hans Hermann August Adolf Stobbe (1860-1938)

Stock system
named for the German chemist Alfred Stock (1876-1946)

stoichiometry
derived from *stoicheion* (Greek: element) and -metry

stokesite
$CaSnSi_3O_9\cdot 2H_2O$, named for the British physicist Sir George Gabriel Stokes (1819-1903)

Stokes' law
named for the British physicist Sir George Gabriel Stokes (1819-1903)

Stokes' rule
named for the British physicist Sir George Gabriel Stokes (1819-1903)

Stollé synthesis
named for the 20th century German chemist R. Stollé

Stoltz oxidative etherification
named for the US chemist Brian Mark Stoltz (born 1970)

stolzite
$PbWO_4$, named for the Czech scientist Joseph Alexis Stolz (1803-1896)

Stoney's electron hypothesis
named for the British scientist George Johnstone Stoney (1826-1911)

Stork enamine reaction
named for the Belgian-US chemist Gilbert Stork (born 1921)

[1]STP
an abbreviation for standard temperature and pressure

[2]STP (DOM)
$C_{12}H_{19}NO_2$, copied from the abbreviation for scientifically treated petroleum, a trademark for a fuel additive, suggesting that this designer drug does the same for your mind as the fuel additive does for your car; also interpreted as an acronym for serenity, tranquillity, peace

Strandberg anion
$Mo_5P_2O_{23}^{6-}$, named for the 20th century Swedish chemist Rolf Strandberg

strass
named for its inventor, the French jeweler G. F. Strass (1700-1773)

strawberry aldehyde
$C_{12}H_{14}O_3$, named after this compound's strawberry and aldehydic aroma notes – contrary to popular belief this compound is not an aldehyde

Strecker amino acid synthesis
named for the German chemist Adolf Friedrich Ludwig Strecker (1822-1871)

Strecker degradation
named for the German chemist Adolf Friedrich Ludwig Strecker (1822-1871)

Strecker sulfite alkylation
named for the German chemist Adolf Friedrich Ludwig Strecker (1822-1871)

strengite
$FePO_4 \cdot 2H_2O$, named for the German mineralogist Johann August Streng (1830-1897)

streptidine
$C_8H_{18}N_6O_4$, derived from streptomycin and -idin(e)

[1]strept(o)-
derived from from *streptos* (Greek: twisted, easy to bend, pliant), from *strephein* (Greek: to twist, to turn)

[2]strept(o)-
derived from the bacterial genus name *Streptococcus*, from [1]strept(o)-

[3]strept(o)-
derived from the bacterial genus name *Streptomyces*, from [1]strept(o)-

streptodornase
coined by contraction of streptococcal deoxyribonuclease

streptokinase
derived from [2]strept(o)- and kinase

streptolydigin
$C_{32}H_{44}N_2O_9$, derived (with contraction) from the bacterial species name *Streptomyces lydicus*, from [1]strept(o)- and *lydicus* (Latin: Lydian), and -in(e)

streptomycin
$C_{21}H_{39}N_7O_{12}$, derived from [3]strept(o)- and -mycin

streptonigrin
$C_{25}H_{22}N_4O_8$, derived from [3]strept(o)-, *niger* (Latin: black), and -in(e) – referring to this antibiotic's black color

streptose
$C_6H_{10}O_5$, derived from streptomycin and -ose

streptovirudin
derived from [3]strept(o)-, virus, and -in(e) – referring to this antibiotic's virucidal action

streptozocin
$C_8H_{15}N_3O_7$, derived from [3]strept(o)-, az(a)-, and -icin(e)

strigol
$C_{19}H_{22}O_6$, derived from the genus name *Striga* (tropical parasitic seed plants), from *striga* (Latin: furrow, windrow, swath), and –ol

strobilurin
derived from the genus name *Strobilurus* (toadstools), from *strobilos* (Greek: whirling or twisted object), from *strephein* (Greek: to turn), and *oura* (Greek: tail), and -in(e)

stromeyerite
CuAgS, named for the German mineralogist and chemist Friedrich Stromeyer (1776-1835)

strontia
SrO, $Sr(OH)_2$, derived from strontianite and -a

strontianite
$SrCO_3$, named after this mineral's locality, the village of Strontian (Srón an t-Sithein), Scotland

strontium
Sr, derived from strontia and -ium

strophanthidin
$C_{23}H_{32}O_6$, derived from strophanthin and -idin(e)

strophanthin
derived from strophanth(o)- and -in(e)

strophanth(o)-
derived from the genus name *Strophanthus* (tropical trees, shrubs, vines), from *strophos* (Greek: twisted band) and anth(o)-

struvite
$(NH_4)MgPO_4{\cdot}6H_2O$, named for the Russian diplomat H. C. G. Baron von Struve (1772-1851)

strychnidine
$C_{21}H_{24}N_2O$, derived from strychn(o)- and -idine

strychnine
$C_{21}H_{22}N_2O_2$, derived from strychn(o)- and -ine

strychn(o)-
derived from the genus name *Strychnos* (trees, vines), ultimately from *strychnos* (Greek: nightshade), from *strychnos* (Greek: acrid, bitter)

Stuart-Briegleb molecular models
named for the Swiss physicist Herbert Arthur Stuart (1899-1974) and the German chemist Günther Briegleb (1905-1991)

stumpflite
$Pt(Sb,Bi)_2$, named for the Austrian mineralogist Eugen Friedrich Stumpfl (born 1931)

STX
$C_{10}H_{17}N_7O_4^{+}$, an abbreviation for saxitoxin

styl(o)-
derived from *stylos* (Greek: stalk, stylus)

stylophorine (chelidonine)
$C_{20}H_{19}NO_5$, derived from the genus name *Stylophorum* (poppies), from styl(o)- and phor(o)-, and -in(e)

stylopine
$C_{19}H_{17}NO_4$, coined by variation of stylo-phorine

styphnic acid
$C_6H_3N_3O_8$, derived irregularly from *stryphnos* (Greek: astringent)

styrene
derived from styr(o)- and -ene

styr(o)-
derived from the genus name *Styrax* (storax trees), ultimately from *styrax* (Greek: storax)

styryl
C_8H_7-, derived from styrene and -yl

suanite
$Mg_2B_2O_5$, named after this mineral's locality, the Hol Kol mine, Suan, North Korea

Suarez reaction
named for the 20th century Spanish chemist Ernesto Suárez

sub- (suc-, suf-, sup-)
derived from *sub* (Latin: under)

suberic acid
$C_8H_{14}O_4$, derived from *suber* (Latin: cork)

[1]sublimate
derived from *sublimatus* (Latin: sublimed), from *sublimare* (Medieval Latin: to refine, to purify, to sublime)

[2]sublimate
$HgCl_2$, derived from [1]sublimate

sublime
derived from *sublimare* (Medieval Latin: to refine, to purify, to sublime)

suboxide
derived from sub- and oxide

subrogation
derived from *subrogare* (Latin: to substitute), from *rogare* (Latin: to reach for, to inquire, to defend, to invite)

substance P (SP)
$C_{63}H_{98}N_{18}O_{13}S$, abbreviated from pain – referring to this peptide's analgesic and hyperalgesic properties; also suggested to be an abbreviation of powder, referring to the state in which this material was isolated

subtilin
derived from the specific epithet of the bacterial species name *Bacillus subtilis*, from *subtilis* (Latin: finely woven, fine, thin, subtle), from sub- and *tela* (Latin: web), akin to *texere* (Latin: to weave), and -in(e)

succimer
$C_4H_6O_4S_2$, coined by anagrammatical contraction of dimercaptosuccinic acid

succinic acid
$C_4H_6O_4$, derived from *suc(c)inum* (Latin: amber)

succinite (amber)
$(C_{10}H_{16}O)_n$, derived from succin(o)- and -ite

succin(o)-
derived from succinic acid

sucrase (invertase)
derived from sucrose and -ase

sucr(o)-
derived from sucrose

sucrose
$C_{12}H_{22}O_{11}$, derived from *sucre* (French: sugar), ultimately from *sarkara* (Sanskrit: sugar), and -ose

sudoite
$Mg_2(Al,Fe)_3Si_3AlO_{10}(OH)_8$, named for the Japanese mineralogist Toshio Sudo (born 1911)

sugar
ultimately derived from *sarkara* (Sanskrit: sugar)

Sugasawa reaction
named for the 20th century Japanese chemist Tsutomo Sugasawa

sugilite
$KNa_2(Fe,Mn,Al)_2Li_3Si_{12}O_{30}$, named for the Japanese petrologist Ken-ichi Sugi (1901-1948)

sulfa-
short for sulfanilyl, derived from sulfanilic acid and -yl

Sulfacid process
coined by contraction of sulfuric acid process

sulfamic acid
H_3NO_3S, derived from sulfuric acid and amide

sulfamide
$H_4N_2O_2S$, derived from sulfuric acid and amide

sulfamoyl
H_2NO_2S-, derived from sulfamic acid and -oyl

sulfane
H_2S, derived from sulf(o)- and -an(e)

sulfanilamide
$C_6H_8N_2O_2S$, derived from sulfanilic acid and amide

sulfanilic acid
$C_6H_7NO_3S$, derived from sulf(o)- and -anilic acid

sulfanuric chloride
$Cl_3N_3O_3S_3$, derived (with contraction) from sulf(o)- and cyanuric chloride

sulfanyl (mercapto)
HS−, derived from sulfane and –yl

sulfatase
coined by contraction of sulfate hydrolase

sulfate
SO_4^{2-}, derived from sulf(o)- and -ate

sulfatide
derived from sulfate and [3]-ide

sulfene
R_2CSO_2, derived from sulf(o)- and -ene; patterned after ketene

sulfenic acid
RSOH, derived from sulf(o)-

sulfide
S^{2-}, derived from sulf(o)- and [2]-ide

sulfine
R_2CSO, derived from sulf(o)- and -in(e)

sulfinic acid
RSO_2H, derived from sulf(o)-

sulfite
SO_3^{2-}, derived from sulf(o)- and -ite

sulfo-
HO_3S-, derived from sulfonic acid

sulf(o)-
derived from sulphur

sulfohalite
$Na_6ClF(SO_4)_2$, derived from sulf(o)-, halo-, and -ite

sulfolane
$C_4H_8O_2S$, coined by contraction of sulfone and thiolane

sulfonamide
RSO_2NR_2, derived from sulfonic acid and amide

sulfone
R_2SO_2, derived from sulf(o)- and -one – expressing an obsolete notion of analogy between sulfones and ketones

sulfonic acid
RSO_3H, derived from sulf(o)-

sulforaphane
$C_6H_{11}NOS_2$, derived from sulforaphen and -ane

sulforaphen
$C_6H_9NOS_2$, derived from sulf(o)- and the genus name *Raphanus* (herbs), from *raphanos* (Greek: radish), akin to *rhapys*, *rhaphys* (Greek: turnip), and -en(e)

sulfoxide
R_2SO, derived from sulf(o)- and oxide

sulfoxylic acid
H_2SO_2, derived from sulf(o)-, ox(o)-, and -yl

sulfur
S, derived from *sulpur* (Latin: sulfur), probably from Oscan

sulfuretin
$C_{15}H_{10}O_5$, derived from the specific epithet of the species name *Cosmos sulphurous* Cav. (orange cosmos), from *sulfureus* (Latin: sulfur-colored), and -etin(e)

sulfuric acid
H_2SO_4, derived from sulfur

sulfurium
S, New Latin name for sulfur, derived from sulfur and -ium

sulfurous acid
H_2SO_3, derived from sulfur

sulfuryl
$-SO_2-$, derived from sulfuric acid and -yl

sultam
coined by contraction of sulfonic acid and lactam

sultim
coined by contraction of sulfonic acid and lactim

sultone
coined by contraction of sulfonic acid and lactone

super-
derived from *super* (Latin: above, over)

superactinoid (superactinide)
elements 122-153, derived from super- and actinoid

superhydride
$C_6H_{16}BLi$, named after its superiority as hydride donor relative to lithium tetrahydridoborate

supermesityl
$C_{18}H_{29}-$, named after this substituent's superior bulkiness relative to ^{2}mesityl

superoxide
O_2^-, derived from super- and oxide

superoxide dismutase (SOD)
derived from superoxide, dismutation, and -ase

superphosphate
derived from super- and phosphate – referring to the notion that this material is a better fertilizer than rock phosphate

supersilyl
$C_{12}H_{27}Si-$, named after this substituent's superior bulkiness relative to trimethylsilyl

supra-
derived from *supra* (Latin: above, over)

suprafacial
derived from supra- and face

suprasterol
derived from supra- and sterol

surinamine (andirine, angeline, geoffroy-ine, ratanhine)
$C_{10}H_{13}NO_3$, derived from the specific epithet of the species name *Geoffroya surinamensis* (a tropical tree), from *surinamensis* (New Latin: Surinamese), and amine

survivin
derived from survival and -in(e)

Süs reaction
named for the German chemist Oskar Süs (1903-1978)

suspension
derived from *suspendere* (Latin: to hang up), from sub- and *pendere* (Latin: to hang)

Suzuki coupling
named for the Japanese chemist Akira Suzuki (born 1930)

Swain-Lupton equation
named for the US chemists Charles Gardner Swain (1917-1988) and Elmer Cornelius Lupton, Jr. (born 1945)

Swain-Scott equation
named for the US chemists Charles Gardner Swain (1917-1988) and Carlton Browne Scott (born 1924)

Swarts reaction
named for the Belgian chemist Frédéric J. E. Swarts (1866-1940)

Swern oxidation
named for the US chemist Daniel Swern (1916-1982)

Swern reagents
named for the US chemist Daniel Swern (1916-1982)

swertiamarin
$C_{16}H_{22}O_{10}$, derived from the genus name *Swertia* (herbs), after the Dutch botanist Emanuel Sweerts (1552-1612), marine, and -in(e)

sydnone
named for the city of Sydney, Australia where these compounds were first prepared

syenite
a rock species, named after the ancient town of Syene, now Assuan, Egypt

sylvanite
$(Au,Ag)_2Te_4$, named after this mineral's locality Transylvania, Romania

sylvite (sylvine)
KCl, derived from *sal digestevus Sylvii* (New Latin: digestive salt of Sylvius), named for the Dutch physician and chemist Franciscus de la Boë (1614-1672), a.k.a. Sylvius

sym-
derived from symmetrical

syn- (syl-, sym-, sys-)
derived from *syn- (syl-, sym-, sys-)* (Greek: with, together with, by means of, at the same time)

synartetic (anchimeric)
derived from *synartan* (Greek: to join)

synchysite
$Ca(Ce,La)(CO_3)_3F$, derived from *synchys*

(Greek: confounding) and -ite – referring to an early misidentification of this mineral as parisite

syndiotactic
derived from *syndyo* (Greek: two together) and tactic

syngenite
$K_2Ca(SO_4)_2 \cdot H_2O$, derived from syn-, -gen, and -ite – referring to this mineral's similarity with polyhalite

synthase (ligase, lyase)
derived from synthesis and -ase

synthesis
derived from syn- and *thesis* (Greek: a placing, an arranging)

synthetase (ligase)
derived from synthesis and -ase

synthon
derived from synthesis and [3]-on

syringaldehyde
$C_9H_{10}O_4$, derived from syringin and aldehyde

syringin
$C_{17}H_{24}O_9$, derived from the genus name *Syringa* (shrubs, trees), from *syrinx* (Greek: panpipe, fistula, tube), and -in(e)

systemin
$C_{85}H_{144}N_{26}O_{28}S$, derived from systemic (action) and -in(e)

szaibelyite (ascharite)
$MgHBO_3$, named for the Hungarian mine surveyor Stephan Szaibely (1777-1855)

Szilard-Chalmers effect
named for the Hungarian-US physicist Leo Szilard (1898-1964) and the 20th century British physicist T. A. Chalmers

T

tabernanthine
$C_{20}H_{26}N_2O$, derived from the genus name *Tabernanthe* (West African shrubs), from the name of the German herbalist Jacobus Theodorus Tabernaemontanus, a.k.a. Jakob Theodor von Bergzabern, (1522-1590), and anth(o)-, and -in(e)

tabun (GA, trilon 83)
$C_5H_{11}N_2O_2P$, an arbitrary code name of undisclosed etymology

tachy-
derived from *tachys* (Greek: swift)

tachysterol
$C_{28}H_{44}O$, derived from tachy- and sterol – referring to the fact that this sterol is a liquid at ambient temperature

tacrine
$C_{13}H_{14}N_2$, coined by contraction of 1,2,3,4-tetrahydroacridin-9-amine

tactic
derived from *taktos* (Greek: ordered), from *tassein* (Greek: to arrange, to order)

TAED
$C_{10}H_{16}N_2O_4$, an abbreviation for *N,N,N',N'*-tetraacetylethylenediamine

taenite
γ-(Fe,Ni), derived from *tainia* (Greek. band) and -ite – referring to this mineral's flattened crystals

Tafel rearrangement
named for the Swiss-German chemist Julius Tafel (1862-1918)

Taft equation
named for the US chemist Robert Wheaton Taft, Jr. (1922-1996)

tagatose
$C_6H_{12}O_6$, anagrammatically derived from galactose

Takata-Ara reaction
named for the Japanese physicians M. Takata (1892-1978) and K. Ara (born 1894)

talc (talcum, steatite, soapstone)
$Mg_3Si_4O_{10}(OH)_2$, ultimately derived from *talq* (Arabic: talc)

talitol (altritol)
$C_6H_{14}O_6$, derived from talose and -itol

tall oil
derived from *tall* (Swedish: pine tree)

tallysomycin
derived from the bacterial genus name *Streptoalloteichus*, from strepto-, all(o)-, teicho-, and -mycin

talonic acid
$C_6H_{12}O_7$, anagrammatically derived from galactonic acid

talose
$C_6H_{12}O_6$, derived from talonic acid and -ose

Tammann's rule
named for the German chemist Gustav Tammann (1861-1938)

tanacetin
$C_{15}H_{20}O_4$, derived from the genus name *Tanacetum* (tansy), ultimately from *athanasia* (Greek: immortality), from a(n)- and *thanatos* (Greek: death), and -in(e)

tannase
derived from tannin and -ase

tannic acid (tannin)
derived from tannin

tannin (tannic acid)
derived from *tannum* (Medieval Latin: oak tree) and -in(e)

tantalite
(Fe,Mn)(Ta,Nb)$_2$O$_6$, derived from tantalum and -lite

tantalocene
$C_{10}H_{10}Ta$, derived from tantalum and -ocene

tantalum
Ta, named for Tantalos, a Greek mythical figure, punished by never being allowed to drink – referring to the fact that Ta_2O_5 does not absorb water nor dissolve in acid

tapiolite (ferrotapiolite)
(Fe,Mn)(Ta,Nb)$_2$O$_6$, named for the Old Finnish god Tapio

TAPS
$C_7H_{17}NO_6S$, an abbreviation for *N*-[tris(hydroxymethyl)methyl]-3-aminopropane-sulfonic acid

tar
derived from *teru* (Old English: tar), from an Indo-European root meaning tree

taraxasterol
$C_{30}H_{50}O$, derived from the genus name *Taraxacum* (perennial herbs), from *tarakhshaqun* (Arabic: wild chicory), and sterol

taraxein
derived from *taraxis* (Greek: confusion), from *tarassein* (Greek: to trouble, to confuse), and -in(e) – referring to the isolation of this protein complex from the blood serum of schizophrenics

taraxerol
$C_{30}H_{50}O$, coined by variation of taraxasterol

tartar
$C_4H_5KO_6$, ultimately derived, via *tartarum* (Medieval Latin: tartar), *tartaron* (Medieval Greek: tartar), and *durdhia* (Arabic: lees), from *durd* (Persian: lees)

tartaric acid (threaric acid)
$C_4H_6O_6$, derived from tartar

tartrate
$C_4H_4O_6^{2-}$, derived from tartaric acid and -ate

tartrazine
$C_{16}H_9N_4Na_3O_9S_2$, derived from tartaric acid, az(o)-, and -in(e)

tartronic acid
$C_3H_4O_5$, derived from tartaric acid and -one; patterned after malonic acid

taurine
$C_2H_7NO_3S$, derived from taur(o)- and -in(e)

taur(o)-
derived from *taurus* (Latin: bull)

taurocholic acid
$C_{26}H_{45}NO_7S$, derived from taur(o)- and cholic acid

tauryl
$C_2H_6NO_2S-$, derived from taurine and -yl

tauto-
derived from *to auton* (Greek: the same)

tautomer
derived from tauto- and -mer

taxane
$C_{20}H_{36}$, derived from tax(o)- and -an(e)

taxicin
derived from tax(o)- and -icin(e)

taxine
$C_{37}H_{51}NO_{10}$, derived from tax(o)- and -in(e)

tax(o)-
derived from the genus name *Taxus* (yews), from *taxus* (Latin: yew), probably of Greek origin

taxodione
$C_{20}H_{26}O_3$, derived from the genus name *Taxodium* (deciduous trees), from tax(o)- and -odium, from *-odes* (Greek: a thing that resembles), di-, and -one

taxol
$C_{47}H_{51}NO_{14}$, derived from tax(o)- and -ol

-taxy
derived from *taxis* (Greek: arrangement), from *tassein* (Greek: to arrange, to order)

tazettine
$C_{18}H_{21}NO_5$, derived from the specific epithet of the species name *Narcissus tazetta* (polyanthus narcissus), from *tazetta*, diminutive of *tazza* (Italian: cup), and -in(e)

TBP
$C_{12}H_{27}O_4P$, an abbreviation for tributyl phosphate

TCDD
$C_{12}H_4Cl_4O_2$, an abbreviation for 2,3,6,7-tetrachlorodibenzo[*b*,*e*][1,4]dioxin

TCDF
$C_{12}H_4Cl_4O$, an abbreviation for tetrachlorodibenzofuran

TCNE
C_6N_4, an abbreviation for tetracyanoethylene

TCNQ
$C_{12}H_4N_4$, an abbreviation for tetracyanoquinodimethane

TDA
an abbreviation for thermodilatometry

[1]TDI
$C_9H_6N_2O_2$, an abbreviation for toluene 2,4-diisocyanate

[2]TDI
an abbreviation for tolerated daily intake

TDP (dTDP)
$C_{10}H_{16}N_2O_{11}P_2$, an abbreviation for thymidine 5'-diphosphate

TEA
$C_6H_{15}NO_3$, an abbreviation for triethanolamine

Tebbe olefination
named for the 20th century US chemist Fred N. Tebbe

Tebbe reagent
$C_{13}H_{18}AlClTi$, named for the 20th century US chemist Fred N. Tebbe

technetium
Tc, derived from *technetos* (Greek: artificial), from *techne* (Greek: art), and -ium – referring to the fact that this element does not occur in nature

technetocene
$C_{10}H_{10}Tc$, derived from technetium and -ocene

Teclu burner
named for the Romanian chemist Nicolae

Teclu (1839-1916)

tecomanine
$C_{11}H_{17}NO$, derived from the genus name *Tecoma* (tropical American shrubs, trees), ultimately from *tecomaxochitl* (Nahuatl: *Tecoma*), from *tecomatl* (Nahuatl: clay pot) and *xochitl* (Nahuatl: flower), and -in(e)

tect(o)-
derived from *tektos* (Greek: melted), from *tekein* (Greek: to melt)

tectorigenin
$C_{16}H_{12}O_6$, derived from the specific epithet of the species name *Iris tectorum* Maxim. (wall iris), from *tectum* (Latin: roof, house, building), -gen, and -in(e)

tectosilicate
derived from tect(o)- and silicate

TEELS
an abbreviation for transmission electron energy loss spectroscopy

teflon
coined by contraction of the name poly[tetrafluoroethene]

Teichmann crystal
named for the Polish physician L. C. Teichmann-Stawiarski (1823-1895)

teicho-
derived from *teichos* (Greek: wall)

teichoic acid
derived from teicho- – referring to the occurrence of these acids in the cell walls and membranes of bacteria

teicoplanin
derived by contraction of the bacterial species name *Actinoplanes teichomyceticus*, from actin(o)-, *planes* (Greek: wanderer), teicho-, and -in(e)

teinochemistry
derived from *teinein* (Greek: to stretch)

TEL
$C_8H_{20}Pb$, an abbreviation for tetraethyllead

tele-
derived from *tele* (Greek: far, far away)

telechele
derived from tele- and chel(e)-

teleocidin
$C_{28}H_{41}N_3O_2$, derived from the subclass name *Teleostei* (fish), from tele- and *osteon* (Greek: bone), -cide, and -in(e) – referring to this compound's toxicity to teleosts

tellane
TeH_2, derived (with contraction) from tellur(o)- and -an(e)

telluric acid (orthotelluric acid)
H_6TeO_6, derived from tellurium

tellurium
Te, derived from *tellus* (Latin: earth) and -ium

tellur(o)-
derived from tellurium

tellurophen
C_4H_4Te, derived from tellur(o)- and phene

tellurous acid
H_2TeO_3, derived from tellurium

tel(o)-
derived from *telos* (Greek: end, consummation, completeness)

telomer
derived from tel(o)- and –mer

telomerase
derived from telomere and -ase

telomere
derived from tel(o)- and -mer(e)

telomycin
$C_{59}H_{77}N_{13}O_{19}$, derived from tel(o)- and -mycin

TEM
an abbreviation for transmission electron microscopy

TEMPO
$C_9H_{18}NO$, an abbreviation for 2,2,6,6-tetramethylpyridin-1-yloxy

tennantite
$(Cu,Fe)_{12}As_4S_{13}$, named for the British chemist Smithson Tennant (1761-1815)

tenorite (melaconite, melachalcite)
CuO, named for the Italian botanist Michele Tenore (1780-1861)

tenuazonic acid
$C_{10}H_{15}NO_3$, derived from the specific epithet of the fungal species name *Alternaria tenuis* Auct., from *tenuis* (Latin: thin, slight, tenuous), az(o)-, and -one

tephro-
derived from *tephros* (Greek: ash gray), from *tephra* (Greek: ashes)

tephroite
Mn_2SiO_4, derived from tephro- and -ite – referring to this mineral's gray color

tephrosin
$C_{23}H_{22}O_7$, derived from the genus name *Tephrosia* (herbs, undershrubs), from tephro-, and -in(e)

TEPP
$C_8H_{20}O_7P_2$, an abbreviation for tetraethyl pyrophosphate

ter-
derived from *ter* (Latin: three times)

tera-
derived from *teras* (Greek: monster)

terbia
Tb_2O_3, derived from terbium and -a

terbium
Tb, derived (with contraction) from Ytterby, Sweden, where the first terbium ores were found, and -ium

terebene
a mixture of terpenes, derived from terebinth, from terebinth tree (*Pistacia terebinthus*), ultimately from *terebinthos* (Greek: terebinth tree)

terebic acid
$C_7H_{10}O_4$, derived from terebene

terephthalic acid
$C_8H_6O_4$, derived from terebene and phthalic acid

termone
coined by contraction of determining hormone

terpene
$(C_5H_8)_2$, derived from *Terpentin* (German: turpentine), ultimately from *terebinthos* (Greek: terebinth tree)

terpenoid
derived from terpene and -oid

terpenylic acid
$C_8H_{12}O_4$, derived from terpene

terpin (terpinol)
$C_{10}H_{20}O_2$, derived from terpene and -in(e)

terpinene
$C_{10}H_{16}$, derived from terpin and -ene

terpineol
$C_{10}H_{18}O$, derived from terpin and -ol

terreic acid
$C_7H_6O_4$, derived from the specific epithet of the fungal species name *Aspergillus terreus*, from *terreus* (Latin: of the earth), from *terra* (Latin: earth) – this compound is not a carboxylic acid

tertiary
derived from *tertius* (Latin: third)

tertiomycin
derived from tertiary and -mycin

terylene
coined by contraction of the name poly[ethylene terephthalate]

TES
$C_6H_{15}NO_6S$, an abbreviation for *N*-[tris(hydroxymethyl)methyl]-2-amino-ethane-sulfonic acid

testosterone
$C_{19}H_{28}O_2$, derived from *testis* (Latin: testicle), steroid, and -one

TETD
$C_{10}H_{20}N_2S_4$, an abbreviation for *N,N,N',N'*-tetraethylthiuram disulfide

tetra-
derived from *tetra-* (Greek: four-), from *tettares, tessares* (Greek: four)

tetracarbane
C_4H_{10}, systematic, but discouraged name for butane, derived from tetra-, carbon, and -an(e)

tetraconta-
derived from *tetrakonta* (Greek: forty)

tetracycline
$C_{22}H_{24}N_2O_8$, derived from tetra-, cycl(o)-, and -in(e)

tetradeca-
derived from *tessareskaideka* (Greek: fourteen)

tetradymite
Bi_2Te_2S, derived from *tetradymos* (Greek: fourfold) and -ite – referring to the twinning of this mineral's crystals in groups of four

tetraglyme
$C_{10}H_{22}O_5$, derived from tetra- and glyme

tetrahedrane
C_4H_4, derived from tetrahedron and -an(e) – referring to this hydrocarbon's tetrahedral molecular shape

tetrahedrite
$(Cu,Fe)_{12}Sb_4S_{13}$, derived from tetrahedron and -ite – referring to this mineral's crystal form

tetrahedro-
derived from tetrahedron

tetrahedron
derived from tetra-, *hedra* (Greek: seat, base), and [1]-on

tetrakis-
derived from *tetrakis* (Greek: four times)

tetralin
$C_{10}H_{12}$, coined by contraction of *1,2,3,4-Tetrahydronaphthalin* (German: tetrahydronaphthalene)

tetralol
$C_{10}H_{12}O$, coined by contraction of 1,2,3,4-tetrahydronaphthalen-2-ol

tetrandrine
$C_{38}H_{42}N_2O_6$, derived from the specific epithet of the species name *Stephania tetrandra* S. Moore (the Chinese medicinal herb *han fang ji*), from *tetrandrus* (New Latin: having four stamens), from tetra- and *aner* (Greek: man = male person, stamen), and -in(e)

tetranectin
derived from tetra-, *nectere* (Latin: to tie), and -in(e)

tetraose
tetrasaccharides, derived from tetra- and -ose

tetraprismane (cubane, quadriprismane)
C_8H_8, derived from tetra-, prism, and -an(e)

tetraterpene
$(C_5H_8)_8$, derived from tetra- and terpene

tetrel
C, Si, Ge, Sn, Pb, derived from tetra- and element – referring to the fourth main group of the periodic system

[1]**tetrin**
derived from tetraene and -in(e)

[2]**tetrin**
derived from tetrapeptide and -in(e)

tetritol
$C_4H_{10}O_4$, derived from tetrose and -itol

tetrodotoxin (tetraodontoxin)
$C_{11}H_{17}N_3O_8$, derived from the family name *Tetraodontidae* (fish), from tetra-, *odontos* (Greek: toothed), from *odon* (Greek: tooth), and toxin

tetronasin
$C_{35}H_{54}O_8$, derived from tetronic acid, -ase, and -in(e)

tetroquinone
$C_6H_4O_6$, coined by contraction of 2,3,5,6-tetrahydroxy-*p*-quinone

tetrose
$C_4H_8O_4$, derived from tetra- and -ose

tetrulose (ketotetrose)
$C_4H_8O_4$, derived from tetra- and -ulose

tetryl (nitramine)
$C_7H_5N_5O_8$, derived from tetra- and -yl – referring to the presence of four nitro groups in this explosive

Teuber reaction
named for the German chemist Hans-Joachim Teuber (born 1918)

Texaco process
named for the energy and petrochemical company Texaco, White Plains, NY, USA

TFA
$C_2HF_3O_2$, an abbreviation for trifluoroacetic acid

TG
an abbreviation for thermogravimetry

TGA
an abbreviation for thermogravimetric analysis

TGS
$C_6H_{17}N_3O_{10}S$, an abbreviation for triglycine sulfate

thalenite
$Y_3Si_3O_{10}(OH)$, named for the Swedish physicist Tobias Robert Thalén (1827-1905)

thalicarpine
$C_{41}H_{48}N_2O_8$, coined by contraction of the species name *Thalictrum dasycarpum* Fisch. & Ave-Lall. (purple meadow rue), from *thaliktron* (Greek: meadow rue), *dasys* (Greek: thick with hair or leaves), and carp(o)-, and toxin

thalidomide
$C_{13}H_{19}N_2O_4$, coined by contraction of 3-(phthalimido)glutarimide

thallane
TlH_3, derived from thallium and -an(e)

thallium
Tl, derived from *thallos* (Greek: young shoot, green stalk), from *thallein* (Greek: to sprout), and -ium – referring to a bright green line in this element's spectrum

THAM (tris, trometamol)
$C_4H_{11}NO_3$, an abbreviation for tris(hydroxymethyl)aminomethane

thaumatin
derived from the species name *Thaumatococcus danielli* Benth. (West African miracle berry), from *thauma* (Greek: miracle, wonder) and *kokkos* (Greek: grain, kernel, kermes berry), and -in(e)

THC
$C_{21}H_{30}O_2$, an abbreviation for tetrahydrocannabinol

the(a)-
derived from the genus name *Thea* (tea plants), from *thea* (New Latin: tea plant)

theaflavine
$C_{29}H_{24}O_{12}$, derived from the(a)- and flavine

thearubigin
derived from the(a)-, rubi-, and -in(e) – referring to these phenolic pigments' orange-brown color

thebaine
$C_{19}H_{21}NO_3$, derived from *thebaia* (New Latin: opium produced in the ancient town of Thebes, Egypt) and -in(e)

theabinone
$C_{18}H_{21}NO_3$, derived from thebaine and -one

-thecium
derived from *thekion*, diminutive of *theke* (Greek: case, chest)

THEED
an abbreviation for transmission high energy electron diffraction

theine (caffeine)
$C_8H_{10}N_4O_2$, derived from the(a)- and -in(e)

Thenard blue
$CoAl_2O_4$, named for the French chemist Louis Jacques Thenard (1777-1857)

thenardite
Na_2SO_4, named for the French chemist Louis Jacques Thenard (1777-1857)

thenoic acid
$C_5H_4O_2S$, ultimately derived from thiophene; patterned after benzoic acid

thenoyl
C_5H_3OS-, derived from thenoic acid and -yl; patterned after benzoyl

thenyl
C_5H_5S-, derived from 2-thienylmethyl; patterned after benzyl

theobromine
$C_7H_8N_4O_2$, derived from the genus name *Theobroma* (cocoa trees), from *theos* (Greek: god) and *broma* (Greek: food), and

-in(e)

theophylline
$C_7H_8N_4O_2$, derived from the(a)-, phyll(o)-, and -in(e)

therm(o)-
derived from *thermos* (Greek: hot)

thermolysin
coined by contraction of the specific epithet of the bacterial species name *Bacillus thermoproteolyticus*, from therm(o)- and lysis, and -in(e)

thermonatrite
$Na_2CO_3{\cdot}H_2O$, derived from therm(o)-, natrium, and -ite – referring to this mineral's formation by thermal drying of natrite

thermorubin
derived from the bacterial genus name *Thermoactinomyces,* from therm(o)-, rubi-, and -in(e)

thetin(e)
coined by contraction of thi(o)- and betain(e)

thevetin
derived from the genus name *Thevetia* (tropical American trees, shrubs), after the French traveler and author Frère André Thévet (1516-1592), and -in(e)

thexyl
$C_6H_{13}-$, coined by contraction of *tert*-hexyl

THF
C_4H_8O, an abbreviation for tetrahydrofuran

thi(a)-
derived from *theion* (Greek: sulfur) of uncertain etymology

thiamine (aneurin)
$C_{12}H_{17}ClN_4OS$, derived from thi(o)- and amine

thianaphthene
C_8H_6S, coined by contraction of thi(a)- and naphthalene

thiarubrine
derived from thi(a)-, rubri-, and *-in* (German: -yne)

thiazole
C_3H_3NS, derived from thi(a)-, az(a)-, and -ol(e)

Thiele reaction
named for the German chemist Johannes Thiele (1865-1918)

Thiele theory
named for the German chemist Johannes Thiele (1865-1918)

Thiele-Winter reaction
named for the German chemist Johannes Thiele (1865-1918) and thc Estonian chemist E. A. Winter (1867-1921)

thienamycin
$C_{11}H_{16}N_2O_4S$, derived from thi(a)-, enamine, and -mycin

thien(o)-
derived from thiophene; patterned after benz(o)-

thienyl
C_4H_3S-, derived from thiophene; patterned after phenyl

thi(o)-
derived from *theion* (Greek: sulfur), of uncertain etymology

thioacid
derived from thi(o)- and acid

thioaldehyde
RCHS, derived from thi(o)- and aldehyde

thiochrome
$C_{12}H_{14}N_4OS$, coined by contraction of thiamine and -chrome

thioctic acid (α-lipoic acid)
$C_8H_{14}O_2S_2$, coined by contraction of 6,8-dithiooctanoic acid

thioester
derived from thi(o)- and ester

thioether
RSR, an obsolete name for sulfide, derived from thi(o)- and [1]ether

thioflavine
derived from thi(o)- and flavine

thioindigo
$C_{16}H_8O_2S_2$, derived from thi(o)- and indigo – referring to the replacement of indigo's two imino groups with sulfur

thioketone (thione)
$R_2C{=}S$, derived from thi(o)- and ketone

thiol
RSH, derived from thi(o)- and -ol

thiolutin
$C_8H_8N_2O_2S_2$, derived from thi(o)-, lute(o)-, and -in(e) – referring to this compound's yellow color

thionalide
$C_{12}H_{11}NOS$, coined by contraction of the obsolete name thioglycolic β-amino-naphthalide

thione (thioketone)
$R_2C{=}S$, derived from thi(o)- and -one

thioneine (ergothioneine)
$C_9H_{15}N_3O_2S$, coined by contraction of ergothioneine

thionine
$C_{12}H_{10}ClN_3S$, derived from thi(o)- and -in(e)

thionyl
$-S(O)-$, derived from thi(o)- and -yl

thiopeptin
derived from thi(o)-, peptide, and -in(e)

thiophene
C_4H_4S, derived from thi(o)- and phene

thiophenine
$C_4H_5N_S$, coined by contraction of thio-phenamine; patterned after aniline

thiophenol
C_6H_6S, derived from thi(o)- and phenol

thiophosgene
CCl_2S, derived from thi(o)- and phosgene

thioredoxin
derived from thi(o)-, redox, and -in(e)

thiosemicarbazide
CH_5N_3S, derived from thi(o)- and semicarbazide

thiosinamine
$C_4H_8N_2S$, derived from thi(o)-, sin(a)-, and amine – this compound is not an amine

thiostrepton
$C_{72}H_{85}N_{19}O_{18}S_5$, derived from thi(o)-, [3]strept(o)-, and -on(e)

thiourea
CH_4N_2S, derived from thi(o)- and urea

thioxanthene
$C_{13}H_{10}S$, derived from thi(o)- and xanthene

thioxanthone
$C_{13}H_8OS$, derived from thioxanthene and -one

THIP
$C_6H_8N_2O_2$, an abbreviation for 4,5,6,7-tetrahydroisoxazolo[5,4-*c*]pyridin-3(2*H*)-one

thiuram
probably coined by contraction of thi(o)-, urea, and amide

thixotropy
derived from *thixis* (Greek: action of touching), from *thinganein* (Greek: to touch), and -tropy

Thomas steel
named for the British metallurgist and inventor Sidney Gilchrist Thomas (1850-1885)

Thomsen-Berthelot principle
named for the Danish chemist Hans Peter Jørgen Julius Thomsen (1826-1909) and the French chemist Marcellin Pierre Eugène Berthelot (1827-1907)

thomsenolite
$NaCaAlF_6 \cdot H_2O$, named for the Danish chemist Hans Peter Jørgen Julius Thomsen (1826-1909)

thomsonite
$NaCa_2(Al_5Si_5)O_{20} \cdot 6H_2O$, named for the Scottish chemist Thomas Thomson (1773-1852)

thoreaulite
$SnTa_2O_6$, named for the Belgian mineralogist Jacques Thoreau (1886-1973)

thorex process
an abbreviation for thorium extraction process

thoria
ThO_2, derived from thorium and -a

thorianite
ThO_2, derived from thoria and -ite

thorite
$(Th,U)SiO_4$, named for Thor, the Old Norse god of thunder

thorium
Th, derived from thorite and -ium

thoron
Rn, unsuccessfully suggested name for radon, derived from thorium and [1]-on

Thorpe reaction
named for the British chemist Sir Jocelyn Field Thorpe (1872-1940)

thortveitite
$(Sc,Y)_2Si_2O_7$, named for the Norwegian engineer Olaus Thortveit (1872-1917)

Thoulet's solution
K_2HgI_4, named for the French mineralogist Julien Thoulet (1843-1936)

threaric acid (tartaric acid)
$C_4H_6O_6$, derived from threose and -aric acid

threitol
$C_4H_{10}O_4$, derived from threose and -itol

threo-
derived from threose

threonic acid
$C_4H_8O_5$, derived from threose

threonine
$C_4H_9NO_3$, derived from threonic acid and -in(e)

threose
$C_4H_8O_4$, anagrammatically derived from erythrose

thrix(o)-
derived from *thrix* (Greek: hair)

thrombin
derived from thromb(o)- and -in(e)

thromb(o)-
derived from *thrombos* (Greek: lump, blood clot)

thromboxane (TX)
$C_{20}H_{40}O$, derived from thromb(o)-, ox(a)- and -an(e)

thuj(a)-
derived from the genus name *Thuja* (thuja), ultimately from *thyia* (Greek: a kind of cedar)

thujane
$C_{10}H_{18}$, derived from thuj(a)- and -an(e)

thujaplicin
$C_{10}H_{12}O_2$, derived from the species name *Thuja plicata* (red cedar), from thuj(a)- and *plicatus* (Latin: plaited, folded), from *plicare* (Latin: to fold, to pleat), and -in(e)

thujic acid
$C_{10}H_{12}O_2$, derived from thuj(a)-

thujone
$C_{10}H_{16}O$, derived from thuj(a)- and -one

thujopsene (widdrene)
$C_{15}H_{24}$, derived from the genus name *Thujopsis* (Japanese evergreen trees), from thuj(a)- and -opsis, and -ene

thulia
Tm_2O_3, derived from thulium and -a

thulium
Tm, named for Scandinavia, derived from *(Ultima) Thule*, the Latin name of the northernmost part of the Earth known in Antiquity, for most purposes equivalent to Scandinavia

thuringite
$(Fe^{II},Fe^{III},Mg,Al)_6(Si,Al)_4O_{10}(O,OH)_8$, named after this mineral's locality Reichmannsdorf, Thuringia, Germany

thymidine (2'-deoxythymidine)
$C_{10}H_{14}N_2O_5$, derived from [1]thym(o)- and -idin(e)

thymine
$C_5H_6N_2O_2$, derived from [1]thym(o)- and -in(e)

thyminose (2-deoxyribose)
$C_5H_{10}O_4$, derived from thymidine and -ose

[1]thym(o)-
ultimately derived from thymus gland, from *thymos* (Greek: thymus gland)

[2]thym(o)-
ultimately derived from *thymon* (Greek: thyme)

thymol
$C_{10}H_{14}O$, derived from [2]thym(o)- and -ol

thymol blue
$C_{27}H_{30}O_5S$, derived from thymol

thymolphthalein
$C_{28}H_{30}O_4$, derived from thymol and -phthalein; patterned after phenolphthalein

thymotic acid (thymotinic acid)
$C_{11}H_{14}O_3$, derived from [2]thym(o)-

thyr(eo)-
derived from thyroid gland, from *thyreoeides* (Greek: shaped like a shield), ultimately from *thyra* (Greek: door)

thyronine
$C_{15}H_{15}NO_4$, derived from thyr(eo)-, -one, and -in(e)

thyrotrophin (thyrotropin)
derived from thyr(eo)- and -trophin

thyrotropin (thyrotrophin)
derived from thyr(eo)- and -tropin

thyroxine (T_4)
$C_{15}H_{11}I_4NO_4$, derived from thyr(eo)-, ox(o)-, and -in(e)

TIC
an abbreviation for total inorganic carbon

Tiemann rearrangement
named for the German chemist Johann Carl Wilhelm Ferdinand Tiemann (1848-1899)

Tiffeneau-Demyanov rearrangement
named for the French chemist Marc Emile Pierre Adolphe Tiffeneau (1873-1945) and the Russian chemist Nikolai Jakovlevich Demyanov (1861-1938)

tiger's eye
SiO_2, named after its appearance

tiglic acid
$C_5H_8O_2$, derived from the specific epithet of the species name *Croton tiglium* (purging croton), perhaps from *tilos* (Greek: diarrhea)

tigloidine
$C_{13}H_{21}NO_2$, derived from tiglic acid and -idin(e)

tigogenin
$C_{27}H_{44}O_3$, derived from tigonin, -gen, and -in(e)

tigonin
$C_{56}H_{92}O_{27}$, derived (by anagrammatical contraction) from the genus name *Digitalis* (foxgloves), from *digitalis* (Latin: belonging to a finger), from *digitus* (Latin; finger, toe), -one, and -in(e)

tiliacorine
$C_{36}H_{36}N_2O_5$, derived from the genus name *Tiliacora* (trees), from *tiliakoru* (Bengal: tiliacora), and -in(e)

Tillmans reagent
$C_{12}H_7Cl_2NO_2$, named for the German chemist Josef Tillmans (1876-1935)

tin
Sn, an indigenous English word

tincal (borax)
$Na_2B_4O_7 \cdot 10H_2O$, derived from *tingkal* (Malay: borax)

tincalconite
$Na_2B_4O_5(OH)_4 \cdot 3H_2O$, derived from tincal, *konia* (Greek: ashes, dust, powder), and -ite – referring to this mineral's composition and powdery appearance

tincture
derived from *tinctura* (Latin: a dyeing), from *tingere* (Latin: to dye)

tiron
$C_6H_4Na_2O_8S_2$, coined by contraction of titanium and iron – referring to the usefulness of this agent in the determination of metal ions

Tishchenko reaction
named for the Russian chemist Vyacheslav Evgenievich Tishchenko (1861-1941)

titania
TiO_2, derived from titanium and -a

titanite (sphene)
$CaTiO(SiO_4)$, derived from titanium and -ite

titanium
Ti, named for the Titans, Greek mythical giants

titanocene
$C_{10}H_{10}Ti$, derived from titanium and -ocene

titin
derived from the Titans, Greek mythical giants, and -in(e) – referring to the 'titanic' size of these proteins

titration
derived from *titre* (French: fineness of, for instance, gold)

TLC
an abbreviation for thin-layer chromatography

TMD
$C_{13}H_{24}O$, an abbreviation for *trans*-1,1,10-trimethyl-2-decalol

TMP (dTMP)
$C_{10}H_{15}N_2O_8P$, an abbreviation for thymidine 5'-monophosphate

TMS
$C_4H_{12}Si$, an abbreviation for tetramethylsilane

TMTD
$C_6H_{12}N_2S_4$, an abbreviation for *N,N,N',N'*-tetramethylthiuram disulfide

TMU
$C_5H_{12}N_2O$, an abbreviation for 1,1,3,3-tetramethylurea

TNAZ
$C_3H_4N_4O_6$, coined by contraction of 1,3,3-trinitroazetidine

TNF
an abbreviation for tumor necrosis factor

TNT
$C_7H_5N_3O_6$, an abbreviation for 2,4,6-trinitrotoluene

TNX
$C_8H_7N3O_6$, an abbreviation for 2,4,6-trinitro-*m*-xylene

TOAC
$C_{10}H_{19}N_2O_3$, coined by contraction of the (unsystematic) name 2,2,6,6-tetramethylpiperidine-*N*-oxyl-4-amino-4-carboxylic acid

tobermorite
$Ca_5Si_6O_{16}(OH)_2{\cdot}4H_2O$, named after this mineral's locality Tobermory, Scotland, UK

Tobias acid
$C_{10}H_9NO_3S$, named for the 19th century German chemist Georg Tobias

tobramycin
$C_{18}H_{37}N_5O_9$, coined by variation of nebramycin

TOC
an abbreviation for total organic carbon

tocol
$C_{26}H_{44}O_2$, coined by contraction of tocopherol

tocopherol
derived from *tokos* (Greek: childbirth), *pherein* (Greek: to bear), and -ol

TOD
an abbreviation for total oxygen demand

TOF-MS
an abbreviation for time-of-flight mass spectroscopy

tolan
$C_{14}H_{10}$, derived from tol(u)- and -an(e)

tolidine
$C_{14}H_{16}N_2$, derived from tol(u)- and -idin(e); patterned after benzidine

Tollens reagent
$[Ag(NH_3)_2]OH$, named for the German chemist Bernhard Christian Gottfried Tollens (1841-1918)

tol(u)-
derived from tolu balsam, named after the town of Santiago de Tolú, Colombia

toluene
C_7H_8, derived from tol(u)- and -ene

toluic acid
$C_8H_8O_2$, derived from tol(u)-

toluidine
C_7H_9N, derived from tol(u)- and -idin(e)

toluoyl (toluyl)
C_8H_7O-, derived from toluic acid and -oyl; patterned after benzoyl

tolyl (toluyl)
C_7H_7-, derived from toluene and -yl

tomatidine
$C_{27}H_{45}NO_2$, coined by variation of tomatine

tomatine
$C_{50}H_{83}NO_{21}$, derived from tomato plant (*Lycopersicon esculentum*), from *tomatl* (Nahuatl: tomato plant), and -in(e)

tombac
(Cu,Zn), ultimately derived from *tembaga* (Malay: copper)

Toms effect
named for the 20th century British physicist B. A. Toms

tonin
ultimately derived from *tonos* (Greek: tension), form *teinein* (Greek: to stretch), and -in(e)

-tonin
ultimately derived from *tonos* (Greek: tension), from *teinein* (Greek: to stretch), and -in(e)

topaz
$Al_2SiO_4(F,OH)_2$, named for Antiquity's mythical Topazos Island in the Red Sea, possibly from *topazein* (Greek: to conjecture)

-topic
derived from *topos* (Greek: place)

topicity (prochirality)
derived from -topic

topo-
derived from *topos* (Greek: place)

topochemistry
derived from topo- – referring to chemistry taking place in the solid state

topoisomerase
derived from topo- and isomerase

topomerism
derived from topo- and isomerism

topotactic
derived from topo- and -tactic

torbernite
$Cu(UO_2)_2(PO_4)_2 \cdot 10H_2O$, named for the

Swedish chemist Torbern Bergman (1735-1784)

torularhodin
$C_{40}H_{52}O_2$, derived from the fungal genus name *Torula*, from *torulus*, diminutive of *torus* (Latin: protuberance, bulge, cushion, couch, torus molding), rhod(o)-, and -in(e)

tosyl
$C_7H_7O_2S-$, coined by contraction of *p*-toluenesulfonyl

tourmaline
$(Ca,Na)(Al,Li,Mg)(Al,Fe,Mn)_6(BO_3)_3(Si_6O_{18})(OH)_4$, ultimately derived from *toramalli* (Singhalese: cornelian), and -in(e)

toxaphene
coined by contraction of tox(o)- and polychlorocamphene

toxiferine
derived from the specific epithet of the species name *Strychnos toxifera* (curare), from tox(o)- and -fer, and -in(e)

toxin
derived from tox(o)- and -in(e)

toxisterol
derived from tox(o)- and sterol

tox(o)-
ultimately derived from *toxikon pharmakon* (Greek: arrow poison), from *toxon* (Greek: bow) and *pharmakon* (Greek: drug)

toxoflavin
$C_7H_7N_5O_2$, derived from tox(o)- and flavin

toxoid
derived from tox(o)- and -oid – referring to the toxophoric group of the original toxin having been removed

toxopyrimidine
$C_6H_9N_3O$, derived from tox(o)- and pyrimidine

toyocamycin
$C_{12}H_{13}N_5O_4$, derived from the specific epithet of the bacterial species name *Streptomyces toyocaensis* and -mycin

tPA
an abbreviation for tissue plasminogen activator

TPMPA
$C_6H_{12}NO_2P$, an abbreviation for (1,2,3,6-tetrahydropyridin-4-yl)phosphinic acid

TPN(NADPH, NADPH$_2$)
$C_{21}H_{30}N_7O_{17}P_3$, an abbreviation for triphosphopyridine nucleotide

TPP
$C_{12}H_{19}ClN_4O_7P_2S$, an abbreviation for thiamine pyrophosphate

trachy-
derived from *trachys* (Greek: rough)

trachylobane
$C_{20}H_{32}$, derived from the specific epithet of the species *Acalypha trachyloba* (a tropical herb), from trachy- and *lobos* (Greek: lobe), and -an(e)

trachyte
a rock species, derived from trachy- and -ite – referring to this material's coarse structure

tragacanth
derived from *tragakanthos* (Greek: tragacanth), from *tragos* (Greek: he-goat) and acanth(o)-

Tramex process
coined by contraction of transition metal extraction process

trans-
derived from *trans* (Latin: on the other side)

transducin (transductase)
derived from trans-, *ducere* (Latin: to lead), and -in(e)

transferrin
derived from trans-, ferr(o)-, and -in(e)

transmethylase (methylase)
coined anagrammatically from methyl-transferase

transmutation
derived from *transmutatio* (Latin: transmutation), from *transmutare* (Latin: to change, to interchange), from trans- and *mutare* (Latin: to change)

transoid
derived from trans- and -oid

Traube cell
named for the German chemist M. Traube (1826-1894)

Traube purine synthesis
named for the German chemist Wilhelm Traube (1866-1942)

Traube's rule
named for the German chemist I. Traube (1860-1943)

traumatic acid
$C_{12}H_{20}O_4$, derived from *trauma* (Greek: wound), akin to *tryein* (Greek: to wear out, to distress)

travertine
a rock species, ultimately derived from *lapis tiburtinus* (Latin: stone from Tibur, Latium)

trehalase
derived from trehalose and -ase

trehalose
$C_{12}H_{22}O_{11}$, derived from trehala (edible pupal covering of the beetle *Larinus maculatus*), ultimately from *tighal* (Persian: trehala), and -ose

trehalostatin (trehazolin)
$C_{13}H_{22}N_2O_{10}$, derived from the specific epithet of the fungal species *Amycolatopsis trehalostatica*, from trehalase and -staticum – referring to this antibiotic's trehalase-inhibiting properties

tremetone
$C_{13}H_{14}O_2$, derived from trembles (a cattle disease caused by tremetone poisoning), -ete, and -one

tremolite (grammatite)
$Ca_2(Mg,Fe)_5Si_8O_{22}(OH)_2$, named after this mineral's mock locality Tremola Valley, Switzerland

tremorine
$C_{12}H_{20}N_2$, derived from tremor and -in(e) – referring to the tremorigenic action of this compound

tretinoin (retinoic acid)
$C_{20}H_{28}O_2$, derived (with contraction) from *all-trans*-retinoic acid, and -in(e)

trevorite
$NiFe^{III}{}_2O_4$, named for the 20th century South African mining inspector Major Tudor Gruffydd Trevor

TRH
$C_{16}H_{22}N_6O_4$, an abbreviation for thyro-tropin-releasing hormone

tri-
derived from *treis* (Greek: three)

triacetin
$C_9H_{14}O_6$, coined by contraction of

Triacetylglycerin (German: triacetyl-glycerol)

triaconta-
derived from *triakonta* (Greek: thirty)

triangulo-
derived from *triangulum* (Latin: triangle)

triaose
trisaccharides, derived from tri-, the arbitrary insert -a-, and -ose – so coined to avoid confusion with triose

tribo-
derived from *tribein* (Greek: to rub)

tribology
derived from tribo- and -logy

triboluminescence
derived from tribo- and luminescence

tributyrin
$C_{15}H_{26}O_6$, coined by contraction of *Tributyrylglycerin* (German: tributyryl glycerol)

tricarballylic acid
$C_6H_8O_6$, derived (with contraction) from tri-, carboxyl, and allyl

tricarbane
C_3H_8, systematic, but discouraged name for propane, derived from tri-, carbon, and -an(e)

trichione
$C_{17}H_{14}O_8$, derived from the genus name *Trichia* (myxomycetes), from trich(o)-, and -one

trich(o)-
derived from *thrix* (Greek: hair)

tricholom-
derived from the genus name *Tricholoma* (mushrooms), from trich(o)- and *loma* (Greek: hem, fringe)

tricholomenyn
derived from tricholom-, -ene, and -yne

tricholomic acid
$C_5H_8N_2O_4$, derived from tricholom-

trichomycin (hachimycin)
derived from the bacterial genus name *Trichomonas*, from trich(o)- and -monas, the fungal genus name *Trichophyton*, from trich(o)- and phyt(o)-, and -mycin – referring to this antibiotic's antiprotozoal and antifungal activity

trichosanthin
derived from the genus name *Trichosanthes* (Asiatic and Australian herbs), from trich(o)- and anth(o)-, and -in(e)

trichostatin
derived from trich(o)- and -statin

trichothecane
$C_{15}H_{26}O$, derived from trichothecine and -an(e)

trichothecine
$C_{19}H_{24}O_5$, derived from the fungal genus name *Trichothecium*, from trich(o)- and -thecium, and -in(e)

tricine
$C_6H_{13}NO_5$, coined by contraction of *N*-[tris(hydroxymethyl)methyl]glycine

triclinic
derived from tri- and -clinic

tricyclene
$C_{10}H_{16}$, derived from tricyclic and -ene

trideca-
derived from *triskaideka* (Greek: thirteen)

tridymite
SiO_2, derived from *tridymos* (Greek: threefold) and -ite – referring to this mineral's common twinning as trillings

triel
B, Al, Ga, In, Tl, coined by contraction of tri- and element – referring to the third main group of the periodic system

trientine
$C_6H_{18}N_4$, coined by contraction of triethylenetetramine

triergolic
derived from tri- and -ergolic

triflic acid
CHF_3SO_3, coined by contraction of trifluoromethanesulfonic acid

triflinic acid
CHF_3SO_2, coined by contraction of trifluoromethanesulfinic acid

trigonal
derived from tri- and -gon

trigonelline
$C_7H_7NO_2$, derived from the genus name *Trigonella* (herbs), from *trigonellum*, diminutive of *trigonum* (Latin: triangle), from tri- and -gon, and -in(e)

trilobine
$C_{35}H_{34}N_2O_5$, derived from the specific epithet of the species name *Cocculus trilobus* DC. (the Chinese medicinal herb *mu fang ji*), from tri- and *lobos* (Greek: lobe), and -in(e)

trilon
an arbitrary code name for German WWII nerve gases, of undisclosed etymology

trimellitic acid
$C_9H_6O_6$, derived from tribasic and mellite

trimer
derived from tri- and -mer

trimesic acid
$C_9H_6O_6$, coined by contraction of tribasic and mesitylenic acid

trimyristin
$C_{45}H_{86}O_6$, derived by contraction of *Trimyristylglycerin* (German: trimyristylglycerol)

trinactin
$C_{43}H_{70}O_{12}$, coined by contraction of trihomononactin, from tri-, homo-, and nonactin

trinor-
derived from tri- and nor-

triolein (olein)
$C_{57}H_{104}O_6$, derived by contraction of *Trioleoylglycerin* (German: trioleoyl glycerol)

triose
$C_3H_6O_3$, derived from tri- and -ose

triostin
derived from the specific epithet of the bacterial species name *Streptomyces triostinicus*, from tri- and *osteon* (Greek: bone), and -in(e)

tripalmitin (palmitin)
$C_{51}H_{98}O_6$, coined by contraction of *Tripalmitoylglycerin* (German: tripalmitoylglycerol)

triphenylene
$C_{18}H_{12}$, derived from tri- and phenylene

triphylite
$LiFe^{II}PO_4$, derived from tri-, phyl(o)-, and -ite – referring to the erroneously assumed presence of three cations (Li^+, Fe^{2+}, Mg^{2+}) in this mineral

triplite
$Mn_2PO_4(F,OH)$, derived from *triploos* (Greek: triplex) and -ite – referring to the threefold cleavage of this mineral

triprenyl (farnesyl)
$C_{15}H_{25}$–, derived from tri- and prenyl

triprismo-
derived (with contraction) from triangular prism

triptycene
$C_{20}H_{14}$, derived from *triptychos* (Greek: threefold) and -ene – referring to this compound's molecular symmetry

tris (THAM, trometamol)
$C_4H_{11}NO_3$, coined by contraction of *N*,*N*,*N*-tris(hydroxymethyl)methanamine

tris-
derived from *tris* (Greek: thrice)

tristearin (stearin)
$C_{57}H_{110}O_6$, coined by contraction of *Tristearoylglycerin* (German: tristearoyl glycerol)

triterpene
$(C_5H_8)_6$, derived from tri- and terpene

tritide
$^3H^-$, derived from tritium and ²-ide

tritium
^3H, derived from *tritos* (Greek: the third) and -ium

triton
$^3H^+$, derived from tritium and ³-on

trituration
derived from *triturare* (Latin: to thresh, to grind), from *terere* (Latin: to rub)

trityl
$C_{19}H_{15}$–, coined by contraction of triphenylmethyl

triuret
$C_3H_6N_4O_3$, derived from tri-, urea, and -et(e)

tRNA
an abbreviation for transfer ribonucleic acid

TRNOE
an abbreviation for transferred nuclear Overhauser effect

Tröger's base
$C_{17}H_{18}N_2$, named for the German chemist Carl Julius Ludwig Tröger (1862-1942)

troilite
FeS, named for the Italian scientist Dominico Troili (1722-1792)

trometamol (THAM, tris)
$C_4H_{11}NO_3$, coined by anagrammatic contraction of tris(hydroxymethyl)amino-methane

Trommer's test
named for the German chemist K. A. Trommer (1806-1879)

Trommsdorf-Norrish effect
named for the German chemist E. Trommsdorf (1905-1996) and the British chemist Ronald George Wreyford Norrish (1897-1978)

Trommsdorf's reagent
named for the German bacteriologist R. Trommsdorf (1874-1944)

trona
$Na_3(HCO_3)(CO_3)\cdot 2H_2O$, derived from *tron* (Arabic: aphaeretic form of *natrun*), from *natrun* (Arabic: natron)

trop(a)-
derived from atropine

tropaeolin
named after the color of the genus *Tropaeolum* (tropical American herbs), from *tropaeum* (Latin: trophy), and -in(e)

tropane
$C_8H_{15}N$, derived from trop(a)- and -an(e)

-trope
derived from *tropos* (Greek: turn), from *trepein* (Greek: to turn)

-trophic
derived from troph(o)-

-trophin
derived from troph(o)- and -in(e)

troph(o)-
derived from *trephein* (Greek: to nourish)

-tropic
derived from *tropos* (Greek: turn), from *trepein* (Greek: to turn)

tropic acid
$C_9H_{10}O_3$, derived from trop(a)-

tropilidene
C_7H_8, derived from trop(a)- and -idene

-tropin
derived from tropo- and -in(e)

tropine
$C_8H_{15}NO$, derived from trop(a)- and -in(e)

tropo-
derived from *tropos* (Greek: turn), from *trepein* (Greek: to turn)

tropolone
$C_7H_6O_2$, derived from trop(a)- and [1]-olone

tropomerism (tautomerism)
derived from tropo- and isomerism

tropomyosin
derived from tropo-, my(o)-, and -in(e)

troponin
coined by contraction of tropomyosin

-tropy
derived from *tropos* (Greek: turn), from *trepein* (Greek: to turn)

tropyl
C_7H_7-, derived (with contraction) from tropilidene and -yl

tropylium
$C_7H_7^+$, derived from trop(a)-, -yl, and -ium

Trost allylation (Tsuji-Trost reaction)
named for the US chemist Barry Martin Trost (born 1941)

Trost desymmetrization
named for the US chemist Barry Martin Trost (born 1941)

trotyl (TNT)
$C_7H_5N_3O_6$, coined by contraction of 2,4,6-trinitrotoluene and (unsystematic) -yl

Trouton constant
named for the Irish physicist F. T. Trouton (1863-1922)

troxerutin
$C_{33}H_{42}O_{19}$, coined by contraction of tris(hydroxethyl)rutin

Truce-Smiles rearrangement
named for the 20th century US chemist William E. Truce and the British chemist Samuel Smiles (1877-1953)

truxillic acid
$C_{18}H_{16}O_4$, derived from truxilline

truxilline
$C_{38}H_{46}N_2O_8$, derived from Truxillo coca (*Erythroxylon truxillense*), named after the town of Trujillo, Peru, and -in(e)

TRXRF
an abbreviation for total reflection X-ray fluorescence spectroscopy

trypan blue
$C_{34}H_{24}N_6Na_4O_{14}S_4$, named to indicate this dye's trypanocidal properties

trypan red
$C_{32}H_{19}N_6Na_5O_{15}S_5$, named to indicate this dye's trypanocidal properties

trypsin
derived from *tryein* (Greek: to wear out, to digest), *pepsis* (Greek: digestion), and -in(e)

tryptamine
$C_{10}H_{12}N_2$, derived from trypt(o)- and amine

tryptazan
$C_{10}H_{11}N_3O_2$, derived from trypt(o)-, az(a)-, and -an(e)

trypt(o)-
derived from tryptophan

tryptophan
$C_{11}H_{12}N_2O_2$, derived from trypsin and -phan(e)

tryptophol
$C_{10}H_{11}NO$, derived (with contraction) from tryptophan and -ol

tsavorite
$Ca_3Al_2Si_3O_{12}$, named after this mineral's locality Tsavo, Kenya

tschermakite
$Ca_2(Mg_3AlFe^{III})Si_6Al_2O_{22}(OH)_2$, named for the Austrian mineralogist Gustav Tschermak von Sessenegg (1836-1927)

TSH
an abbreviation for thyroid-stimulating hormone

Tscherniac-Einhorn reaction
named for the German chemists J. Tscherniac and Alfred Einhorn (1857-1917)

tsuduranine (tuduranine)
$C_{18}H_{19}NO_3$, derived (with contraction) from *tudurafuzi* (Japanese: Chinese moonseed, *Sinomenium acutum* (Thunb.) Rehd. & Wils.) and -in(e)

Tsuji-Trost reaction (Trost allylation)
named for the 20th century Japanese chemist Jiro Tsuji and the US chemist Barry Martin Trost (born 1941)

T-2 toxin
$C_{24}H_{34}O_9$, an abbreviation of trichothecene mycotoxin

TTF
$C_6H_4S_4$, an abbreviation for 1,4,5,8-tetrathiafulvalene

TTP (dTTP)
$C_{10}H_{17}N_2O_{14}P_3$, an abbreviation for thymidine 5'-triphosphate

tuaminoheptane (tuamine)
$C_7H_{17}N$, derived from the unsystematic

name 2-aminoheptane

tuber-
derived from *tuber* (Latin: hump, knob, tumor, truffle, tuber)

tuberactinomycin
derived from the subspecific epithet of the bacterial species name *Streptomyces griseoverticillatus* var. *tuberacticus*, from *tuberacticus* (New Latin: antitubercular), and -mycin

tubercidin
$C_{11}H_{14}N_4O_4$, derived from the specific epithet of the bacterial species name *Streptomyces tubercidicus*, from *tubercidicus* (New Latin: antitubercular), and -in(e)

tuberin
$C_{10}H_{11}NO_2$, derived (with contraction) from antitubercular and -in(e)

tubocuraran
$C_{32}H_{30}N_2O_2$, derived from tubocurarine and -an(e)

tubocurarine
$C_{37}H_{41}N_2O_6^+$, derived from *tubus* (Latin: tube), curare, and -in(e) – referring to the earlier common shipment of curare in sections of hollow bamboo

tubulin
derived from *tubulus*, diminutive of *tubus* (Latin: tube), and -in(e) – referring to this protein's role in the function of microtubules of eukaryotic cells

tubulos-
derived from *tubulosus* (Latin: resembling or having the form of a tube), from *tubus* (Latin: tube)

tubulosan
$C_{27}H_{33}N_3$, derived from tubulosine and -an(e)

tubulosine
$C_{29}H_{37}N_3O_3$, derived from the specific epithet of the species name *Pogonopus tubulosus* (DC.) Schumann (a plant of the *Rubiacaea* family), from tubulos-, and -an(e)

tuftsin
$C_{21}H_{40}N_8O_6$, named for Tufts University, Boston, MA, USA – in turn named for the US businessman and philanthropist Charles Tufts (1781-1876) – where this tetrapeptide was discovered, and -in(e)

tugtupite
$Na_4BeAlSi_4O_{12}Cl$, named after this mineral's locality Tugtup agatakorfia, Tunugdliarfik, Greenland

tundrite
$Na_2Ce_2TiO_2(SiO_4)(CO_3)_2$, named after this mineral's localities in the tundra

tungsten
W, derived from *tungsten* (Swedish: scheelite, literally heavy stone), probably via *tungstène* (French: tungsten), via regulus of tungsten - ironically the Swedish name for tungsten is volfram

tungstenite
WS_2, derived from tungsten and -ite

tungstocene
$C_{10}H_{10}W$, derived from tungsten and -ocene

tunicamycin
derived from *tunica* (Latin: tunic), of Semitic origin, and -mycin – referring to this antibiotic's action on membrane synthesis

tunichrome
derived from tunicate, from *tunica* (Latin: tunic), of Semitic origin, and -chrome

tunicin
derived from tunicate, from *tunica* (Latin: tunic), of Semitic origin, and -in(e)

turanose
$C_{12}H_{22}O_{11}$, derived from *Turan* (Persian: Turkestan) and -ose – referring to this sugar's occurrence in a manna found in Turkestan

turgorin
derived from turgor and -in(e)

turmerone
$C_{15}H_{22}O$, derived from turmeric, ultimately from *terra merita* (Latin: deserving or deserved earth), and -one

Turnbull's blue, insoluble
$Fe^{III}[Fe^{III}Fe^{II}(CN)_6]_3$, named for the chemical company Arthur & Turnbull, Camlachie, Scotland, UK

Turnbull's blue, soluble
$K[Fe^{III}Fe^{II}(CN)_6]$, named for the chemical company Arthur & Turnbull, Camlachie, Scotland, UK

turpentine
ultimately derived from *terebinthos* (Greek: turpentine tree)

turquoise
$CaAl_6(PO_4)_4(OH)_8{\cdot}4H_2O$, derived from *turquois* (French: Turkish) – named after Turkey where this mineral was first found

tutin
$C_{15}H_{18}O_6$, derived from the Maori name *tutu* of the plant *Coriaria japonica* A. Gray (a shrub), and -in(e)

Tutton salt
$M^{I}M^{II}(SO_4)_2{\cdot}6H_2O$, named for the British chemist A. E. H. Tutton (1864-1938)

twistane
$C_{10}H_{16}$, derived from twist and -an(e) – referring to this hydrocarbon's twisted structure

Twitchell process
named for the 19th century US chemist Ernst Twitchell

TX
$C_{20}H_{40}O$, an abbreviation for thromboxane

tyl(o)-
derived from *tylos* (Greek: knob, lump, callus, pad)

tylocrebrine
$C_{24}H_{27}NO_4$, derived (with contraction) from the species name *Tylophora crebriflora* S. T. Blake (coast tylophora), from tyl(o)- and *creber* (Latin: crowded, numerous, full), and -in(e)

tylophorine
$C_{24}H_{27}NO_4$, derived from the genus name *Tylophora* (milkweeds), from tyl(o)- and -phor, and -in(e)

tylosin
$C_{46}H_{77}NO_{17}$, a name coined as a trademark without any stated reason

tyramine
$C_8H_{11}NO$, derived (with contraction) from tyrosine and amine

tyr(os)-
derived from *tyros* (Greek: cheese)

tyrocidine
coined by variation of tyrothricin

Tyrode's solution
named for the US pharmacologist M. V. Tyrode (1878-1930)

tyrosinase
derived from tyrosine and -ase
tyrosine
$C_9H_{11}NO_3$,derived from tyr(os)- and -in(e)

tyrosol
$C_8H_{10}O_2$, derived (with contraction) from tyrosine and -ol

tyrothricin
derived from the bacterial group name *Tyrothrix*, from tyr(os)- and *thrix* (Greek: hair), and -in(e)

U

ubiquinone (coenzyme Q)
derived from *ubique* (Latin: everywhere) and quinone

ubiquitin
derived from *ubique* (Latin: everywhere) and -in(e)

UDP
$C_9H_{14}N_2O_{12}P_2$, an abbreviation for uridine 5'-diphosphate

Uemura oxidation
named for the 20th century Japanese chemist Sakae Uemura

Ugi reaction
named for the Estonian-German chemist Ivar Karl Ugi (1930-2005)

ugrandite
coined by contraction of uvarovite, grossular, and andradite

-uide
an arbitrary suffix coined by variation of ^{2}-ide

uintaite (gilsonite)
named after this mineral's locality, the Uinta Valley, UT, USA

ulexite
$NaCa[B_5O_6(OH)_6]\cdot 5H_2O$, named for the German chemist Georg Ludwig Ulex (1811-1883)

uliginosin
derived from the specific epithet of the species name *Hypericum uliginosum* HBK (a South American weed), from *uliginosus* (Latin: growing in wet or swampy ground), ultimately from *udus*, *uvidus* (Latin: damp, moist), and -in(e)

ullmannite
NiSbS, named for the German chemist and mineralogist Johann Christoph Ullmann (1771-1821)

Ullmann reaction
named for the German chemist Fritz Ullmann (1875-1939)

-ulose
an arbitrary suffix derived from levulose

Ultée cyanohydrin method
named for the 20th century Dutch chemist A. J. Ultée

ultra-
derived from *ultra* (Latin: on the other side, beyond)

ultramarine
$(Ca,Na)_8(AlSi_3O_4)_6(Cl,S,SO_4)$, derived from *ultramarinus* (Latin: from beyond the sea), from ultra- and *mare* (Latin: sea)

ulvöspinel
Fe_2TiO_4, named after this mineral's locality, the island of Ulvön, Sweden

umami
derived from *umai* (Japanese: something delicious)

umbelliferone
$C_9H_6O_3$, derived from the family name *Umbelliferae* (fragrant or aromatic plants), from *umbella* (Latin: parasol, umbrella, umbel), diminutive of *umbra* (Latin: shadow), and -fer, and -one

UMP
$C_9H_{13}N_2O_9P$, an abbreviation for uridine 5'-monophosphate

umpolung
derived from *Umpolung* (German: pole reversal), from *umpolen* (German: to reverse poles), from *Pol* (German: pole)

undeca-
derived from *undecim* (Latin: eleven)

uni-
derived from *unus* (Latin: one)

unimolecular
derived from uni- and molecular

univalent
derived from uni- and -valent

Upjohn reaction
named for the chemical company The Upjohn Company, Kalamazoo, MI, USA

UPS
an abbreviation for ultraviolet photoelectron spectroscopy

uracil
$C_4H_4N_2O_2$, derived from urea and acrylic acid – referring to this compound's laboratory preparation from these starting materials

uramil
$C_4H_5N_3O_3$, derived from urea, amide, and -il(e)

uranediol
$C_{21}H_{36}O_2$, coined by contraction of urine and (3β,5α,17α,17aβ)-17-methyl-*D*-homo-androstane-3,17a-diol

uraninite
UO_2, derived from uran(o)-, -in(e), and -ite

uranium
U, named for the planet Uranus, named for the Greek god of the sky Uranos, ultimately derived from *ouranos* (Greek: sky)

uran(o)-
derived from uranium

uranocene
$C_{16}H_{16}U$, derived from uran(o)- and -ocene

uranophane (uranotile)
$Ca(UO_2)_2(SiO_3,OH)_2 \cdot 5H_2O$, derived from uran(o)- and -phan(e)

uranyl
UO_2^{2+}, derived from uran(o)- and -yl

urazole
$C_2H_3N_3O_2$, derived from urea, az(a)-, and -ol(e)

urea (carbamide)
CH_4N_2O, derived from *urée* (French: urea), from ur(o)-

urease
derived from urea and -ase

Urech cyanohydrin method
named for the 19th century Spanish chemist F. Urech

Urech hydantoin synthesis
named for the 19th century Spanish chemist F. Urech

urethan
$C_3H_7NO_2$, derived from urea, eth(a)- and -an(e)

-urgy
ultimately derived from *ergon* (Greek: work)

uric acid
$C_5H_4N_4O_3$, derived from ur(o)-

uricase
coined by contraction of uric oxidase

uridine
$C_9H_{12}N_2O_6$, derived (with contraction) from uracil and -idin(e)

ur(o)-
derived from *ouron* (Greek: urine)

urobilin
derived from ur(o)- and bilin

urochloralic acid
$C_8H_{11}Cl_3O_7$, derived from ur(o)- and chloral

urocortin
derived from ur(o)- and corticotropin

urogastrone (epidermal growth factor)
derived from ur(o)-, gastr(o)-, and -one

urokinase
derived from ur(o)- and kinase

-uronic acid
an arbitrary suffix derived from glucuronic acid

urothion
$C_{11}H_{11}N_5O_3S_2$, derived from ur(o)-, thi(o)-, and -on(e)

urotropin (methenamine)
$C_6H_{12}N_4$, coined as a trademark, derived from ur(o)-, -tropic, and -in(e) – referring to this compound's antiseptic action on the urinary tract

ursane
$C_{30}H_{52}$, derived from ursolic acid and -an(e)

urso-
derived from *ursus* (Latin: bear)

ursocholic acid
$C_{24}H_{40}O_5$, derived from urso- and chol(e)-

ursodiol
$C_{24}H_{40}O_4$, derived from urso-, di-, and -ol

ursolic acid
$C_{30}H_{48}O_3$, derived from urso- and -ol – referring to this compound's isolation from bearberry (*Arctostaphylos uva-ursi* L.)

Urushibara's catalyst
named for the Japanese chemist Y. Urushibara (1901-1973)

urushiol (urushic acid)
$C_{21}H_{32}O_2$, derived from *urushi* (Japanese: lacquer) and -ol

uscharidin
$C_{29}H_{38}O_9$, coined by variation of uscharin

uscharin
$C_{31}H_{41}NO_8S$, derived from *uschari*, a native African word for the arrow poison obtained from the African plant rubberbush, apple of Sodom (*Calotropis procera*), ultimately probably from *usher* (Arabic: rubberbush, apple of Sodom, *Calotropis procera*), and -in(e)

usnic acid
$C_{18}H_{16}O_7$, derived from the genus name *Usnea* (mosses), from *ushnah* (Arabic: moss)

ussingite
$Na_2AlSi_3O_8(OH)$, named for the Danish mineralogist Niels Viggo Ussing (1864-1911)

ustilagic acid
derived from the fungal genus name *Ustilago*, from *ustilago* (Late Latin: a thistle), from *ustus* (Latin: burnt), from *urere* (Latin: to burn) – referring to the scorched appearance of these smut fungi

UTP
$C_9H_{15}N_2O_{15}P_3$, an abbreviation for uridine 5'-triphosphate

UV
an abbreviation for ultraviolet

uv(a)-
derived from *uva* (Latin: grape)

uvarovite
$Ca_3Cr_2(SiO_4)_3$, named for the Russian statesman and amateur mineral collector Count Sergei S. Uvarov (1765-1855)

uvitic acid
$C_9H_8O_4$, derived from uv(a)- and -ite – referring to the possible formation of this acid from tartaric acid

uvitonic acid
$C_8H_7NO_4$, coined by variation of uvitic acid – referring to the fact that this acid is a pyridine analog of the benzene derivative uvitic acid

UV/VIS
an abbreviation for ultraviolet/visible

uzarigenin
$C_{23}H_{34}O_4$, derived from uzarin, -gen, and -in(e)

uzarin
$C_{35}H_{54}O_{14}$, derived from uzara root (*Xysmalobium undulatum* (L.) R. Brown or *Dicoma anomala*), possibly of Turkic origin, and -in(e)

V

vaccenic acid $C_{18}H_{34}O_2$, derived from *vacca* (Latin: cow) and -ene – referring to the occurrence of this acid in bovine fat

valence
derived from *valentia* (Latin: strength, capacity), from *valere* (Latin: to be strong, to thrive)

valentinite
Sb_2O_3, named for the fictitious 15th century German alchemist Basilius Valentinus, supposedly a Benedictine monk, regarded as an alias for several later alchemistic writers

valeric acid
$C_5H_{10}O_2$, derived from the genus name *Valeriana* (valerian), after the Roman province of Pannonia Valeria, now part of Hungary

valer(o)-
derived from valeric acid

valine
$C_5H_{11}NO_2$, derived (with contraction) from isovaleric acid and -in(e)

valinomycin
$C_{54}H_{90}N_6O_{18}$, derived from valine and -mycin

vanadinite
$Pb_5(VO_4)_3Cl$, derived from vanadium and -ite

vanadium
V, named for Vanadis, an Old Norse goddess of beauty, also called Freya – referring to the beautiful colors of vanadium compounds

vanad(o)-
derived from vanadium

vanadocene
$C_{10}H_{10}V$, derived from vanad(o)- and -ocene

van Arkel-de Boer process
named for the Dutch chemists Anton Eduard van Arkel (1893-1976) and Jan Hendrik de Boer (1899-1971)

vancomycin
$C_{66}H_{75}Cl_2N_9O_{24}$, coined as a trademark without any stated reason, derived from -mycin

van der Waals equation of state
named for the Dutch chemist Johannes Diderik van der Waals (1837-1923)

vanillic acid
$C_8H_8O_4$, derived from vanillin

vanillin
$C_8H_8O_3$, derived from the genus name *Vanilla* (vanilla), ultimately from *vanilla* (Spanish; vanilla), diminutive of *vaina* (Spanish: sheath, pod)

vanilmandelic acid
$C_9H_{10}O_5$, derived from vanillin and mandelic acid

van Slyke method
named for the US biochemist Donald Dexter van Slyke (1883-1971)

vanthoffite
$Na_6Mg(SO_4)_4$, named for the Dutch chemist Jacobus Henricus van't Hoff (1852-1911)

van't Hoff's law
named for the Dutch chemist Jacobus Henricus van't Hoff (1852-1911)

van Urk reaction
named for the Dutch pharmacist H. W. van Urk (1886-1945)

vapor
derived from *vapor* (Latin: steam, vapor)

variscite
$AlPO_4 \cdot 2H_2O$, named after *Variscia* (Medieval Latin: Vogtland, Germany), the locality of this mineral

Varrentrapp reaction
named for the German chemist Franz Varrentrapp (1815-1877)

Varrentrapp-Will method
named for the German chemists Franz Varrentrapp (1815-1877) and H. Will (1812-1890)

Vasella reaction
named for the Swiss chemist Andrea Vasella (born 1943)

vasicine
$C_{11}H_{12}N_2O$, derived from the specific epithet of the species name *Adhatoda vasica* Nees (Malabar nut), from *vasika, vasaka* (Sanskrit: Malabar nut), from *vasayati* (Sanskrit: it perfumes, it makes fragrant), and -in(e)

Vaska compound
$M^{I}Cl(CO)(PPh_3)_2$, named for the Estonian-US chemist Lauri Vaska (born 1925)

vasopressin (adiuretin, antidiuretin)
$C_{46}H_{64}N_{14}O_{12}S_2$, derived from *vas* (Latin: vessel), pressure, and -in(e)

vaterite
$CaCO_3$, named for the German mineralogist and geologist Heinrich Vater (1859-1930) who synthesized this mineral

Vauquelin's salt
$[Pd(NH_3)_4][PdCl_4]$, named for the French chemist Nicolas Louis Vauquelin (1763-1829)

VB
an abbreviation for valence bond

veatchine
$C_{22}H_{33}NO_2$, derived from the specific epithet of the species name *Garrya veatchii* Kellogg (canyon silktassel), after the US naturalist John Allen Veatch (1808-1870) and/or his son, the US mining engineer Andrew Allen Veatch (1832-1871), and -in(e)

Vedejs reagent
$C_{11}H_{23}MoN_4O_6P$, named for the Latvian-US chemist Edwin Vedejs (born 1941)

VEELS
an abbreviation for vibrational electron energy loss spectroscopy

vellosimine
$C_{19}H_{20}N_2O$, derived from the specific epithet of the species name *Geissospermum vellosii* Allem. (pereira tree) and amine

venturicidin
derived from the fungal genus name *Venturia*, after the Italian 19th century botanist A. Venturi, -cide, and -idin(e) – referring to this antibiotic's activity against apple scab (*Venturia inaequalis*)

Venturi tube
named for the Italian physicist Giambattista Venturi (1746-1822)

veralkamine
$C_{27}H_{43}NO_2$, derived (with contraction) from the species name *Veratrum album* (white hellebore), from veratr(o)- and *albus* (Latin: white), and amine

veratraman
$C_{27}H_{43}N$, derived from veratr(o)-, amine, and -an(e)

veratramine
$C_{27}H_{39}NO_2$, derived from veratr(o)- and amine

veratric acid
$C_9H_{10}O_4$, derived from veratr(o)-

veratridine
$C_{36}H_{51}NO_{11}$, derived from veratr(o)- and -idin(e)

veratr(o)-
derived from the genus name *Veratrum* (hellebore), from *verus* (Latin: true) – referring to the ancient belief that subsequent sneezing confirms the truth of a statement

veratrole
$C_8H_{10}O_2$, derived from veratr(o)- and -ol(e)

veratroyl
$C_9H_9O_3-$, derived from veratric acid and -oyl

verbascose
$C_{30}H_{52}O_{26}$, derived from the genus name *Verbascum* (mullein), from *verbascum* (Latin: mullein), and -ose

verbenalin
$C_{17}H_{24}O_{10}$, derived from verben(o)-, -al, and -in(e)

verben(o)-
derived from the genus name *Verbena* (medicinal plants), from *verbenae* (Latin: sacred boughs of laurel, olive, or myrtle), from *verber* (Latin: whip)

verbenone
$C_{10}H_{14}O$, derived from verben(o)- and -one

verdigris
$Cu(C_2H_3O_2)_2{\cdot}H_2O$, ultimately derived from *vert de Grice* (Old French: green of Greece)

vermiculite
$Al,Fe,Mg)_3(Al,Si)_4O_{10}(OH)_2{\cdot}4H_2O$, derived from *vermiculus*, diminutive of *vermis* (Latin: worm) and -ite – referring to the fact that this mineral's scales, when heated, open out into wormlike forms

Verneuil process
named for the French chemist Auguste L. V. Verneuil (1856-1913)

verno(l)-
derived from the genus name *Vernonia* (tropical herbs, shrubs), after the British botanist William Vernon (deceased 1711)

vernolepin
$C_{15}H_{16}O_5$, derived (with contraction) from the species name *Vernonia hymenolepis* (sweet bitterleaf), from verno(l)- and *lepis* (Greek: flake, scale), and -in(e)

vernolic acid
$C_{18}H_{12}O_3$, derived from verno(l)-

veronal
$C_8H_{12}N_2O_3$, coined as a trademark, probably named for the city of Verona, Italy

verrucarin
derived from the specific epithet of the fungal species name *Myrothecium verrucaria* (Albertini & Schweinitz) Ditmar ex Fries, from *verruca* (Latin: wart), and -in(e)

verticillin
derived from the fungal genus name *Verticillium*, from *verticillatus* (New Latin: arranged in verticils), from *verticillus* (Latin: whorl of a spindle), diminutive of *vertex* (Latin: whirl), and -in(e)

verticine
$C_{27}H_{45}NO_3$, derived from the specific epithet of the species name *Fritillaria verticillata* Willd. (fritillary), from *verticillatus* (New Latin: arranged in verticils), from *verticillus* (Latin: whorl of a spindle), diminutive of *vertex* (Latin: whirl), and -in(e)

vesuvianite (idocrase)
$(Ca,Na)_{19}(Al,Mg,Fe)_{13}(SiO_4)_{10}(Si_2O_7)_4(OH,F,O)_{10}$, named after this mineral's locality Mount Vesuvio, Italy

vetiv-
derived from the genus name *Vetiveria* (perennial grasses), ultimately from *vettiver* (Tamil: vetiver, *Andropogon zizamoides*), from *vetti* (Tamil: khuskhus, *Andropogon zizamoides*) and *ver* (Tamil: root)

vetivazulene
$C_{15}H_{18}$, derived from vetiv- and azulene

vetivone
$C_{15}H_{22}O$, derived from vetiv- and -one

vic-
derived from vicinal

vicinal
derived from *vicinalis* (Latin: neighborly), from *vicinus* (Latin: neighboring), ultimately from *vicus* (Latin: village, farm, town quarter)

vicine
$C_{10}H_{16}N_4O_7$, derived from the genus name *Vicia* (vetch), from *vicia* (Latin: vetch), ultimately from *vincire* (Latin: to bind, to tie), and -in(e)

Vickers hardness scale
named for the British industrialist Edward Vickers (1804-1897)

Viehe's salt
$C_3H_6Cl_3N$, named for the German-Belgian chemist Heinz Günter Viehe (born 1929)

Vigreux column
named for the 20th century French glassblower M. Henri Vigreux

villiaumite
NaF, named for the 20th century French artillery officer Maxime Villiaume who discovered this mineral

Vilsmeier reaction
named for the German chemist Anton Vilsmeier (1894-1962)

Vilsmeier-Haack reaction
named for the German chemists Anton Vilsmeier (1894-1962) and A. Haack (1898-1976)

vinblastine (vincaleukoblastine)
$C_{46}H_{58}N_4O_9$, coined by contraction of vincaleukoblastine

vinca-
derived from the genus name *Vinca* (periwinkle), from *pervinca* (Latin: periwinkle), from *vincire* (Latin: to bind)

vincaleukoblastine (vinblastine)
$C_{46}H_{58}N_4O_9$, derived from vinca-, leukoblast, and -in(e)

vincamine
$C_{21}H_{26}N_2O_3$, derived from vinca- and amine

vincane
$C_{19}H_{24}N_2$, derived from vinca- and -an(e)

vincennite
($HCN,AsCl_3,SnCl_4,CHCl_3$), named after the town of Vincennes, France

vincristine
$C_{46}H_{56}N_4O_{10}$, derived from vinca-, *crista* (Latin: crest), and -in(e)

vindoline
$C_{25}H_{32}N_2O_6$, derived (with contraction) from vinca-, indole, and -in(e)

vinic acid
$C_4H_6O_4$, an obsolete name for racemic acid, derived from *vinum* (Latin: wine)

vinic alcohol
C_2H_6O, obsolete name for ethanol, derived from spiritus vini

vinyl
C_2H_3-, derived from *Vine*, *Vinegas* (obsolete and uncharacteristic German: ethene) and -yl

vinyl alcohol
C_2H_4O, obsolete name for ethenol, derived from vinyl and alcohol – in contemporary terminology vinyl alcohol is not considered as an alcohol, but as an enol

vinylogy
coined by contraction of vinyl homology

violacein
$C_{20}H_{13}N_3O_3$, derived from the specific epithet of the bacterial species name *Chromobacterium violaceum*, from *violaceus* (Latin: bluish), from *viola* (Latin: the flower violet), and -in(e)

violaxanthin
$C_{40}H_{56}O_4$, derived from violet and xanthin

viologen
derived from violet and -gen

violuric acid
$C_4H_3N_3O_4$, derived from violet and urea

viomycin (enviomycin)
$C_{25}H_{43}N_{13}O_{10}$, derived from violet and -mycin – referring to the color of the *Streptomyces* species producing this polypeptide

VIP
an abbreviation for vasoactive intestinal polypeptide

viquidil
$C_{20}H_{24}N_2O_2$, coined by contraction of vinyl, quinoline,and -il(e)

virginiamycin
derived from the specific epithet of the bacterial species name *Streptomyces virginiae*, after the state of Virginia, USA, and -mycin

virginium
Fr, a name unsuccessfully suggested for francium, derived from the state of Virginia, USA and -ium

viridicatin
$C_{15}H_{11}NO_2$, derived from the specific epithet of the fungal species name *Penicillium viridicatum* Westling, from *viridicatus* (Latin: colored green), from virid(o)-, and -in(e)

viridin
$C_{20}H_{16}O_6$, derived from the specific epithet of the fungal species name *Gliocladium virens*, from *virens* (Latin: greening), from *virere* (Latin: to green), and -idin(e)

virid(o)-
derived from *viridis* (Latin: grass green)

virucide
derived from virus and -cide

VIS
an abbreviation for visible (spectroscopy)

viscose
derived from viscous and -ose

viscosin
$C_{54}H_{95}N_9O_{16}$, derived from the specific epithet of the bacterial species name *Pseudomonas viscosa*, from *viscosus* (Late Latin: viscous), from *viscum* (Latin: mistletoe, birdlime), and -in(e)

visnadine
$C_{21}H_{24}O_7$, derived (with contraction) from the specific epithet of the species name *Ammi visnaga* L. (toothpick plant), from *ammi* (Greek: an umbelliferous plant), possibly from *ammos* (Greek: sand), and *bisnaga* (Spanish: toothpick plant), from *pastinaca* (Latin: parsnip), ultimately from *pastum* (Akkadian; adze, ax cleaver), and -idin(e)

vis vitalis
derived from *vis* (Latin: power) and *vitalis* (Latin: vital), from *vita* (Latin: life) – referring to the alchemistic belief that organic compounds could only be generated by living organisms

Vitali-Morin reaction
named for the Italian pharmacist D. Vitali (1832-1917) and the Swiss pharmacist A. Morin (1800-1879)

vitamin
derived from *vita* (Latin: life) and amine – referring to the fact that the first vitamin to be discovered, thiamine, was an amine

vitellin
derived from vitell(o)- and -in(e)

vitell(o)-
derived from *vitellus* (Latin: egg yolk)

vitreous
derived from *vitrum* (Latin: glass)

vitriol
an archaic name for metal sulfate, derived from *vitriolum* (Latin: something glassy), from *vitrum* (Latin: glass), from *vitrum* (Latin: woad) – referring to the blue-green color of antiquity's glass

vitriolic acid
H_2SO_4, an archaic name for sulfuric acid, derived from vitriol

vitriolic acid air
SO_2, an archaic name for sulfur dioxide, derived from vitriolic acid and air

vitriolum
New Latin name for vitriol

vivianite
$Fe_3(PO_4)_2{\cdot}8H_2O$, named for the British mineralogist J. H. Vivian (1785-1855) or the British mine manager W. Vivian (1817-1879)

VLDL
an abbreviation for very low density lipoprotein

Vleminckx' solution
named for the Belgian physician J.-F. Vleminckx (1800-1876)

voacamine
$C_{43}H_{52}N_4O_5$, derived from the genus name *Voacanga* (voacanga), from *voacanga* (Malagasy: voacanga), and amine

vobasan
$C_{20}H_{26}N_2$, derived from vobasine and -an(e)

vobasine
$C_{21}H_{24}N_2O_3$, derived (with contraction) from the genus name *Voacanga* (voacanga), from *voacanga* (Malagasy: voacanga), base, and -in(e)

vobtusine
$C_{43}H_{50}N_4O_6$, coined by contraction of the species name *Voacanga thouarsii* var. *obtusa* (K. Schum.) (a voacanga subspecies), from *voacanga* (Malagasy: voacanga) and *obtusus* (Latin: dulled), from *tundere* (Latin: to beat), and -in(e)

VOC
an abbreviation for volatile organic compound(s), volatile organic chemical(s), or volatile organic carbon

Voight amination
named for the 19th century German chemist K. Voight

Volhard-Erdmann cyclization
named for the German chemists Jacob Volhard (1834-1910) and H. Erdmann

Volhard titration
named for the German chemist Jacob Volhard (1834-1910)

Volhard-Wolff titration
named for the German chemists Jacob Volhard (1834-1910) and N. Wolff

volicitin
$C_{23}H_{38}N_2O_5$, coined by contraction of elicitor of volatile biosynthesis and -in(e)

voltammetry
coined by contraction of voltametry and amperometry

vomicine
$C_{22}H_{24}N_2O_4$, derived from the specific epithet of the species name *Strychnos nux vomica* L. (strychnine tree), from *vomica* (Latin: disaster), from *vomere* (Latin: to vomit), and -in(e)

vomitoxin
$C_{15}H_{20}O_6$, derived from *vomere* (Latin: to vomit) and toxin

Vorbrüggen glycosylation
named for the German chemist Helmut Vorbrüggen (born 1930)

Vortmann's sulfate
$Co_6H_{79}N_{27}O_{33}S_6$, named for the 20th century German chemist Georg Vortmann

vrbaite
$Tl_4Hg_3Sb_2As_8S_{20}$, named for the Czech mineralogist Karel Vrba (1845-1922)

VSEPR model (Gillespie-Nyholm model)
an abbreviation for valence shell electron pair repulsion

vulcanization
named for Vulcanus, the Roman god of fire and metalworking, probably of Etruscan origin

vulgaxanthin
derived from the specific epithet of the species name *Beta vulgaris* (common beet), from *vulgaris* (Latin: of the mob), from *vulgus* (Latin: mob, common people), xanth(o)-, and -in(e)

vulpinic acid (vulpic acid)
$C_{19}H_{14}O_5$, derived from the specific epithet of the species name *Letharia vulpina* (wolf lichen, wolf moss), ultimately from *vulpinus* (Latin: of the fox), from *vulpes* (Latin: fox) – referring to the use of this plant as a poison for wolves and foxes

vuonnemite
$Na_{11}Nb_2TiSi_4O_{17}(PO_4)_2F_2$, named after this

mineral's locality, the Vuonnemi River, Kola Peninsula, Russia

VX (A 4)
$C_{11}H_{26}NO_2PS$, an arbitrary US military code name

W

Wackcnrodcr's liquid
named for the German chemist Heinrich Wilhelm Ferdinand Wackenroder (1798-1854)

Wacker oxidation
named for the chemical company Wacker-Chemie GmbH, Munich, Germany, founded by the German chemist Alexander Wacker (1846-1922)

Wacker-Tsuji oxidation
named for the chemical company Wacker-Chemie GmbH, Munich, Germany, and the 20th century Japanese chemist Jiro Tsuji

Wade's rule
named for the British chemist K. Wade (born 1932)

Wadsworth-Emmons reaction
named for the US chemists William Steele Wadsworth, Jr. (born 1927) and William David Emmons (1924-2001)

wagnerite
$Mg_2PO_4(F,OH)$, named for the German mining official F. M. von Wagner (1768-1851)

Wagner-Jauregg reaction
named for the German-Swiss chemist Theodor Wagner-Jauregg (1903-1992)

Wagner-Meerwein rearrangement
named for the Russian chemist Georg Egorovich Wagner (1849-1903) and the German chemist Hans Leberecht Meerwein (1879-1965)

Walden inversion
named for the German chemist Paul Walden (1863-1957)

Walden's rule
named for the German chemist Paul Walden (1863-1957)

Wallach degradation
named for the German chemist Otto Wallach (1847-1931)

Wallach reaction
named for the German chemist Otto Wallach (1847-1931)

Wallach rearrangement
named for the German chemist Otto Wallach (1847-1931)

Walsh diagram
named for the Scottish chemist Arthur Donald Walsh (1916-1977)

warburganal
$C_{15}H_{22}O_3$, derived from the genus name *Warburgia* (tropical trees), named for the German botanist Otto Warburg (1859-1938), -an(e), and -al

Warburg-Dickens-Horecker pathway
named for the German biochemist Otto Heinrich Warburg (1883-1970), the British biochemist Frank Dickens (1899-1986), and the US biochemist Bernard Leonard Horecker (born 1914)

Warburg's respiratory enzyme (cytochrome oxidase)
named for the German biochemist Otto Heinrich Warburg (1883-1970)

warfarin
$C_{19}H_{16}O_4$, coined by contraction of the abbreviation WARF, for Wisconsin Alumni Research Foundation, and coumarin

water
H_2O, an indigenous English word

water glass
named after these alkali silicates' water solubility and glassy consistency

Watson-Crick model
named for the US biochemist James Dewey Watson (born 1928) and the British biochemist Francis Harry Compton Crick (1916-2004)

wavellite
$Al_3(PO_4)_2(OH,F)_3 \cdot 5H_2O$, named for the British physician William Wavell (deceased 1829) who discovered this mineral

Wawzonek-Yeakey rearrangement
named for the US chemists Stanley Wawzonek (1914-1998) and Ernest Leon Yeakey (born 1934)

Weakly anion
$EW_{10}O_{36}{}^{n-12}$ (n = valency of E), named for the 20th century British chemist Timothy J. R. Weakly

Weerman degradation
named for the 20th century Dutch chemist R. A. Weerman

Weidenhagen synthesis
named for the German chemist Rudolf Weidenhagen (1900-1979)

Weinhold vessel
named for the German physicist Adolf Ferdinand Weinhold (1841-1917)

Weiss reaction
named for the Czech-US chemist Ulrich Weiss (1908-1989)

Weisz ring oven
named for the Austrian chemist Herbert Weisz (born 1922)

Weitz-Scheffer reaction
named for the 20th century German chemists E. Weitz and A. Scheffer

Weldon process
named for the British chemist Walter Weldon (1832-1885)

Werner complex
named for the Swiss chemist Alfred Werner (1866-1919)

Wessely oxidation
named for the Austrian chemist Friedrich Wessely (1897-1967)

Wessely-Moser rearrangement
named for the Austrian chemists Friedrich Wessely (1897-1967) and G. H. Moser

Western blot (Western blotting)
jocularly so named with reference to Southern blot

Weston element
named for the British chemist and inventor Edward Weston (1850-1936)

Westphalen-Lettré rearrangement
named for the 20th century German chemists T. Westphalen and H. Lettré

West's solution
(P_4,S_8,CH_2I_2), named for the US mineralogist C. D. West (1903-1993)

Wharton reaction
named for the British-US chemist Peter

Stanley Wharton (born 1931)

Wheland intermediate
named for the US chemist George Willard Wheland (1907-1972)

whewellite
$CaC_2O_4 \cdot H_2O$, named for the British mineralogist, crystallographer, and natural philosopher William Whewell (1794-1866)

Whiting reaction
named for the 20th century British chemist M. C. Whiting

Wichterle reaction
named for the Czech chemist Otto Wichterle (1913-1998)

Wickbold method
named for the German chemist R. Wickbold (1910-1991)

Widman-Stoermer synthesis
named for the 19th century Swedish chemist Oskar Widman and the 20th century German chemist R. Stoermer

Widmark method
named for the Swedish physiologist Erik Matteo Prochet Widmark (1889-1945)

Widmer column
named for the 20th century Swiss chemist Gustav Widmer

Wieland-Gumlich aldehyde
$C_{19}H_{22}N_2O_2$, named for the German chemists Heinrich Wieland (1877-1957) and Walter Gumlich

Wieland-Miescher ketone
$C_{11}H_{14}O_2$, named for the Swiss chemist Peter Wieland (born 1920) and the Swiss-Italian chemist Karl Miescher (1892-1974)

Wiesner reaction
named for the Austrian chemist Julius Ritter von Wiesner (1838-1916)

Wigner rule
named for the Hungarian-US physicist Eugene Paul Wigner (1902-1995)

Wijs' reagent
(ICl,CH_3COOH), named for the Dutch chemist J. J. A. Wijs (1864-1942)

Wilke process
named for the German chemist Günther Wilke (born 1925)

Wilkinson's catalyst
$C_{54}H_{45}ClP_3Rh$, named for the British chemist Sir Geoffrey Wilkinson (1921-1996)

willemite
Zn_2SiO_4, named for the king of the Netherlands Willem I (1772-1843)

Willgerodt-Kindler reaction
named for the German chemists Conrad Heinrich Christoph Willgerodt (1841-1930) and Karl Kindler (1891-1967)

Williamson ether synthesis
named for the British chemist Alexander William Williamson (1824-1904)

Willstätter reaction
named for the German chemist Richard Martin Willstätter (1872-1942)

wiluite
$Ca_{19}(Al,Mg,Fe,Ti)_{13}(B,Al,[])_5Si_{18}O_{68}(O,OH)_{10}$, named after this mineral's locality, the Wilui River, Yakutia, Russia

Wilzbach technique
named for the US chemist K. E. Wilzbach (born 1920)

Winkler burette
named for the German chemist Clemens Winkler (1838-1904)

Winkler generator
named for the German chemist F. Winkler (1888-1950)

Winstein-Holness equation
named for the Canadian-US chemist Saul Winstein (1912-1969) and the US chemist N. J. Holness

Wiswesser line notation
named for the US chemist W. J. Wiswesser (1914-1989)

withaferin
coined by contraction of the species name *Withania somnifera* Dun. (ashwagandha), named for the British naturalist Henry Thomas Maire Witham (1779-1844), and -fer, and -in(e)

witherite
$BaCO_3$, named for the British physician and mineralogist William Withering (1741-1799)

Witt diazotization method
named for the German chemist Otto Nikolaus Witt (1853-1915)

wittichenite
$Cu_6Bi_2S_6$, named after this mineral's locality Wittichen, Germany

Wittig reaction
named for the German chemist Georg Wittig (1897-1987)

[1,2]-Wittig rearrangement
named for the German chemist Georg Wittig (1897-1987)

[2,3]-Wittig rearrangement
named for the German chemist Georg Wittig (1897-1987)

WLN
an abbreviation for Wiswesser line notation

Wohl degradation
named for the German chemist Alfred Wohl (1863-1939)

wöhlerite (wohlerite)
$NaCa_2(Zr,Nb)Si_2O_7(O,OH,F)_2$, named for the German chemist Friedrich Wöhler (1800-1882)

Wöhler's urea synthesis
named for the German chemist Friedrich Wöhler (1800-1882)

Wohl-Ziegler reaction
named for the German chemists Alfred Wohl (1863-1939) and Karl Ziegler (1878-1973)

Wolffenstein-Böters reaction
named for the German chemists Richard Wolffenstein (1864-1929) and O. Böters

Wolff-Kishner reduction
named for the German chemist Ludwig Wolff (1857-1919) and the Russian chemist Nikolai Matveevich Kishner (1867-1935)

Wolff rearrangement
named for the German chemist Ludwig Wolff (1857-1919)

wolframite
$(Fe,Mn)WO_4$, derived from *Wolfram* (German: wolframite), probably from *Wolf* (German: wolf) and *Rahm* (German: froth, dirt), akin to *spuma lupi* (Medieval Latin: wolframite, literally wolf's froth) – referring to the medieval miners' observation that wolframite makes molten tin turn to slag the

same way as a wolf eats a sheep – and -ite

wolframium
W, New Latin name for tungsten, derived from *Wolfram* (German: tungsten) and -ium

wollastonite
$CaSiO_3$, named for the British mineralogist and chemist William Hyde Wollaston (1766-1828)

Wood-Bonhoeffer method
named for the US physicist R. W. Wood (1868-1955) and the German chemist Karl Friedrich Bonhoeffer (1899-1957)

Wood's metal
(Bi,Pb,Sn,Cd), named for the US chemist B. Wood

Woodward-Hoffmann rules
named for the US chemist Robert Burns Woodward (1917-1979) and the Polish-US chemist Roald Hoffmann (born 1937)

Woodward *cis*-hydroxylation
named for the US chemist Robert Burns Woodward (1917-1979)

Woodward's reagent K
$C_{11}H_{11}NO_4S$, named for the US chemist Robert Burns Woodward (1917-1979)

Wood-Werkman reaction
named for the US biochemists H. G. Wood (1907-1991) and C. H. Werkman (1893-1962)

wortmannin
$C_{23}H_{24}O_8$, derived from the specific epithet of the fungal species name *Penicillium wortmanni* Klocker, after the German botanist Julius Wortmann (1856-1925), and -in(e)

Woulfe flask
named for the British chemist Peter Woulfe (1727-1803)

wulfenite
$PbMoO_4$, named for the Austrian mineralogist Franz Xaver von Wulfen (1728-1805)

Wurster's blue
$C_{10}H_{16}N_2$, named for the German chemist Casimir Wurster (1854-1913)

Wurster's red
$C_8H_{12}N_2$, named for the German chemist Casimir Wurster (1854-1913)

Wurster's salt
named for the German chemist Casimir Wurster (1854-1913)

Wurtz coupling reaction
named for the French chemist Charles Adolphe Wurtz (1817-1884)

Wurtz-Fittig reaction
named for the French chemist Charles Adolphe Wurtz (1817-1884) and the German chemist Rudolph Fittig (1835-1910)

wurtzite
ZnS, named for the French chemist Charles Adolphe Wurtz (1817-1884)

Wurtz reaction
named for the French chemist Charles Adolphe Wurtz (1817-1884)

Wurzschmitt method
named for the German chemist Bernhard Wurzschmitt (1895-1975)

wüstite
FeO, named for the German geologist and paleontologist Ewald Wüst (1875-1934)

X

XAES (XEAES)
an abbreviation for X-ray induced Auger electron spectroscopy

XANES
an abbreviation for X-ray absorption near edge

xanthan gum
derived from the bacterial genus name *Xanthomonas*, from xanth(o)-, and -an

xanthate
derived from xanthic acid and -ate

xanthatin
$C_{15}H_{18}O$, derived from the genus name *Xanthium* (herbs), from *xanthion* (Greek: a plant used to color the hair yellow), from xanth(o)-, and -in(e)

xanthene
$C_{13}H_{10}O$, derived from xanth(o)- and -ene

xanthic acid
$C_3H_6OS_2$, derived from xanth(o)-

xanthine (xanthin)
$C_5H_4N_4O_2$, derived from xanth(o)- and -in(e) – referring to the yellow residue left after evaporation of xanthine with nitric acid

xanthine oxidase
derived from xanthine and oxidase

xanth(o)-
derived from *xanthos* (Greek: yellow)

xanthone
$C_{13}H_8O_2$, derived from xanthene and -one

xanthophyll (lutein)
$C_{40}H_{56}O_2$, derived from xanth(o)- and -phyll

xanthopterin
$C_6H_5N_5O_2$, derived from xanth(o)- and pterin

xanthosine
$C_{10}H_{12}N_4O_6$, derived (with contraction) from xanthine, ribose, and -in(e)

xanthotoxin (ammoidin)
$C_{12}H_8O_4$, derived from xanth(o)- and toxin – referring to this compound's yellow color

xanthoxyletin
$C_{15}H_{14}O_4$, derived from the genus name *Xanthoxylum* (shrubs, trees), from xanth(o)- and xyl(o)-, and -etin(e)

xanthoxylin
$C_{10}H_{12}O_4$, derived from the genus name *Xanthoxylum* (shrubs, trees), from xanth(o)- and xyl(o)-, and -in(e)

xanthurenic acid
$C_{10}H_7NO_4$, derived from xanth(o)- and -urenic; patterned after kynurenic acid

xanthydrol
$C_{13}H_{10}O_2$, coined by contraction of xanthene and -hydrol

xanthyletin
$C_{14}H_{12}O_3$, derived from the genus name *Xanthoxylum* (shrubs, trees), from xanth(o)- and xyl(o)-, and -etin(e)

XDP
$C_{10}H_{14}N_4O_{12}P_2$, an abbreviation for xantho-

sine 5'-diphosphate

xenate
$HXeO_4^-$, derived from xenic acid and -ate

xenic acid
H_2XeO_4, derived from xenon

xeno-
derived from *xenos* (Greek: strange)

xenobiotic
derived from xeno- and bio-

xenon
Xe, derived from xeno- and [1]-on

xenonate (perxenate)
XeO_6^{4-}, derived from xenonic acid and -ate

xenonic acid (perxenic acid)
H_4XeO_6, derived from xenon

xenotime
YPO_4, derived from *kenos* (Greek: empty) and *time* (Greek: honor) – referring to the fact that this mineral was erroneously assumed to contain an unknown metal

xenyl (biphenylyl)
$C_{12}H_9$–, derived from xeno- and -yl

xer(o)-
derived from *xeros* (Greek: dry)

xerocomic acid
$C_{18}H_{12}O_8$, derived from the genus name *Xerocomus* (mushrooms), from xer(o)- and *kome* (Greek: hair, crest)

xerogel
derived from xer(o)- and gel

xerulin
$C_{18}H_{14}O_2$, derived from the genus name *Xerula* (mushrooms), from xer(o)-, and -in(e)

XMP
$C_{10}H_{13}N_4O_9P$, an abbreviation for xanthosine 5'-monophosphate

XOD
an abbreviation for xanthine oxidase

xonotlite (eaklyite, jurupaite)
$Ca_6Si_6O_{17}(OH)_2$, named after this mineral's locality Tetela de Xonotla, Puebla, Mexico

XPES (XPS)
an abbreviation for X-ray photoelectron spectroscopy

XRD
an abbreviation for X-ray diffraction

XRF
an abbreviation for X-ray fluorescence (spectroscopy)

XRS
an abbreviation for X-ray spectroscopy

XTP
$C_{10}H_{15}N_4O_{15}P_3$, an abbreviation for xanthosine 5'-triphosphate

xylan
derived from xylose and -an

xylanase
derived from xylan and -ase

xylene
C_8H_{10}, derived from xyl(o)- and -ene – referring to this hydrocarbon's formation in wood tar

xylenol
$C_8H_{10}O$, derived from xylene and -ol

xylidine
$C_8H_{11}N$, derived from xylene and -idin(e)

xylitol
$C_5H_{12}O_5$, derived from xyl(o)- and -itol

xyl(o)-
derived from *xylon* (Greek: wood)

xylose
$C_5H_{10}O_5$, derived from xyl(o)- and -ose

xylulose
$C_5H_{10}O_5$, derived from xyl(o)- and -ulose

xylyl
C_8H_9-, derived from xylene and -yl

[1]**xylylene**
C_8H_8, derived from xylyl and -ene

[2]**xylylene**
$-CH_2C_6H_4CH_2-$, derived from xylyl and -ene

Y

YAG
$Al_5Y_3O_{12}$, an abbreviation for yttrium-aluminum garnet

yangonin
$C_{15}H_{14}O_4$, derived from *yangona* (Fijian: kava, *Piper methysticum*) and -in(e)

YIG
$Fe_5Y_3O_{12}$, an abbreviation for yttrium-iron garnet

yingzhaosu
derived from *yingzhao* (Chinese: ylang-ylang vine, *Artabotrys uncinatus* (L.) Merr.)

-yl
derived from *hyle* (Greek: matter) – referring to organic chemistry's radicals as 'ultimate matter'; first used in benzoyl

ylangene
$C_{15}H_{24}$, derived from ylang-ylang, from *ylang-ylang* (Tagalog: ylang-ylang, *Canangium odoratum*), and -ene

ylide
derived from -yl and ^{2}-ide

ylo-
derived from -yl

-yne
an arbitrary suffix patterned after -in(e)

yohimb(a)-
derived from the specific epithet of the species name *Coryanthe yohimbe* (yohimbé), from Bantu, and -in(e)

yohimban
$C_{19}H_{24}N_2$, derived from yohimb(a)- and -an(e)

yohimbine
$C_{21}H_{26}N_2O_3$, derived from yohimb(a)- and -in(e)

Young method
named for the British chemist S. Young (1857-1937)

yperite (lost, mustard gas)
$C_4H_8Cl_2S$, French military code name, after the town of Ypres (Ieper), Belgium

ytterbia
Yb_2O_3, derived from ytterbium and -a

ytterbium
Yb, named for the village of Ytterby, Sweden where the first ytterbium ores were found

yttria
Y_2O_3, derived from yttrium and -a

yttrium
Y, derived (with contraction) from Ytterby, Sweden where the first yttrium ores were found, and -ium

yuehchukene
$C_{26}H_{26}N_2$, derived from *yueh-chu* (Chinese: orange jessamine, *Murraya paniculata* (L.) Jack), and -ene

Z

z
derived from *zusammen* (German: together)

Zachariasen's rule
named for the Norwegian-US crystallographer F. W. H. Zachariasen (1906-1979)

Zaitsev-Rosenmund reaction
named for the Russian chemist M. M. Zaitsev (1845-1904) and the German chemist Karl Wilhelm Rosenmund (1884-1965)

Zaitsev's (Saytsev's, Saytzeff's) rule
named for the Russian chemist Aleksandr Mikhailovich Zaitsev (1841-1910)

zapato-
derived from *zapato* (Spanish: shoe)

zarzissine
$C_5H_5N_5$, a name coined without any stated reason, maybe derived from a lay name of the Mediterranean sponge *Anchinoe paupertas* which produces this alkaloid

ZDO
an abbreviation for zero differential overlap

ze(a)-
derived from the genus name *Zea* (corn), from *zea* (Greek: single-grained wheat)

zearalenone
$C_{18}H_{22}O_5$, derived from ze(a)-, -al, -ene, and -one

zeatin
$C_{10}H_{13}N_5O$, derived from ze(a)- and -in(e)

zeaxanthin
$C_{40}H_{56}O_2$, derived from ze(a)- and xanthin

zein
derived from ze(a)- and -in(e)

Zeisel method
named for the Austrian chemist Simon Zeisel (1854-1933)

Zeise's salt
$C_2H_4Cl_3KPt{\cdot}H_2O$, named for the Danish chemist William Christopher Zeise (1789-1847)

ZEKE
an abbreviation for zero kinetic energy

Zemplén degradation of sugars
named for the Hungarian chemist Géza Zemplén (1883-1956)

Zemplén saponification
named for the Hungarian chemist Géza Zemplén (1883-1956)

zeolite
derived from *zein* (greek: to boil) and -lite – referring to these minerals' expulsion of boiling water upon heating

Zerewitinoff method
named for the Russian chemist T. Zerewitinoff (1874-1947)

zero
ultimately derived from *sifr* (Arabic: empty, zero)

Ziegler method
named for the German chemist Karl Ziegler (1898-1973)

Ziegler-Natta catalyst
$TiCl_4/C_6H_{15}Al$, named for the German chemist Karl Ziegler (1898-1973) and the Italian chemist Giulio Natta (1903-1970)

Ziegler-Natta polymerization
named for the German chemist Karl Ziegler (1898-1973) and the Italian chemist Giulio Natta (1903-1970)

Zimmermann reaction
named for the German chemist W. Zimmermann (1910-1982)

zinc
Zn, ultimately derived from *Zinke* (German: barb) – referring to the molten metal's formation of jags at certain temperatures

zincite
ZnO, derived from zinc and -ite

Zincke disulfide cleavage
named for the German chemist Theodor Ernst Carl Zincke (1843-1928)

Zincke-König reaction
named for the German chemists Theodor Ernst Carl Zincke (1843-1928) and W. König (1878-1964)

Zincke nitration
named for the German chemist Theodor Ernst Carl Zincke (1843-1928)

Zincke-Suhl reaction
named for the German chemists Theodor Ernst Carl Zincke (1843-1928) and R. Suhl

zincocene
$C_{10}H_{10}Zn$, derived from zinc and -ocene

zincum
Zn, New Latin name for zinc, derived from zinc and –um

ZINDO
an abbreviation for Zerner intermediate neglect of differential overlap, named for the US chemist Michael Charles Zerner (1940-2000)

zingerone
$C_{11}H_{14}O_3$, derived (with contraction) from zingiber- and -one

zingiber-
derived from the genus name *Zingiber* (ginger), from *zingiber, gingiber* (Latin: ginger), ultimately from *srnga-vera* (Sanskrit: antler-shaped)

zingiberene
$C_{15}H_{24}$, derived from zingiber- and -ene

Zinin reduction
named for the Russian chemist Nikolai Nikolaevich Zinin (1812-1880)

zinkenite (zinckenite)
$Pb_9Sb_{22}S_{42}$, named for the German mineralogist J. K. L. Zincken (1798-1862)

zinnwaldite
$K(Al,Fe,Li)_3(Si,Al)_4O_{10}(OH)F$, named after its locality Zinnwald, now Cinovec, Bohemia, Czech Republic

Zintl phases
named for the German chemist Eduard Zintl (1898-1941)

Zintl rule
named for the German chemist Eduard Zintl (1898-1941)

zircon
$(Zr,Hf)SiO_4$, derived from *giargone* (Italian: zircon), from *zargun* (Arabic: gold color)

zirconia
ZrO_2, derived from zirconium and –a

zirconium
Zr, derived from zircon and -ium

zirconocene
$C_{10}H_{10}Zr$, derived from zirconium and -ine

zirconyl
−ZrO−, derived from zirconium and -yl

zoapatanol
$C_{20}H_{34}O_4$, derived from zoapatle (*Montanoa tomentosa*), from *zoapatl* (Nahuatl: zoapatle), -an(e), and -ol

zoisite
$Ca_2Al_3(SiO_4)_3(OH)$, named for the Austrian natural scientist Sigismund Zois von Edelstein (1747-1819)

zoo-
derived from *zoon* (Greek: animal), from *zoe* (Greek: life)

Zosel method
named for the German chemist K. Zosel (1913-1989)

ZSM
an abbreviation for zeolite Socony (Standard Oil Company of New York) Mobil

zwieselite
$Fe_2PO_4(F,OH)$, named after this mineral's locality Rebenstein, Zwiesel, Germany

Zwikker reaction
named for the Dutch 20th century pharmacist J. J. L. Zwikker

zwitterion
derived from *Zwitter* (German: hermaphrodite, hybrid), from *zwei* (German: two), and ion

zymase
derived from zym(o)- and -ase

zym(o)-
derived from *zyme* (Greek: yeast)

zymogen (proenzyme)
derived from zym(o)- and -gen

zymosterol
$C_{27}H_{44}O$, derived from zym(o)- and sterol